THE SCIENCES IN THE AMERICAN CONTEXT: NEW PERSPECTIVES

Edited by Nathan Reingold

SMITHSONIAN INSTITUTION PRESS
WASHINGTON, D.C.
1979

© 1979 by Smithsonian Institution. All rights reserved.

Library of Congress Cataloging in Publication Data
American Association for the Advancement of Science.
  The sciences in the American context.
  Papers presented at a bicentennial conference at the Feb. 1976 Boston meeting of the American Association for the Advancement of Science; cosponsored by Section L of the Association and the History of Science Society.
  1. Science—United States—History—Congresses.
I. Reingold, Nathan, 1927-   II. American Association for the Advancement of Science. Section on History and Philosophy of Science. III. History of Science Society. IV. Title.
Q127.U6A69 1979   509'.73   79-607059
ISBN 0-87474-798-8
ISBN 0-87474-797-X pbk.

Printed in the United States of America
Designed by Stephen Kraft

TABLE OF CONTENTS

1　Preface

9　Reflections on 200 Years of Science in the United States
Nathan Reingold

21　Paradigm Lost
William H. Goetzmann

35　Americans Abroad: Science and Cultural Nationalism in the Early Nineteenth Century
Bruce Sinclair

55　Astronomy in Antebellum America
Deborah Jean Warner

77　Science, Public Policy, and Popular Precepts: Alexander Dallas Bache and Alfred Beach as Symbolic Adversaries
Robert Post

99　The American Scientist in Higher Education, 1820-1910
Stanley M. Guralnick

143　Rationalization and Reality in Shaping American Agricultural Research, 1875-1914
Charles E. Rosenberg

165 From the Grand Canyon to the Marianas Trench: The Earth Sciences After Darwin
Steve Pyne

193 Industrial Research Laboratories
Kendall Birr

209 The Rise and Spread of the Classical School of Heredity, 1910-1930: Development and Influence of the Mendelian Chromosome Theory
Garland E. Allen

229 American Foundations as Patrons of Science: The Commitment to Individual Research
Stanley Coben

249 Warren Weaver and the Rockefeller Foundation Program in Molecular Biology: A Case Study in the Management of Science
Robert E. Kohler, Jr.

295 The Physics Business in America, 1919-1940: A Statistical Reconnaissance
Spencer R. Weart

359 Science Agencies in World War II: The OSRD and Its Challengers
Carroll Pursell

379 Academic Science and the Military: The Years Since the Second World War
Harvey M. Sapolsky

PREFACE

With the exception of the editor's introduction, all the papers in this volume were presented at a bicentennial conference at the February 1976 Boston meeting of the American Association for the Advancement of Science. Section L of the Association and the History of Science Society were cosponsors; the extensive program was possible only because of support from the National Endowment for the Humanities and the National Science Foundation.

In putting together the program, the editor was not merely rising to a patriotic occasion, for these papers can be viewed on several levels. In the last ten years, there has been a notable upsurge in research in the history of the sciences in the United States; what we give here are both summaries of the state of knowledge and new facts and interpretation. More than anything else, the upsurge in research reflects the persistent belief that the pure and applied sciences are particularly important in understanding the course of American history. Even contributions narrowly focused, in the proper manner of academic specialization, display subtle resonances with themes in the general literature of American history. Yet, if that belief animates research, there is little here or elsewhere to define just what is particularly important about the sciences in American history. The belief turns out to be a puzzle still awaiting solution.

On another level these essays—so different in subject and style—show a long-term trend underway in the history of science in general. People still talk of internal and external histories. Increasingly, the explicit intention is to join the two—to have a complete historical picture of the life of science and of scientists. On both sides of the Atlantic are desires to see the past in the round, not merely idealized fragments. From these desires one can sense tensions, ambitions, and pretenses as historian after historian seeks analytic explanations, synthetic themes, and heuristic models.

The essays are, then, contributions to the national history, plus something else. Not every major topic is covered; there are too many for one conference and one volume. But enough is here to suggest the fascinating problems awaiting solution. The essays are modest but unmistakable harbingers of a vastly altered history of science in the process of emerging. The altered history will be committed to a merger of the intellectual and social histories of the sciences, the story of people and their environments. Unlike the older history of science, concepts and data will serve as important elements of human environments, not as the principal objects of study justifying the endeavor. The new history will be one of styles and attitudes, of the lay public as well as the great investigators. This new development is stirring just over our horizon. We see shafts of light, we hear sounds of harmonies. And as we go about the tasks set by professional inertia, we can only hope our labors are hastening the day.

After the Boston sessions, I was visited by an editor of *Nature* magazine who wanted to discuss them. When I complained that Europeans were largely ignorant of America's part in the recent history of the sciences, he challenged me to write a piece for *Nature*'s bicentennial issue on why his readership should know this story. I could not refuse an opportunity for a patriotic fulmination aimed at a largely British audience, but the resulting essay, included in this volume, was quite different from our original intentions. Instead of a recital of intellectual triumphs and institutional genealogies, an irresistible impulse produced a historiographic polemic in capsule form. Instructions to the Boston participants had suggested either new findings, summaries of knowledge, or both. My "Reflections" piece is best described as a modest survey of future research topics.

Quite unexpected was the particular thrust of William Goetzmann's essay. Given his intellectual history, a new look at explorations and surveys was not unexpected, particularly with the recent interest in space. A Pulitzer Prize winner, Goetzmann is interested in how people of the past reacted to new specifics of time and place. His vision takes in a broader cultural perspective than the specialized fields of science. From that vantage point, Goetzmann protests about the ahistoric vision of a T. S. Kuhn, which retrospectively limits past perception to those individuals and topics analogous to the current situation. Paradoxically, Goetzmann's complaint is similar to that of historians calling for history from the bottom. He wants to extend "science" to a larger number of actors than an invisible college of savants. Dr. Goetzmann is at the University of Texas in American Studies.

Bruce Sinclair's essay is in a familiar genre, an account of Americans abroad in the Old World. Henry James the elder was with Joseph Henry in 1837; his son would raise the theme of Americans abroad to high art. But somehow the younger Henry James never quite caught all the nuances of this encounter of cultures. Mark Twain did better. The scientists of Sinclair's essay are not innocents abroad, no matter how impressed and wide-eyed they seem at times. They showed a shrewd realism, a cool appraisal of what might travel well across the ocean. Perhaps Japanese scientists of later generations displayed a similar concern. Sinclair is at the University of Toronto and specializes in the history of technology.

A quite different and militant note appears in Deborah Warner's work on antebellum astronomy. Despite a strong interest in minorities in science and in the lesser practitioners, she draws back from any downgrading of the essential roles of internal factors in the development of science. This approach is a credit, no doubt, to her Harvard training, and the result is a fresh and spirited look at the astronomy of that day. Not everyone will agree with Warner's conclusions, but they must be seriously considered in all future studies. Her paper is, perhaps unintentionally, a genuine contribution to the merger of internal and external histories of science. Ms. Warner is curator of astronomy at the National Museum of History and Technology, Smithsonian Institution.

To the historian, the clash between men of theory and men of practice is a recurrent issue, sometimes assuming epic proportions. Robert Post gives a classic instance of this clash in antebellum America, and he writes from the standpoint of many involved in the history of technology. There is a coolness, if not hostility, to the scientist's claim to deference, especially in terms of practical results. Exalting basic research implicitly downgrades the men of practice and their achievements. With some awkwardness, Post's school of thought postulates a complete split between theory and practice, especially in terms of the past. Consequently, Post and like-minded scholars have an antipathy to certain historical scientists (Alexander Dallas Bache in this case) and also to policies reflecting the viewpoints of pure scientists. An air of unreality hovers around this argument; nevertheless, Post has performed a sparkling act of debunking. He is at the National Museum of History and Technology, Smithsonian Institution.

A revisionist of a different sort is Stanley Guralnick of the Colorado School of Mines. Although interested in the physical sciences in the last century, he is best known for taking a fresh look at science in the traditional liberal arts college. The results of that research overturned much conventional wisdom. In this piece, Guralnick has extended his work in both time and scope into the early years of this century. This is a kind of no-nonsense, hard-boiled look at the college professor who happened to be a scientist. Because the setting is institutional, the scientists are unfamiliar to those accustomed to the older literature but familiar to observers of the present. This essay subtly but pointedly counters the myth that federal money after World War II corrupted higher education.

A large problem is just how scientists motivated by a pure science ideology could effectively function in settings dominated by utility. The *Nature* paper I wrote briefly suggests that scientists often succeeded in defining applied missions broadly, to include theoretical concerns. For many years, Charles Rosenberg of the University of Pennsylvania has studied and written about agriculture and medicine in the United States. Like many others in this volume, Rosenberg came to these concerns from American history, not one of the sciences or philosophy, and his attention is firmly fixed on a social situation in which the contents of scientific fields are simply one element. The article here is a continuation of Rosenberg's look at agricultural experiment stations. By uncovering the complexities of their social setting, Rosenberg avoids the kind of sterile polarization of other writings: pure versus applied, the little men versus the big men of science. He shows how the differing views of the agricultural scientists and their client constituencies produced a compromise interaction satisfactory to both.

Pathbreaking and breathtaking are the words elicited by Steve Pyne's essay. It will not be to everyone's taste, this ambitious attempt to compare two differing world views in geology. There is an almost Whitmanesque drive to encompass here, a barely suppressed passion to list and to see connections. A student of Goetzmann's, Pyne is now writing a history of wildland fire protection in the United States. "From the Grand Canyon to the Marianas Trench" is a byproduct of his graduate work. Until recently, historians of geology went slightly beyond Darwin to show how evolution was verified by the fossil record. This approach resulted in a subtle, unintended diminution of the field, as if geology existed only to provide footnotes on the *Origin of Species*. Pyne has moved into the next century, where the study of the earth becomes an intellectual battleground for competing word views with far-reaching ramifications.

Everyone with any interest in modern science and technology knows that the industrial research laboratory is very important. One of the few actually to study the laboratory as a type is Kendall Birr of the State University of New York (Albany). His is a summation of the state of our knowledge on this subject. Until very recently, corporate reticence has limited scholarly access to the records of industrial research. It is not enough simply to look at lab notebooks or a few files of the corporate director. Decisions at the top of the hierarchy are crucial, as are the viewpoints of engineering and marketing people in the corporation. The corporate laboratory remains a major target for future historical research.

Garland E. Allen has contributed the only piece on a single individual. On the surface, this paper is the most resolutely conventional, in the sense of being internalistic. A professional biologist at Washington University of St. Louis, Allen has produced a stream of very high-quality articles on a historical moment of great importance—the transition from the older tradition of morphological research to the new experimental and quantitative biology. Some of the principal actors and events in this transition were in the United States. Allen is not simply concerned with a succession of technicalities; he too looks for larger consequences in broader settings. The work on Thomas Hunt Morgan is a needed preliminary to an ongoing, large-scale investigation of the historical relations of genetics and racial beliefs and policies in twentieth-century America. What differentiates Allen from others with similar backgrounds is a concern and involvement with social issues.

The culture of the twenties is a speciality of Stanley Coben of the University of California (Los Angeles). Within this period, he has noted the rise of various research traditions. Here he considers the contributions of the private sector via the mechanism of the foundation. In contrast to the mission-oriented federal or corporate sectors, Coben finds in the foundations he studied a rather generous, open support of individual investigators. To a contemporary reader, this essay has a faint aroma of the lost cause, a nostalgic look backwards from an academia buffeted by demands for relevance. Enamored by his leading characters, Coben hardly says anything about the general trends of the period, nor does he analyze in depth the motivations of the learned communities benefiting from philanthropic largesse. But he does demonstrate that the Jazz Age was much more than gilded youths and bootleggers.

Patronage also attracts Robert Kohler of the University of Pennsylvania, but there is a difference in his approach. The previous piece conveyed a sense of high-minded disinterest; here we encounter a high-minded sense of mission. Kohler's piece is not simply on one individual, Warren Weaver, but on how the great foundations deliberately went about achieving self-defined goals. This aspect of the foundations is well known to specialists but hardly appreciated by many. A foundation need not be a passive recipient of requests for support, nor does it necessarily rely upon peer review—by now one of the sacraments of the learned world.

Because Weaver's molecular biology program succeeded, there is little inclination today to question the effort. But Kohler's essay does so implicitly. Believers in peer review might wonder at the value judgments of a mathematician in this field; others might mull over roads not taken outside of biology. Still others might assess someday the specific effects on the biological sciences of such targeted support. Did Weaver's tilt hobble other specialties in an era when federal funds were minuscule?

Spencer Weart of the American Institute of Physics does a superb job of describing the state of a great field, physics, during the years between the two wars. The description is wholly in terms of social and economic parameters, not of intellectual advances. In its own way, yet so different from Goetzmann's, this essay is another instance of the tacit repudiation of T. S. Kuhn's views and of older modes in the history of science. Of all the information conveyed by Weart, the most important concerns the effects of the Depression. Weart clearly shows that these effects were both minor and of limited duration: Physics did very well indeed during the years of the Depression.

Taken in conjunction with his other statistics, Weart's work has an unspoken message. It repudiates the image of a golden age, strongly and sincerely held by many physicists. According to this image, before the corruption of the Bomb and defense moneys, physicists were humble, threadbare seekers after the truths of the universe. Because of work by Weart and others, we now know physics to have been a rising field attracting considerable support, both relatively and absolutely, as far back as 1900. The golden age image is a way of disclaiming responsibility for later developments in science policy.

The World War II experience had great impact on public policy and on the national perception of the sciences. Carroll Pursell of the University of California (Santa Barbara) looks at that experience from a position of dissent with many of the resulting clichés.

Like Post, he views the basic science ideology as wrong in its historic foundations and pernicious in many of its consequences. This look at alternatives to the policies associated with the Wartime Office of Scientific Research and Development is tinged with concern for the possible antidemocratic role of expertise and the rise of the corporate state solidly anchored on the military-industrial complex. The essay is tinged with nostalgia for lost causes, what-might-have-beens, and roads not taken. Because the issues involved are still with us, Pursell's words will evoke varying responses from readers.

What did come out of the war was the great role of national security as a justification of research and the source of funds on a scale unimaginable a few years before. This large, complex story so far has eluded the historian's grasp. Not only were many documents closed to impartial scrutiny, but national security issues produced tendencies to one-dimensional or black-and-white treatments. Still, time is on the side of Clio. Harvey Sapolsky of the Massachusetts Institute of Technology (MIT) is a political scientist concerned with institutional behavior. He has written a fresh account, based upon original sources, of the early years of the Office of Naval Research (ONR). For a brief period, ONR was the most effective supporter of basic research outside of medicine and atomic energy. Sapolsky's account of policy formulation has many virtues, not the least of which is to introduce many readers to Adm. Harold Bowen. One result of that introduction is the realization that the history of science has to include more than scientists and their creations.

By way of acknowledgments, James M. Hobbins helped materially in the early planning of the bicentennial conference. He and my other colleagues at the Joseph Henry Papers, notably Arthur P. Molella and Marc Rothenberg, acted as referees of the papers submitted for this work. Beverly Jo Lepley prepared the manuscript. Joyce Latham served as copy eiditor. They all deserve credit for the volume's merits.

The introduction and a condensed version of the Weart study appeared earlier in *Nature* magazine; Rosenberg's paper was printed in *Social Studies of Science*; the Coben piece was in *Minerva*, as was a shorter version of Kohler's paper. Permission to use them in this volume is gratefully acknowledged.

N.R.

REFLECTIONS ON 200 YEARS OF SCIENCE IN THE UNITED STATES*

NATHAN REINGOLD
Smithsonian Institution

EARLIER THIS YEAR, A BRITISH HISTORIAN OF SCIENCE WROTE ME that he was giving a few lectures in a course on the institutional history of American science before 1914. I was astonished. Hardly any non-American historians do research or display interest in science in the United States; presumably their courses reflect this. An East German has a few pieces dealing with American astronomy; a Dutch historian this year published an article on a nineteenth-century American physicist; a few years ago, British geologists did a multivolume history of geomorphology, a field they defined as overwhelmingly derived from U.S. geologists.

In contrast, there is no shortage of writings and courses outside the United States on science policy or sociology of science, with strong emphasis on recent events in the American republic. Nobel Prizes are counted, statistics marshaled, and very current trends analyzed. In this dispensation, science in the United States usually starts with a contemporary occurrence—the Mansfield Amendment, Sputnik overhead, the atom bomb, or the arrival of refugees from Hitlerism. But research and development in the United States are important on a global scale. They could not spring up overnight but are the result of a complex and fascinating historical metamorphosis.

In 1776, the British North Americans south of former French

---

* Support of the U.S. National Endowment for the Humanities is gratefully acknowledged.

Canada had one great investigator, Benjamin Franklin, and a handful of others we can denote as scientists. By 1900, the new nation had a vigorous scientific community, with many of the basic structural and functional characteristics so evident today. These included strong commitments to fundamental research in various fields. In the present century, these characteristics have produced what is now so visible. This is a meaningful historical movement, not simply a banal success story, and it deserves critical scrutiny by scholars. Rather than recite great names, notable contributions, or the genealogies of institutions, let us consider the sciences in America as a set of historiographic problems particularly worth studying in comparative prospect.

AMERICANS AND EUROPEANS ARE STILL influenced by de Tocqueville's *Democracy in America* (1835). The work is best described as a hortatory essay to the French; it is an account of a mythical country perversely given the name of a real republic on the continent of North America. Noting the absence of the usual landmarks of European history—kings, churches, nobles, castles, dynastic squabbles, and sectarian wars—de Tocqueville concluded that the mechanism of American history was different, a vast arena for the smooth workings of impersonal, dynamic laws. He approved this lack of romance, drama, and great trends as a way to avoid the disruption of revolutions.

This view reoccurs in the influential writings of Henry Adams, the descendant of presidents, who cast a jaundiced eye on both his society and the consequences of science. The fact that Adams thought America and China were alike in their lack of drama says much about his insight into both histories. Like the immigrant myth of gold in American streets to explain prosperity, this Newtonian line of reasoning aborted analysis; change simply occurred, avoiding any need for study of the prior actions of people and institutions.

De Tocqueville noted the presence of energetic go-getters intent on acquiring wealth; he met no savants or devotees of abstract thought who required a niche immune from daily pressures in a hierarchical, aristocratic society. He saw a democratic society that was technologically oriented and indifferent to basic research. In fact, though, there were scientists in antebellum America. Someone like Nathaniel Bowditch, the translator of Laplace, or Joseph Henry at Princeton might have dented de Tocqueville's certitude. Or going forward in time to a school chum of Henry Adams, we have Charles Sanders Peirce, the mathematical physicist turned idealist philosopher. He was every bit a grand savant, and there were others.

De Tocqueville did not say there would be no science nor great pure scientists in the United States. He thought that applied science would yield bits of basic science—that is, he was of an older view that did not see basic science as logically and chronologically coming before technology. And by chance a great pure scientist might arise in the American republic. De Tocqueville believed that pure science ideally required the individualism and the drama of an aristocratic order. Although he was wrong in predicting a steady state country (as was Henry Adams in seeing the heat death of entropy), de Tocqueville had grasped an important truth. The western European ideal of a value-free science, independent of society, would take different forms in the United States. One can argue the extent to which this ideal was realized in Europe; the structured society of its nations, at the very least, permitted the illusion of realization. There great men, ideas, and events existed, at least figuratively, in isolation. In the United States, their counterparts were often bathed, if not immersed, in social, political, economic, and ideological realities.

To this day, many American scientists incline to the view that their society is indifferent, if not hostile, to basic science, despite ample evidence of the growth over time of support for research, even research of the most abstruse nature. A proper perspective on this peculiar attitude is a comparison with literature, music, and art. Once the arts also were presumed blighted by the American environment. A democratic society could support popular or vernacular cultures but not high cultures. No one makes this accusation any more about the arts, which now flourish in symbiosis with the popular culture. Only in science is there uneasiness and sensitivity, obviously a phenomenon of great importance.

Early in the 1850s, the aged Alexander von Humboldt wrote approvingly of American efforts in a wide range of physical and biological topics related to geography. Because he wanted Americans to receive greater international visibility, Joseph Henry of the Smithsonian Institution suggested preparing what eventually became the Royal Society's catalog of scientific literature of the last century. Henry originally limited the work to the exact sciences, so he clearly had more than scientific geography in mind. Although the first series of this catalog, up to 1863, disclosed that roughly 5 percent of the authors were from the United States, Europeans were not impressed.

By the end of the century, American activity in the earth sciences loomed large. Though an extension of Humboldt's view, geology reeked of the romance of the Old West, so attractive to generations of man-boys on both sides of the Atlantic. It did not hurt that the leader of the geological community in America was John Wesley Powell, whose hand was lost at Shiloh in the Civil War and who had first explored the untamed Colorado. Powell's other role as a pioneer ethnologist only heightened the effect. Although very creditable, the scientific exploration of the West was like the rumored gold in the streets: something waiting there. The effort did not impress those interested in mathematical and experimental fields. Only by 1951–52, did a French geologist, Emanuel de Margerie, produce his two volume *Etudes américaines*. Earlier, a few works noted that astronomy had become a very active discipline in the United States by 1900.

Shortly after that date, a significant act of recognition occurred. Reacting to the founding of the Carnegie Institution of Washington and the Rockefeller Institute, the Germans founded the Kaiser Wilhelm Gesellschaft. In the years before World War I, a number of German observers commented favorably on the quality and scale of the sciences in the United States, some warning against the threat to German hegemony in research. Apparently, the familiar competition of Britain and France was less worrisome.

One of the Carnegie beneficiaries, astrophysicist George Ellery Hale, added Rockefeller funds and built the great 200-inch telescope at Mount Palomar. That great instrument and E. O. Lawrence's pre-World War II cyclotron symbolized science in the United States to many on both sides of the Atlantic. Such large-scale efforts reflected the nation's natural endowment; great instrumentation reflected the supposed American propensity for "practical" devices rather than the abstruse theorizing of true savants. The attributes of Hale and Lawrence reinforced these stereotypes. Hale was a fabulous promoter and organizer, a veritable Morgan or Rockefeller of science; Lawrence, also a manipulator, went on to invent a television system.

Usually overlooked were scientists of a different mold. A contemporary of Hale's, the great geneticist Thomas Hunt Morgan, used the fruit fly in a series of austere, rigorous experiments. With a passion for scientific exactitude, Morgan shunned empire building and applications. When Hale and Robert A. Millikan invited Morgan to the California Institute of Technology, they initially envisioned a program of science applied to medicine, but Morgan insisted on pure research. Although the United States was a world power in the sciences when Morgan went to Pasadena, a generation passed before sustained, serious historical research began, and it did so then largely because of World War II and its consequences.

THIS 200-YEAR HISTORY IS NOT explicable by any assumptions of American singularity. Like the nation, the sciences in the United States are part of western civilization; the thirteen rebellious colonies were a provincial outpost of this civilization. It is not wholly accidental that the first college in this country was founded in 1636 in Massachusetts Bay Colony in a village later named Cambridge. Modern science originated in western Europe and developed an eastern wing in Russia and a western one in the United States of America. Despite the obvious influence of British (and French) science, the most interesting comparisons and relations of the research scene in these former colonies of the United Kingdom are with Germany and the Soviet Union, a point to which I will return.

Because of its origins as a provincial outpost of western Europe, the United States was never properly described as a young country, that is, a country lacking the kind of history associated with the cultures of the European nation states. More precisely, the United States had never experienced a youth consisting of dynastic and sectarian violence, a rigid class structure, and an all-powerful established church. This country was born to adulthood in the sense of having many of the trappings of modernity. To conservatives, the lack of the usual traditions and their detritus was a basic flaw; to liberals and radicals their absence was a glorious opportunity. With some exceptions, both groups were skeptical about developing a high culture of science in a democratic environment.

In 1776, the United States was not a blank slate. Selectively, patterns of life and thought in seventeenth and eighteenth-century Europe flourished and persisted; in time they became Americanized, diverging from a western Europe undergoing its own reactions to intellectual and economic changes. At the same time, a kind of feedback from these European developments prevented any great cultural split between the populations on both sides of the Atlantic.

In the United States, there was an avid pursuit of the European printed word. Even modest cities of the Mississippi valley in the antebellum era had a good number of books and journals. Young scientists visited Europe, and many went on to study at overseas universities. In turn, a Louis Agassiz came in 1846, as did many others in the last century. The German emigrés of the 1930s were simply an instance in a two-century-old process. In 1973, about 6,000 scientists and engineers emigrated to the United States. At all times in our history, many of our scientists have been foreign born.

The colonists who revolted insisted they were only asking for their just rights as Britons, even though most in the United Kingdom probably did not see those rights in the same way. Facing western Europe, American investigators regarded science as part of their heritage; ruthlessly they brought over the ocean whatever they wanted of this scientific heritage, disregarding any plaints about the European nature of the best of science. Being provincials in the eyes of western Europe, the colonists often were infuriated by patronizing treatment. In fact, American scientists were still vexed by patronizing Europeans as late as the interwar years. Yet, from the very beginning, American investigators wanted nothing as much as the good opinions of their peers in Europe. At the same time, many strongly desired to present their results as a symbolic return to Europe for the great gift of science.

Because of this ambiguous relationship with European science, the American scientific community developed a different pattern of nationalism and internationalism than that of its overseas counterparts. Disregard the fatuous cliché that "the sciences were never at war." It is a meaningless truism applicable to sets of concepts and data, not communities of humans. Real, live scientists usually rise and salute when the flag is raised, especially during wartime. Even in peace, national differences and rivalries influence the behavior of scientists. Being outside the rivalry between the French and Germans after Sedan and not sharing the British anguish that Perkin discovered the dye but Germany garnered the industry, American scientists could work very comfortably in an international setting. At the same time, national interests were never totally absent.

ANOTHER DIFFERING PATTERN ARISING FROM colonial roots is the development of a cult of the people, substituting for king, nobility, and gentry. In a post-World War II movie about a hack senator trying for the presidency, William Powell in the title role makes a campaign promise to give a Harvard degree to every American at birth. Mass culture, like dreams, sometimes discloses suppressed beliefs. Just as scientists (and conservative critics) were uneasy about the lack of a proper hierarchy and widely agreed-upon norms, the American public has had qualms about status based on education. Contrary to popular belief, the nation inherited a class system from Europe and still displays wealth-based social stratification. Contrary to de Tocqueville, quite a number of families retained high status over many generations, in some instances even constituting a patriciate. This is true despite a significant degree of upward mobility in the past for all groups, with the notable exception of those of African descent. Yet classlessness is widely assumed in the form of crowding to a middle position.

Education can provide a status to counter classlessness, and traditionally expertise is highly valued here. At the same time, one of the glories of American history is that the learned have much power and honor but lack the certainty of an assured niche. When historians turned to the history of the sciences in America, they noted that scientists wanted an elite, meritocratic basis for their group and often rejected the validity of public intrusion into their affairs. Filled with such historiographic problems, I spent a year in Britain in 1964–65 studying British counterparts to my favorite nineteenth-century American scientists. An acquaintance in the civil service, now deceased, invited me to lunch at the Athenaeum Club. As I looked around, he said, "This is like the Cosmos Club in Washington." But I knew differently. Just a few days before, I had read in manuscript a letter from an officer of the Royal Society (RS) to an RS Council member, then little known. Charles Darwin was invited to the Athenaeum for lunch to discuss a matter before the Royal Society. From my readings of many other manuscripts, I had already concluded that in comparison to British scientists, my American "elitists" were raving egalitarians. Even if they had not been infected with republican or democratic principles, control by one establishment was not feasible, given the size and complexity of the United States.

Just as wealth conferred status over generations, so did education and intellectual attainments. Family groupings of scientists appeared in the last century; for example, modern geodesy was brought to the United States by the Swiss-American, Ferdinand Rudolph Hassler. A generation later, the Nova-Scotia-born mathematical astronomer Simon Newcomb married a Hassler. Their daughter became the wife of W. J. McGee, the geologist and anthropologist who was one of John Wesley Powell's lieutenants. Charles Sanders Peirce's father and brother were professors of mathematics at Harvard. J. Willard Gibbs's father (of the same name) was a professor of languages at Yale. William James's father was a Swedenborgian philosopher, undoubtedly a surprise to his teacher, the physicist Joseph Henry. A present foreign Royal Society fellow is a third-generation scientist.

These family groupings have never constituted an intellectual aristocracy because of the rapid growth of the scientific community over the last two centuries. The father of the mathematician Oswald Veblen studied physics at Johns Hopkins during its golden age; he later taught at a midwestern state university. Oswald's uncle Thorstein became a great social theorist; Oswald was a founder of the Institute for Advanced Study. They were a family of Norwegian immigrant intellectuals. A small town in South Dakota produced two notable physicists, E. O. Lawrence and Merle Tuve. The so-called "new immigration" from eastern and southern Europe by now has yielded scientists of Jewish, Italian, Slavic, and other ancestries. Despite the persistence and importance of a few families as sources for recruits, the scientific community has never been a closed group. The newest immigrants from Asia are entering the various disciplines; by now about 5 percent of U.S. recipients of science and engineering doctorates are of Asiatic origin. Moreover, fifteen years of consciousness raising may very well result in an inflow of blacks, women, Chicanos—even more American Indians.

In the United States, scientists are overwhelmingly a middle-class group. Despite the presence of individuals from the upper class and some from the lower, becoming a scientist until recently was a means of going from the lower middle to the upper middle class. At present, such social mobility seems on the wane; if so, it is all the more important to study that part of the middle class producing those in the professional classes who provide an essential element of stability in the United States. Relatively little is known about such families, in contrast to concern for the genteel culture of the patriciate or the pains and traumas of the poor and the outcast. Because scientists are less numerous than other professional groups, they are more accessible to research. Where family influences are lacking, we still do not know why a small minority of middle-class youngsters in the past opted for scientific careers—until very recently a choice of limited remuneration and power. Perhaps the youngsters acted because they, like most Americans, adhered to a cult of knowledge.

Here the comparisons between the United States, Germany, and the Soviet Union are important. For more than half of its history, the United States was unknowingly in a race with Germany. Early in the eighteenth century, nearly 150 years before the United Kingdom, the Prussian state made elementary schooling compulsory and universal. With the later expansion of German universities, the British were fated to be overtaken. The Germans trained large numbers for particular niches in their society, each niche with a specific place in an elaborate hierarchy. Members of the professoriate were an elite of elites, having gone beyond the doctorate in their habilitation, a point the Americans did not quite grasp when graduate schools arose here under the influence of

German higher education. What American scientists did note was the high status of the German professoriate. In Prussia, each chair holder was a higher civil servant. American scientists often hungered for the same kind of assured high place in their nation.

Throughout the nineteenth century, Americans expanded all levels of their educational system; in retrospect, they were trying to match Germany. The statistics are not very clear, but apparently around 1900, the time Germans became conscious of an American threat, the United States had surpassed Germany, at least in numbers. Quality was another question. American universities were viewed largely as an extension of the democratic thrust to provide a responsible electorate. Some institutions viewed themselves in a patrician sense, as training cultured gentlemen for service to the society. Typically, even the mass schools regarded themselves as engaged in both cultural uplift and a service role. The sciences served both purposes. Mass plus quality tended to become the ideal.

Some examples come to mind. First, in the middle of the large state University of Illinois at Urbana, before a library with nearly four million books, is a plot of corn planted to remind all of the university's origin as a land grant school for agriculture and the mechanical arts.

Second, when the University of Chicago opened its doors in 1891, it was already a great university; this oft-repeated judgment refers to its scholarly research strength. But its president, W. R. Harper, at the same time launched an extension program of correspondence and evening courses. The mathematics professor at Chicago, Eliakim H. Moore, was the first pure mathematician elected to the National Academy of Sciences, and he received an honorary doctorate from Göttingen. However, he conceived his authority to extend from research in pure mathematics to supervising the training of elementary school teachers of arithmetic. Moore, Harper, and the University of Illinois—like many other American individuals and institutions—wanted theory and practice, culture and service, the elite and the masses somehow combined or in juxtaposition.

The Soviet Union and the United States are probably the only remaining countries still believing in some form of the idea of progress, a Marxist version in one, an Enlightenment variant in the other. The two countries have curiously parallel histories in the sciences. Both were Humboldtian because of the need to manage their large land masses, and both turned engineering into mass professions of high status. The Soviets have won the numbers race in that category, a more serious reality than mere heft of armed might. The United States has lagged behind Russia in pure mathematics and may still. Russian universities play a much different role in the sciences. Centralization is great in the Soviet Union, whereas egalitarian instincts limit centralization in the United States. Our society has a primal urge to replicate anything deemed desirable. In addition, strong anachronistic sentiment for the rustic ultimately works against the big. By comparison, being rustic is something the Soviets shun.

BESIDES THE IDEA OF PROGRESS, THE CULT of knowledge in the United States has other traces of the seventeenth and eighteenth centuries now largely absent from western Europe. By establishing much of the broad form and content of what is now designated science, the Scientific Revolution in Europe defined some concerns as nonknowledge and started segregating the high culture of the sciences from the vernacular culture. But the useful arts partially escaped this exclusion until the nineteenth century. At the same time, the casual amateur and the modest practitioner were increasingly displaced by the savant. (Significant scientific amateurism probably survived longer in the United Kingdom and the United States than on the Continent.) The Scientific Revolution gave form to a previously rather loose, amorphous conglomeration, and the Industrial Revolution hardened the new pattern. Much of this process had bypassed the British North American colonies. There the vernacular, the nonscience, the amateur, and the applied survived and flourished after independence, when the high culture of science began a slow, steady growth.

Europeans established a scheme of things in which there was a congruence between social, intellectual, and institutional hierarchies—not a perfect congruence, but enough to avoid many of the problems Americans had. On the western side of the Atlantic, in comparison, the heritage of the past and the thrust of historical development did not neatly separate grand savant and practitioner, theoretician and earnest mechanic, the abstruse and the vernacular. They were scrambled together. Thus, a persisting tension developed between the scientific elite, with its perception of fundamental research, and the mass community, with its thrust for diffusion of knowledge (often older and sometimes applied). As a result, the cult of knowledge in the United States encompassed a research ideal but not a basic research ideal, despite all the exertions of many generations of scientists.

In this research ideal was a blurred boundary between theory and practice. Originally, religion provided a basis for pure research. The idea of design presupposing the existence of a Designer readily yielded a drive for research untrammeled by sordid motives. Secu-

Reingold

larization purged the theological mission, leaving a pure research ideal. At the same time, the old, general notion of the utility of knowledge was elevated by the association with a designing deity. All research carried out His purpose and disclosed His will. Only overly rationalistic, sensitive scientists worried about the problems of defining discrete forms of research styles and goals.

Although this approach was a happy compromise, even before the Civil War American scientists launched a counterstrategy to split off basic research by establishing enclaves largely immune to the national tendency to blend theory and practice. Higher education often served as the site of these enclaves. A notable recent example is the Institute for Advanced Study. As it were, the scientists levied an overhead charge for theoretical science against the gross national product. Many were uneasy about the more common mode of supporting theoretical science, which also appeared before the Civil War. This mode was to define applied missions so as to include necessarily a basic research component. In the 1920s, Harvard's E. H. Hall (of the Hall Effect) corresponded with C. J. Davisson of Bell Telephone Laboratories, a future Nobel Laureate. Impressed by Davisson's work, Hall sent the head of the laboratories, H. D. Arnold, a letter praising the company's enlightened support of basic research. Arnold replied that nothing of the sort was intended, that Bell was interested in ways to increase the supply of electrons.

Sweeping mandates like this are widely present in so-called mission-related research. Believing sincerely in their approach, scientists still argue in favor of broad definitions of applied missions. But many also favor protected enclaves for basic research—enclaves not subject to the vagaries of economics, international affairs, public health policies, and the like. Yet, it is naive to hope for and to expect immunity from the ebb and flow of history.

Despite $32 billion for research and development in 1974, despite more than a half million persons engaged in research and development, despite $4 billion for basic research, U.S. scientists are still uneasy about their place in this society. There is an overwhelming cult of the nation, with such dominating symbols as George Washington, the flag, the national anthem, Abraham Lincoln, the Statue of Liberty. Science is not represented among those symbols; yet even technology once had the iron horse and Thomas Edison. And our country has just gone through ten terrible years raising nagging doubts about progress. Largely because of the anti-science quirks of Lyndon Johnson and Richard Nixon, inflation has eroded the real strength of the national research budget.

From my ivory tower vantage, I find prospects encouraging, starting with former President Ford's increase in the budget to make up some of the slippage from the Nixon and Johnson years. Moreover, American scientists have always functioned well while complaining loudly. Even before the Civil War, they demonstrated great talents as entrepreneurs, improvisers, and artful dodgers. Nor am I nonplussed by the predicted numbers of surplus Ph.D.'s. Except for about fifteen post-World War II years and perhaps a few years before the Great Depression, there has always been a surplus of scientists on paper. A surplus can act as a stimulus, both for the nation and for individuals.

Whatever happens, the course of the sciences in the United States will remain an arena of tensions and ambiguities, a historical movement in which prosaic reality turns out to be a symbol, often exhilarating, sometimes grotesque, occasionally a black comedy. Ours has been a history of trying to have your cake and eat it, but perhaps that is true of all of the past 200 years of our national life.

PARADIGM LOST

WILLIAM H. GOETZMANN
University of Texas

IN 1962, PROF. THOMAS KUHN PUBLISHED HIS IMPORTANT WORK, *The Structure of Scientific Revolutions*. His concepts of "paradigm," "normal science," "scientific community," "anomaly," and even "revolution" generated significant controversy and a large body of literature. In 1970, Professor Kuhn revised his book in response to legitimate criticism. His most significant addendum was an attempt to clarify his use of the term "paradigm," which he said should be taken in two senses:

> On the one hand it stands for the entire constellation of beliefs, values, techniques, and so on shared by members of a given community. On the other, it denotes one sort of element in that constellation, the concrete puzzle-solutions which, employed as models or examples, can replace explicit rules as a basis for the solution of the remaining puzzles of normal science.[1]

It is not the purpose of this paper to add to the list of critical literature and medieval pinhead dancing related to Kuhn's hypothesis. Rather, I am inclined to accept his two-level scenario, but at the same time to suggest that perhaps because his definition of "scientific community" was too precise—too restricted and limited to the so-called "internal history of science"—a whole paradigm shift in western civilization itself has been lost. A very large community of scientific investigators has been relatively ignored by historians of science. A world event comparable in importance to the Scientific Revolution remains indistinct in the tapestry of history. As a result, we have come up short in our knowledge of an important element in the nature of scientific change.

The "lost paradigm," I would assert, is a Second Great Age of Discovery, founded on the world-spanning work of at least three centuries of scientific exploration. This Second Great Age differed in one marked respect from the age of Columbus. It had the benefit of modern science, with all its attendant points of view and techniques. In this second age, knowledge for its own sake, rather than in reference to some medieval typology or classical formulation, began to be important, as signified by the founding of societies and institutions dedicated to knowledge. People began to develop careers focused on organizing knowledge about nature. Then, in the mid-seventeenth century, a "scientific revolution" took place.

Much of the nature of this revolution is already known, especially in the area of Newtonian physics. Less is known about the developments in natural history, and virtually no connection has been made between the emergence of these sciences and the role of the explorer, who formed an important part of a vastly larger scientific community than has previously been delineated. The great number of explorers of a serious scientific bent during the period from 1600 to 1900 throws into question whether or not, as is often asserted, 90 percent of the scientists who ever lived are alive today.

Clearly, I have chosen a large topic—one much too large for a brief paper, and one that wanders far afield from the main subject of our session, "two hundred years of American science." My purpose, however, is to suggest, to sketch out a horizon or field for historical research, some of which I hope to undertake myself in the future. I am also concerned with the world matrix of American science (if a national science can be said to exist), and I am further intrigued by how a focus on the explorer as scientist affects or possibly can make use of the concept of paradigm. Thus, in this paper I will range widely, as befits a historian of exploration; however, by the same token, I hope to add a new dimension or perhaps give a different shape to the contours of scientific time during the period when America began to loom as a field for the study of nature.

CLEARLY, A SECOND GREAT AGE OF DISCOVERY did not spring immediately into being. The striking exhibits in Hugh Honour's *The European Vision of America*, on display at the National Gallery of Art from late 1975 to early 1976, indicate this. The flora and fauna of the New World fascinated Europeans. They fantasized about strange animals, and they were endlessly curious about the Indians. Cartographers, printmakers, draftsmen, artisans, and artists continually represented the emerging New World. Some of them, such as John White and Thomas Heriot of England and Albert Eckhout, Frans Post, Willem Piso, and Georg Maregrave of Holland, attempted exact renditions or descriptions of New World phenomena. The Dutchmen, in their *Historiae Rerum Naturalism Brasiliac*, edited by Johannes de Laet and published in 1648, provided the first printed systematic account of the natural history of any part of the Americas.

Gradually, more books and pictures of a systematic, descriptive nature were produced from New World experience. By the late seventeenth century, the *Wunderkammer* or collection of natural curiosities began to make its appearance among the fashionable of Europe, as did the private garden of exotic plants. But as yet, little in the way of organized or systematic scientific exploration had taken place. In part, this was true because there was no adequate system of natural history classification. Exotic plants were fitted into Renaissance or classical herbals or left out when they did not fit. Animals were hung on the great chain of being, and arguments persisted well into the eighteenth century as to whether New World animals, and people for that matter, were physically inferior to European specimens. But gradually the quest for organized knowledge about the exotic parts of the globe began to take on a more systematic form, thanks to the work of John Ray, Karolus Linnaeus, and others.

Due to the impact of exploration, however, a central focus of scientific activity became that of terrestrial space, including measuring the earth, charting the oceans and continental interiors with precision, and locating patterns of terrestrial distribution for rocks, minerals, fossils, plants, animals, and men. Because of the seeming perfection of the Newtonian mechanistic model and the great chain of being, because of the emergence of seemingly rational systems for biological classification and the assumption that the earth was somehow a finite system, and because of a belief in the mechanistic psychology of John Locke (which seemed to confirm the tenets of Francis Bacon), much of the Second Great Age of Discovery seemed to be taken up with system building, in a drive for completeness in human knowledge. Alexander von Humboldt's statement in this regard will forever remain striking. He wrote from the heart of the Andes:

I have conceived the mad notion of representing . . . the whole
of the physical aspect of the universe in one work, which is to
include all that is at present known . . . from the nature of the
nebula down to the geography of the masses clinging to a granite
rock. . . .[2]

The result of this "mad notion" was, of course, Humboldt's attempted systematic condensation of all knowledge in a work modestly entitled *Kosmos* which, as might be expected, was left unfinished at his death. Humboldt added something to Baconian or Lockean empiricism, however. By the early nineteenth century, he had absorbed Kantian and post-Kantian German philosophy to the extent that nature for him involved two systems: the telluric or empirical, and the phenomenological or causal connections constructed in the mind. He could organize—or so he believed—the enormously variegated profusion of nature's data according to Kantian categories without losing the sense of nature's rich heterogeneity. Humboldt was a romantic scientist.

By and large, however, the drive for system building so characteristic of the Second Great Age of Discovery was not the result of a confidence in civilized man's ability to control nature, or even to control his knowledge of nature. That drive was the result of a continuing series of crises stemming from an information explosion. The emergence of advanced or complex civilizations in western Europe, and to a lesser extent in North America, meant that due to increased and new forms of communication, these societies were absorbing information about the world at a tremendous rate.

The fact that both the eighteenth and nineteenth centuries were ages of imperialism in which territories, dominance of continents, and control of trade and the oceans were of paramount importance meant that governments, full of national pride, and merchant-adventurers, seeking always to expand their opportunities, were prime movers in the global information-gathering process. In 1788, for example, a group of British lords and wealthy merchants turned their Saturday eating and drinking club into the African Association dedicated to "Promoting the Discovery of the Inland Parts of that Quarter of the world." They declared that "no species of information is more ardently desired, or more generally useful than that which improves the science of Geography."[3]

Such instances might be multiplied many times in the western world during the Second Great Age of Discovery. Thus, contrary to Professor Kuhn's rather restricted hypothesis, the community of information seekers was not limited to scientific specialists. Rather, entire nations and whole societies, merchants, dilettantes, and even state legislatures launched exploring expeditions to discover nature's wonders. But the institutions for processing the resulting avalanche of new data, as well as the philosophical structures for ordering that data, were invariably inadequate to the task. No system provided in advance for the unexpected on a vast scale. So the world itself became a large "anomaly," with the result that knowledge gathering did not undergo one dramatic scientific revolution but a sense of continual evolving crises. This result occurred because of an information overload generated by exploring expeditions as well as stay-at-home scientific investigators.

Some investigators, notably Oken and Schelling in Germany, Von Baer in Russia, and Emerson and Thoreau in America, gave up coping with the data problem on a systems level and settled for synecdoche—i.e., any part of nature could stand for the whole. This was an implosion reaction that proved to be central to the Romantic movement and eventually to the self-oriented perceptions of modernism. But for most thinkers in western Europe and America, the Second Great Age of Discovery was a crises-laden world event in which they found themselves, and they faced it with zest and the excitement of possibility. For them, the explorer was the hero. He also should be thought of as integral to any history of science. The makers of modern geography, for example, were not only the Ritters and the Herders, they were the Humboldts, the Cooks, the Condamines, the Mungo Parks, the Lewis and Clarks, the Fremonts and the Wilkes, who spread out over ocean and continent discovering, mapping, describing, and characterizing the unexpected that they set out to find.

To this long list of explorers must of course be added the self-declared scientists they took with them—first generalists or natural philosophers, then more and more specialists, as information piled up in all directions. We must also add the artists and illustrators whose visual representations were essential tools for scientific understanding and classification and who also generated whole new ways of looking at nature that conditioned scientific thinking. Finally, we must add the map makers, who found ingenious ways to digest and incorporate vast amounts of data in a convenient, portable, and integrated way. In large measure, it was the map as a system that made possible the development of geography, which became the key science during the Second Great Age of Discovery.

I hope that at this point the immensely varied pattern or paradigm of the Second Great Age, which spans most of American history, has begun to be apparent, and that the simplistic image of a carefully circumscribed and self-conscious "scientific community" has been broadened to all its complex dimensions. Perhaps it is not necessary in this paper to do more than recall some of the names of explorers in this age who were bent on scientific discovery, although they also may have had other missions relating to the economics of imperialism. Among the oceanic explorers, recall

Cook, Byron, Beecher, Bligh, Fitzpatrick, Vancouver, Ross, Bering, Chirikov, Bellinghausen, Kotzebue, Krusenstern, d'Urville, La Perouse, Bougainville, Darwin, Palmer, Fanning, Wilkes, Rodgers and Ringgold, Peary, Kane, and the whole succession of men who followed in the wake of Sir John Franklin searching for a Northwest Passage. This oceanic activity continued well through the nineteenth century, resulting not only in the discovery of continents, the charting of coasts and islands, the mapping of the oceans and to some extent the ocean floor, but also the discovery of primitive peoples, exotic animals and plants, and a rich sense of the variety of nature and its growth at the ends of the earth.

In the meantime, scientific explorers investigated the interiors of the continents. Mackenzie, Hearne, Thompson, Palliser, and Peter Pond helped to open up Canada to modern scientific civilization. Lewis and Clark, Pike, Long, Fremont, Emory, Ives, Whipple, Stevens, Gunnison, Sitgreaves, Simpson, Macomb, Powell, King, Hayden, Wheeler, Dutton, and a whole coterie of scientific followers brought science to western America, but they were following in the tradition of the Bartrams, Clayton, Audubon, Rafinesque, Mark Catesby, and many others of the eighteenth century.

If one thinks of Africa, the names of Norden, Bruce, Park, Lander, Barth, Champollion, Burton, Speke, Grant, Lugard Baker, Livingston, and Stanley spring immediately to mind. Also in the celebrated "search for the Nile," one sees a whole cross-section of interests and motives: Burton with his interest in ethnology, Speke the geographer and topographer, Baker the curious adventurer, Livingston the man of religion, and Stanley the newspaperman and adventurer who suddenly found geography fascinating.

One could go on in this vein, continent by continent, and call the roll of Russian and English and German explorers in central Asia; the men who opened up South America, following in the wake of the mighty Humboldt and La Condamine; those intrepid but inevitably disappointed souls who crossed over into the interior of Australia; the marchers in the Gobi Desert and the mappers of the Himalayas; the Amundsens and Shackletons and Pearys on the frozen polar continents. The list is long and the harvest for science so rich as to be virtually incalculable.

In addition, such repeated exploring activity had marked impact on the history of scientific institutions. Prof. William Stanton has shown recently that the Great United States Exploring Expedition, with its tremendous accumulation of charts and collections, was crucial in the formation of the Naval Observatory and the founding of the Smithsonian Institution.[4] This expedition, plus the Pacific

Railroad Surveys of 1853, the U.S. Mexican Boundary Survey, the North Pacific Exploring Expedition, and countless other lesser efforts, forced the specialization of science in America where it had not existed before and also made it possible for science to become a paying profession. At the same time, military commanders of U.S. exploring expeditions from Wilkes to Emory to Dutton considered themselves members of the scientific community and were founders of such ventures as the Washington Academy of Sciences, the National Botanical Garden, the Smithsonian, the National Academy of Science, the Cosmos Club, and the National Geographic Society.

Similar institutional patterns emerge in the major countries of Europe during the same period, for the world had become an enormous museum to the cognoscenti of both western Europe and America. In countless instances, too, scientific institutions linked with exploration received acclaim and support from a public dazzled by natural wonders brought from the ends of the earth. Every man became his own tastemaker as South Sea wallpaper, complemented by chinoiserie bric-a-bric, Saracenic furniture, and parlor table bird books, seemed perfectly appropriate in Moorish, Gothic, Egyptian, Palladian, or Tuscan villas. For many, it became the height of fashion to visit George Catlin's Indian show and then ape the noble savage in all manner of life styles. Thus, the "lost horizon" of the Second Great Age of Discovery emerges as part of the Romantic movement.

THE INNER DYNAMISM OF THE PARADIGM—the evolution of the series of "concrete puzzle-solutions" referred to by Professor Kuhn—is equally interesting. Essentially, the Second Great Age of Discovery began with a focus on space—global dimensions and the distribution and location of things—and eventually discovered time, only to lose it again at the end of the period in a maze of competing times and a bewildering sense of simultaneity that intellectual historians call modernism. Darwin's very first words in *The Origin of Species* demonstrate the shift:

When on board H.M.S. "Beagle," as naturalist, I was much struck with certain facts in the distribution of the inhabitants of South America, and in geological relations of the present to the past inhabitants of that continent. These facts seemed to me to throw some light on the origin of species. . . . (Introduction, p. 1)

The inner paradigm development begins with epistemology. The inductive method ascribed to Francis Bacon and carried forward by John Locke formed the basis of science and one reason for exploration in the Second Great Age of Discovery. Just as truly as the heart was a pump for Harvey, man was a data gatherer for Bacon and an information-processing machine for Locke. Men went out into nature, observed it at first hand, took impressions into the brain which, due to its powers of reason, classified nature's data into appropriate categories from the simple to the complex. The problem, as both Hume and Kant and later Humboldt were shortly to observe, was where did the appropriate categories come from and what relationship did they bear to the datum of nature?

At the very time that the philosophers were wrestling with this problem and Kant was fashioning his grand system, the scientific community, which I have already suggested was very broad and ill-defined, was wrestling with the problem of categories on a more down-to-earth, methodological level. The explorer-discoverer was going far afield, collecting vast quantities of new information in good Baconian or Lockean fashion. The scientist or natural philosopher back home in Europe and America was doing the same thing, whether in the Paris chalk basins, the Alpine meadows, the English countryside, or the back country of America.

These individuals were perforce concerned with the order and meaning of the new information. Newtonian physics had demonstrated that the world was orderly and worked by laws, but it said little about the tangled world of natural history. Rather, the traditional framework within which the naturalist worked was the great chain of being, a very old concept. The avalanche of new data severely challenged the utility of the chain of being, first infinitely multiplying its categories and causing, as Lovejoy has observed, shifts, changes, splitting, and movement until the categories were almost meaningless. Then the chain itself had to be expanded until it figuratively burst asunder with the plentitude of new data.

Naturalists like John Ray, who inspired the African explorer Mungo Park, and Karolus Linnaeus, who was in close touch with the eighteenth-century explorers of America (especially the Swede, Peter Kalm), struggled to create a new order for the natural world. The resulting system of class, family, genus, and species unleashed the scientific lumpers and splitters upon the world. Naturalist explorers did two important things. They searched assiduously for every new species in nature, whether it be desert cacti or lush tropical orchids from the Amazon. They also plotted the location of both new and known species on maps. Distribution over space thus became an important question for the naturalist.

As further aid in this quest, the age produced new technological developments. Improved scientific instruments enabled cartographers and chartmakers to make much more accurate maps, and the barometer that measured altitude enabled distribution to be calculated in three dimensions. Thus the map served as a basic organizing device for plotting distribution through global space and was fundamental to the development of geography as a systematized science. Moreover, the map allowed for correlation of plants, animals, climate, altitude, landforms, and ultimately geological formations. Thus, maps inevitably stressed interrelationships and the organic connection of everything in nature. Alexander von Humboldt stood forth as the great master and symbol of the organic-geographic approach to nature as he used all the tools of scientific geography to trace out and put down in his reports and maps the world of teeming varied life he saw on his South American expeditions.

The cartographic-organic image was reinforced by the work of the scientific artist—men such as Captain Cook's Sydney Parkinson and Georg Forster, Humboldt's Bonplan, Mark Catesby, John Bartram, and even Captain Bligh who, though a naval officer, was also an accomplished artist. These scientific illustrators not only stressed the brilliantly colored and exotic specimens of nature; they also portrayed ecological relationships. Very often these artists were well in advance of more specialized scientists in this area of endeavor. They depicted animals, plants, and people in their "natural surroundings" or "native habitats" in relationship to one another because the artists, like the mapmakers and geographers, were interested in the interrelationship of all terrestrial phenomena. Thus the artists and the mapmakers, in collaboration with the explorer-scientists, helped make the growing, proliferating, changing "tree of life" the central metaphor of the age. By the latter quarter of the eighteenth century, even Peter Pallas, scientific explorer of the remotest reaches of Russia, had pictured a tree as representing all the species of plants and animals. Thus, in following Lockean or Baconian empiricism to its logical extreme on all fronts, both naturalist and artist found themselves in the growing, changing, protean world of Romanticism.

THEY ALSO FOUND THEMSELVES IN A WORLD of time. Individual growth could well be cyclical, but distribution of species required time, not to mention causal factors. Professors Edward Lurie and A. Hunter Dupree have ably recounted just how this question came to a head in the Gray-Agassiz debates on Darwin in 1859. It is important for our purposes to note two things about the debates. Data on plant distribution, indicating identity of species between

East Asia and eastern North America, were collected by George Wright on the North Pacific Exploring Expedition, and Gray's triumph over Agassiz depended upon the change of species over time. James Dwight Dana, who served on the Wilkes expedition, also noted the factor of time in the formation of organically composed coral atolls. The study of distribution led to the study of migration and time and eventually, of course, *origin*. It is significant that Darwin and Wallace, both explorer-scientists interested in species distribution, eventually produced the great treatises on time and the evolution of species.

Time also figured importantly for students of the human species. Because of their increasing contact with new people everywhere, explorer-scientists in the eighteenth century reopened the field of social science, which for many years had been laid to rest by Aristotle. The two main questions of the age related to time. First, of course, was what might be called "the diffusion question." If all the lost tribes in forests, jungles, and on remote continents such as newly-discovered Australia, all the scattered evidences of high civilization in "new worlds" had been created simultaneously, then what became of the Adam and Eve story and the 4,000-year time line suggested by the Bible? How could Adam and Eve have been the "first parents" of mankind?

To be a Christian required that one address himself to this question with a view to discovering a history for each tribe and civilization that got them somehow from Eden to their own peculiar remote corner of the globe and accounted for the distinctive differences in their stages of growth. Americans of every persuasion had answers to these questions—from the amateur archeologist Josiah Priest, who traced new world inhabitants back to Romans, Hebrews, Phoenicians, and Greeks to Louis Agassiz, who espoused separate creationism. Joseph Smith of Palmyra, New York, derived a new religion out of the long prehistory of America. But then, his story was not told to him by mere mortals. He had magic spectacles, the Urim and Thummim, and he was way up there on the chain of being conversing with an angel named Moroni.

Differences in the peoples encountered brought up a second question for the social scientists of the age—the question of progress. Most Americans took their cue from an Englishman, Herbert Spencer. Trained as an observer of natural history, Spencer compounded his philosophy of progressive evolutionism out of Lamarck, the Russian *Naturphilosopher*, Von Baer, and Lord Kelvin's laws of thermodynamics. All life was progressing through time as Spencer put it in "a continuous change from incoherent homogeneity, illustrated by the lowly protozoa, to coherent heterogeneity manifested in man and the higher animals." In other words, progress was over a long time span. It was linear and nonreversible and showed increasing specialization of organisms as evidence of progress. Hence, the more specialized the societies and cultures encountered on the globe, the more progressively advanced and very probably the older they were.

Spencer's most important American disciple was Lewis Henry Morgan, whose *Ancient Society* traced the progress of mankind from savagery to barbarism to civilization in stages related to technology. Thus the natural or life sciences in the second age began with space and distribution and climaxed with the discovery of time and nice distinctions of progress. They made history possible.

Darwin, as was previously noted, provided one sort of climax to the age of space turned into time. Charles Lyell, with his *Principles of Geology* (3 vols. 1830–1833) embodying the concept of uniformitarianism, provided another. Uniformitarianism meant that one observed the processes at work on the earth at present and then projected them back into time in a reconstruction of the earth's history. Because observable earth processes, like erosion, worked very slowly, it became apparent that the history of the earth was much longer than anyone suspected.

Operating within the older Wernerian or Huttonian theories of geology, it was possible to cling still to the biblical age of the earth or between four thousand and six thousand years. After Lyell and uniformitarianism, this was not really feasible unless, like Clarence King, one was a catastrophist. As late as 1877 King, like Louis Agassiz, still believed that sudden catastrophic changes had shaped the earth. To him, sudden ice ages; ages of vulcanism, fracture, and fire; the melting of great glaciers; enormous floods and crashing icebergs; and overnight climatic changes could account for the sudden shaping of the earth and its creatures.

But even catastrophists were forced into discussing the earth in terms of its history. For geologists, most major questions had become historical. This view was reinforced by the discovery, made by William Smith in England and Thomas Say in America, that fossils denoted time. Each layer or horizon of the earth's history was marked by characteristic fossils. Thus paleontology, another branch of geology as history, came into being.

In America, James Hall of the New York State Natural History Survey published thirteen volumes on the *Paleontology of New York State* in which, with the aid of Lardner Vanuxem, he fashioned a stratigraphic column for North America that was independent from that of Europe. Hall's proteges carried his column into the American West, which became a great world laboratory for geology. In 1857, John Strong Newberry, while with the Ives Expedition, descended to the floor of the Grand Canyon and identified another stratigraphic column, which he verified later with the Macomb Expedition to the junction of the Green and Grand Rivers. Ferdinand V. Hayden and Fielding B. Meek, working in the Dakota Badlands, traced out with very fine distinction the several layers of the cretaceous horizon. Hall, Marcou, and W.P. Blake somewhat presumptuously drew geological maps of the whole American West, based upon little or no experience with any exploring expeditions. By midcentury, European geologists were doing much the same thing, with the work of Sir Roderick Murchison in the Ural Mountains being the most famous. The dominant paradigm in geology was historical, even in America, despite persistent demands for practical discoveries of mineral deposits by citizens at the state and local levels.

Geologists began not only the making of "time lines" or stratigraphic columns, but they also traced ancient horizons over spatial areas in the new geologic maps they were developing. Thus, time or history could be spatially represented and related to other data collected in other branches, such as those of terrestrial magnetism and the life sciences. Moreover, in finding rich, fossil-laden horizons in the ancient past, geologists contributed to the organic image of the earth that so dominated the Romantic period. Space had turned into time that was linear, seemingly nonreversible, and gradual. Even Kelvin's experiments with the cooling of the earth and his work on the laws of thermodynamics eventually seemed to confirm this theory. Thus not in one striking breakthrough or sudden revolution, but gradually in confrontation with the new all across a whole range of scientific activities (often characteristically associated with exploration and the Baconian or Lockean epistemology), one scientific paradigm dissolved and turned into another.

As this happened, a new cultural paradigm also emerged in the same way. There were indeed instances of striking anomalies encountered along the way: Humboldt's discovery of marine fossils high in the Andes, Smith and Say's discovery that fossils could be used to date strata, Cuvier's discovery of rich fossil deposits in the Paris Basin, Wright's notice of plant species correspondences between eastern Asia and eastern North America, Agassiz's discovery of an Ice Age, and of course Darwin's observations in the Galapagos.

Yet, to me, no one of these developments was sufficient to cause a great leap from the spatially oriented static system to the time-oriented organic and evolutionary paradigm. In fact, of course, during much of the Second Great Age of Discovery both spatial and temporal systems existed at the same time and still do to the present day, depending upon the problem or circumstance. Nonetheless, it is clear that the basic model that lay at the heart of the scientific paradigm did change during the period at hand from a closed, balanced, Newtonian system related to expected finite space to a time-oriented, linear, asymmetrical model that first appeared to have teleological possibilities, then seemed open ended. In America, the adjustment was virtually complete, and two widely adopted social philosophies were based upon it—Social Darwinism (in its many varieties) and Pragmatism, the chance universe philosophy of open-ended limitless possibilities.

Towards the end of the nineteenth century, however, a number of other events took place that, in a sense, might be considered evidence of the end of the Second Great Age of Discovery and the beginning of a third age. This change appears to be bound up in the effacement of linear time and the recognition of the possible finiteness of the earth and its environment. Recognition of this shift roughly corresponds to Turner's perception, in a far different context, of the end of the open, continuous frontier. Empty, unknown terrestrial space was, of course, rapidly becoming a vanishing commodity, although the race for the poles was getting into high gear. Even more, the straightforward concept of time was disappearing before relativity of times, streams of consciousness, a pluralistic universe and hence multiple perspectives, cultural relativism in anthropology, and the resurgence of systems thinking and equilibrium in geology and ecology.

The new age of discovery would seem to be an age of *simultaneity*, where primitive cultures would be equated with highly developed technological cultures. Times would be relative and space conceptually altered, as in the Dymaxion map projections of Buckminster Fuller. Highly experimental science, rather than descriptive and inductive science, would come to the fore. Science as history would seem to be virtually dead.

The Third Great Age of Discovery, within which we are now living, appears once again to be an age of spatial relations wherein science seeks to orient itself within systems and structures, not the least of which is the universe. The new explorer looks for the relations of things and out of this generates new information, unless of course one cares to equate the discovery of a "moon rock" with the

Paradigm Lost

vast rich spaces of the New World. And yet, perhaps history as scientific paradigm is not dead. Is the moon rock the key to the moon's origin and hence to that of the earth and the solar system? Is it the key to a much grander historical process? Perhaps structures and systems are the inevitable reflex action to discovery as man tries to turn new data into meaningful information.

At any rate, much of American history and the history of American science is spanned by the *Second* Great Age of Discovery, wherein terrestrial exploration was a paramount activity—a heightened, more dramatic version of "normal science." This age formed a paradigm unto itself and yet included a significant change of subparadigms, from a focus on space and systems to a focus on evolution and time. Thanks to an age of discovery that will probably never end and perhaps is in continuous evolution, first western Europe, then America, now ever-increasing parts of the globe and civilization itself, become ever more efficient information-processing systems. The human story of science, exploration, and scientific discovery for much of American history can be found imbedded in that vast matrix that was the Second Great Age of Discovery.

1. Thomas S. Kuhn, *The Structure of Scientific Revolutions*, 2nd ed. enlarged (Chicago: University of Chicago Press, 1970), p. 175.

2. Victor W. Von Hagen, *South America Called Them* (New York and Boston: 1955), p. 145.

3. Quoted in Sanche de Gramont, *The Strong Brown God* (Boston: Houghton-Mifflin Co., 1976), p. 17.

4. William Stanton, *The Great United States Exploring Expedition* (Berkeley, Los Angeles, London: University of California Press, 1975), p. 357 ff.

AMERICANS ABROAD:
SCIENCE AND CULTURAL NATIONALISM
IN THE EARLY NINETEENTH CENTURY

BRUCE SINCLAIR
University of Toronto

EUROPE WAS A SUBJECT OF SPECIAL FASCINATION FOR THOSE AMERicans of the early nineteenth century who were concerned about their country's culture. Both as a matter of historical development and because the Old World was better provided with the means and setting, it seemed that creative achievements in the United States must be defined in terms of European standards. Yet political independence called for intellectual independence, and Americans searched for native foundations upon which to build a distinctive identity.

The cultural nationalism of the Jacksonian era presented particular problems for scientists. They shared in the general conviction that science and democracy had a special affinity, and furthermore that the country's needs might legitimately give a certain shape to its scientific inquiry.[1] But at the same time, the rhetoric of science described an international community of seekers after truth, undifferentiated by geography. Thus, unlike those who dreamed of an American style of art or literature, scientists were not usually interested in devising an alternate botany or physics. Scientists might have imagined that open access to knowledge would ultimately lift American science to preeminence, but they recognized Europe's hegemony in teaching, research, and publication. These tensions between nationalism and internationalism were especially acute for Americans abroad.

European travel was not novel to the Jacksonian period, of course. Since the eighteenth century, American students in significant numbers had gone to Edinburgh or to other centers of teaching and research in medicine and the sciences. Benjamin Silliman's tour in 1805–1806 was simply the first of many nineteenth-century visits taken by newly-appointed professors before they assumed their teaching duties.[2] But the 1830s—and notably the years between 1837 and 1839—were unusual, both for the number of travelers and the extent to which their visits reflected an emerging nationalism in American science.

A remarkable number of American scientists went abroad in the last years of the 1830s. The best known among them were Alexander Dallas Bache, Asa Gray, and Joseph Henry, but during the same period Elias Loomis, Robert Hare, John Locke, A. D. Stanley, Lewis Gibbes, and John Farrar were also in Europe, as was Lt. Charles Wilkes of the U.S. Navy and "Prof. Davies, late of the U.S. Military Academy."[3] For Bache, Gray, and Henry, particularly, the first trip abroad was a crucial watershed in their scientific careers. Furthermore, their experiences in Europe provided the basis upon which they defined the special needs of American science.

Two factors help account for the coincidental nature of these several pilgrimages. The first has to do with finance. Money was readily available in the flush times immediately before the Panic of 1837. Industry was expanding rapidly, commerce flourished, and widespread projects of internal improvement provided job opportunities. States like Michigan planned magnificent new universities, and the federal government contemplated a great exploring expedition to the South Seas. Large-scale, publicly supported surveys in geology, meteorology, and terrestrial magnetism reflected the levels of funding available to science, as well as a popular faith in the utility of its applications.

Asa Gray, recently appointed professor of botany at the University of Michigan, went abroad with the promise of $5,000 for the purchase of books and scientific apparatus. In somewhat the same spirit, Princeton's president, John McLean, was concerned that his college excel all others in science teaching; while on an inspection tour of neighboring colleges he wrote Joseph Henry, "If you are willing, arrangements must be made for you to proceed next spring for Europe, to spend there as much time as you desire, & procure whatever apparatus we need."[4] Bache, who had just been named president of the newly founded and richly endowed Girard College in Philadelphia, was supported with similar generosity for a two-year study of European educational institutions; he also had funds for the purchase of books and scientific equipment.[5]

These new levels of support for science coincided with the appearance of America's first generation of professional scientists, and they went to Europe with specific objectives. Henry, Bache, and

Gray were self-conscious about their own careers in science, and self-consciously they also assumed the burden of America's scientific reputation. Asa Gray's primary objective, as his biographer has pointed out, was "to wrest from Europe control of the source material on North American plants."[6] Joseph Henry's aims, defined in a letter to a friend, were to establish personal connections with European men of science, to study their teaching methods, and to learn their research techniques. He sought, as he put it, that "which cannot be learned in books," and he characteristically stated his purposes in terms of his own improvement and that of his country.[7]

All three of these men envisioned their experiences in Europe as a crucial part of the process by which science in America would be advanced. Contact would diminish that sense of remoteness from the world's creative centers of scientific investigation which Bache had felt so sharply upon learning of Gauss's research efforts. "How long shall we . . . remain ignorant," he had written to Henry, "of what is doing in one of the most active portions of the Continent?"[8] And all three were convinced, implicitly or explicitly, that one of the most important results of their visits should be that Americans receive proper recognition abroad for their scientific achievements.

Unlike many Americans who went overseas to learn something of European science and its practitioners, Gray, Henry, and Bache were unusually well-prepared to participate directly in scientific activities there. Circumspect at the beginning, they armed themselves with the customary letters of introduction. Gray carried so many, in fact, that an alarmed French border inspector seized them for investigation.[9] But in practice, these letters were little required beyond the elementary dictates of convention. Each of the scientists already had some reputation abroad. They also benefited directly from the contacts other Americans had made in previous visits, and their objectives were further advanced through the aid of William and Petty Vaughan—two London members of a family remarkable for years of quiet usefulness to the cause of American science.[10]

Joseph Henry, for instance, went abroad with letters from the highest of U.S. political figures, but he could also depend on a network of relationships bequeathed from John Torrey's 1833 tour of Europe, Robert Hare's attendance at the meetings of the British Association for the Advancement of Science in 1836, and upon his own reputation in science. For example, Henry reported to his wife that his letter of introduction proved unnecessary in his visit to Sir David Brewster because Brewster had just recently described Henry's Albany electrical experiments in an encyclopedia article, and "when my name was announced, he gave me a cordial recognition and devoted himself during my stay entirely to me."[11] In addition, more or less as they had done for Bache, Hare, Loomis, Locke, and others, the Vaughans provided Henry with introductions to people he wished to meet. They also located accommodations, helped with banking arrangements, collected and forwarded his mail, booked his ship passage to the continent, and then sat up half the night with him until its early morning departure.[12]

Bache and Henry were connected to British interests in terrestrial magnetism through James Renwick's links to Edward Sabine and by a correspondence Henry had begun in 1834 with James David Forbes of Edinburgh. Henry and Forbes had exchanged magnetic needles and after some observations of his own, Henry loaned Forbes's needle to Bache, who subsequently used it for a set of experiments in a vacuum apparatus he had designed.

In correspondence with Forbes, Henry pointed out the importance of terrestrial magnetism in the U.S. He also suggested both New York State's willingness to obtain apparatus from Europe and the likelihood that Henry and Bache would have charge of the New York survey. When Bache arrived in Britain, bringing with him his vacuum device for Forbes's use, he had the means of access to a British network that included Forbes, Humphrey Lloyd at Dublin, John Stevelly at Belfast, and Sabine, who was soon to command a worldwide magnetic survey.[13]

Bache was preceded by his publications, too. Henry had sent Forbes a copy of Bache's report on the Franklin Institute's boiler explosion investigation, as well as other of Bache's scientific papers. Bache was pleasantly surprised to find some of his work known in London's scientific circles, but the fact that he was Benjamin Franklin's great grandson gave him more immediate and favorable recognition in Europe than anything else.[14]

Asa Gray also entered European scientific circles via a network of relationships and a correspondence of his own, plus evidences of his own talents. Although he was heir to a long tradition of European and American contact in the exchange of plant specimens, through his mentor and colleague John Torrey, Gray also was directly connected to the most active group of British botanists—among them William Hooker, George Bentham, and John Lindley.

Torrey was the American editor of Lindley's *Introduction to the Natural System of Botany*, and when he went to Europe in the spring of 1833, he spent five weeks living with Hooker in Glasgow while the latter was engaged in writing his *Flora borealiamericana*. Gray had begun a correspondence of his own with Hooker in 1834,

Sinclair

and on his arrival in Great Britain he went first to Glasgow to work in Hooker's collections. In a manner that recapitulated Torrey's experience, Hooker later ensured close working relations for Gray with the London botanists, especially Lindley and Bentham, who in turn passed him on to their continental colleagues.[15]

These varied connections gave the Americans unusual opportunities to meet Britain's men and women of science. In a letter to Stephen Alexander, Joseph Henry claimed, "My reception thus far has been cordial and kind beyond my most sanguine expectations and every facility has been given me to prosecute the object of my tour." In a reception that went beyond politeness, he said, Wheatstone had spent a week with him and Bache. And Babbage had not only shown his calculating machine in lengthy detail but also had arranged a party so that Americans could meet the renowned astronomer, Mary Somerville. Bache, too, discovered his European counterparts "much more accessible than I had supposed they would be," and Gray found himself absorbed in the Hooker household as a family member.[16]

The lack of reserve the Americans met meant, among other things, that they could learn which foreign scientists to value and, conversely, that they could teach the Europeans which Americans not to value. With respect to the latter point, Bache recorded in his diary:

With Prof. P[hillips] I had a most interesting conversation
about science and its interests. He informed me that pains had
really been taken to excite a prejudice against Rodgers (H.D.)
by the circulation of that vile attack upon his unpublished report:
that at first it had done harm but now that the report was in hand
all was right. . . . As he was in the stream about our science in
America I took the opportunity to let him know how much harm
was done by encouraging such quacks as Featherstonehaugh by
membership of the Royal Soc. of London.[17]

Charles Wheatstone emerged as one of the valuable friendships for both Bache and Henry. Henry thought him the most "talented" scientist in London, and Bache wrote "The person with whom I have spent most time here is Prof Wheatstone, he is disposed to be very kind & his original views & instruments require to be heard & seen to be fully appreciated."[18]

Most crucial to their objectives, it was through such contacts as Wheatstone that Bache and Henry were able to conduct experiments with their British counterparts. Two days after the Royal Society meeting at which Wheatstone had talked of novel experiments with the thermoelectric spark by "an Italian Prof.," Bache and Henry, along with John F. Daniell, joined Wheatstone at his laboratory in Kings College to repeat the Italian Santi-Linari's discoveries.

First, however, Henry was called upon to repair the spiral coil that Daniell had constructed according to the plan Henry had published in the *American Journal of Science*. The coil was to be used in the thermoelectric experiments, but had failed to function. Henry examined it and thought he recognized the problem. It took nearly two hours to take the coil apart, adjust it, and put it back together again. Much to his mortification, it still failed to work, and he almost regretted having "attempted to experiment in a strange laboratory and among those to whom I had always looked up as masters in Science."[19]

To his great relief, the spiral conductor performed successfully on the second trial, encouraging Henry to use it in an experiment to obtain a spark from the thermoelectric pile. Daniell thought the experiment would not work, but he and Wheatstone managed the apparatus for Henry's attempt. The first two trials ended in failure. Henry, as he recounted the episode in his diary, readjusted the parts more carefully for another effort, "Prof. Wheatstone as before applying the ice and Prof. Daniell the poker."[20]

During the third trial of the experiment, Henry's first attempt to get a spark produced no effect, but on the second a small spark was produced each time Henry broke the connection. Bache had been in an adjoining room, but, Henry said, "The noise we made called him forth. He afterwards made much sport with our enthusiasm. Prof. Daniell flourishing the poker Wheatstone with the ice and I jumping as he said in extacy (*sic*)."[21]

Although Henry had the satisfaction of knowing that he and his collaborators were the first in England to have performed this experiment, he also realized that, with the proper equipment, he could have done it himself two years earlier in Princeton. After all, scientific priority was continually in the minds of the Americans. Direct involvement in science while abroad made clearer what the needs of science were back home and proved one of the best ways for Bache and Henry to estimate their own abilities.

For Asa Gray, the general process of adjusting to a foreign environment was similar, although his task was somewhat more clearly defined. As William Darlington had pointed out in his review of part I of Torrey and Gray's *Flora of North America*, Old World botanists had "reaped a rich harvest of discovery in our forests," but the fruits of it lay in foreign hands. Therefore, it was of utmost importance, Darlington argued, that whoever might undertake to prepare a North American Flora, should be thoroughly acquainted with the labors of preceding botanists; and, by consulting their collections, as far as practicable, be competent to detect their errors, adjust their discrepancies, and determine their various synonyms.[22]

From Torrey's previous visit and from their collaboration on *The Flora of North America*, Gray already knew of the British herbaria that contained significant numbers of North American specimens. He had recognized that a knowledge of those collections was crucial to the completion of the *Flora*, and he made it a condition of the Michigan position that he first have a year abroad.[23] Gray's main job, therefore, was to work through the foreign collections, and his letters to William Hooker hint at the singlemindedness of his labors. He had hoped already to have left for the Continent, Gray wrote to Hooker from London early in March 1838, "but I am not yet ready, although I have worked as hard as possible. I can not work at botany from nine in the morning until twelve o'clock at night as at Glasgow, but my evenings have been fully occupied with Scientific societies, *Microscopic people*, dinner parties, &c."[24]

From his assiduous study of the British herbaria, Gray was increasingly able to recognize faulty identifications. P. Barker Webb, a wealthy English botanist living in Paris, wrote Hooker on that subject: "You may easily believe how much I was pleased to make the personal acquaintance of Dr. Asa Gray, he has set us right here on many American matters."[25] The same skills made it possible for Gray to identify unpublished genera and also to perceive which geographic areas of North America would be worth further exploration.

The ease with which Gray, Henry, and Bache worked with their foreign colleagues, and the friendships that sprang from collaborative investigation, gave substance to the conception of science as an intellectual pursuit that transcended national boundaries. For example, Gray and Torrey were connected with Hooker, John Lindley, and George Bentham in a botanical enterprise that seemed "a kind of freemasonry," as Torrey put it.[26] Indeed, were it not for the headings on their letters to each other, there would be little cause to imagine a difference in nationality. "I have more letters from Dr. Torrey," Gray wrote Hooker in 1839, "who be sure will be surprised to hear of our having such a time together in London, and would have given I know not what to have been with us."[27] And in a letter to Gray, Hooker spoke of their mutual friendship with Torrey in exactly the same spirit: "I quite long to have a chat with him about you & to tell him of our frolicking together in London."[28]

These close relations inevitably diminished nationalistic feelings. Thus, although Gray's ambition was to master the foreign sources of American botany and thereby to repatriate the subject, he imagined his future research in the context of an international peer group. It had occurred to him, he wrote Hooker from Germany, to devote himself to a "detailed treatise on the Geographical Botany of North America." It would take some years, he said, "and I shall not undertake it unless I think I can do the subject justice and myself credit. What do you think of the plan? If you approve will you give me some suggestions?"[29]

Bache and Maj. Edward Sabine were immediately drawn to each other. "He is a man of remarkably pleasing address," Bache wrote in his diary after meeting Sabine. "In a very friendly way he invited himself to breakfast with me on Monday which will be very agreeable." Henry joined them for that breakfast and recorded the event in his diary as "much conversation on the subject of america, magnetism, & &"[30] In fact, Sabine had been stationed in Canada during the War of 1812 and had been responsible for the construction of defensive works at Ft. Erie that were taken by the Americans during the war. However, the discussion moved easily from his military service to their mutual scientific interests.

In Scotland, Henry was invited by the Commissioners of Northern Lighthouses to join an official party touring those lighthouses in which Fresnel's new lens system had been installed. Moreover, in what seemed a marvelous example of the internationalism of science, at a formal dinner the American was invited by the British to speak on the French invention.[31]

Yet Bache, Gray, and Henry were Americans, and their concern with elevating their country's science suggested some limits to the concept of internationalism. Bache and Henry were especially sensitive to the issue of America's scientific reputation. On one hand, it came as something of a shock to realize how little the affairs of the U.S. mattered abroad. Echoing de Tocqueville's observation, Joseph Henry wrote to his wife, "We have too exalted an opinion of our influence and the share of attention we occupy in the minds of Europe."[32] On the other hand, what Europe did learn of American science was unflattering. Audubon was a source of continual embarrassment, for example, and Henry was told there was "no naturalist of any reputation in the U.S."[33] For the Americans, therefore, direct participation in the life of British science—attending lectures and scientific meetings, experimenting or analyzing specimens with foreign colleagues, and publishing in British journals—provided important opportunities for correcting foreign misapprehensions of the level of scientific work in the U.S.

Unfortunately, the activities that yielded the greatest sense of membership in an international fraternity were also those with the greatest potential for arousing nationalist sensitivities. The degree to which it was easy to slide from one feeling to the other is well illustrated by two controversies in which Joseph Henry became engaged.

The first incident involved an abstract of Bache's report on the investigation of steam boiler explosions at Philadelphia's Franklin Institute—an abstract that Henry had been asked to prepare for the British publication, the *Magazine of Popular Science*. Initially, it seemed to Henry just the opening he was seeking, and he wrote his wife that he "was very much engaged in preparing an analysis of the experiments and reports . . . to be published in one of the Scientific Journals of this city."[34] The investigation, to discover the cause of steamboat boiler explosions, had involved a sophisticated set of experiments, and the chance to write an abstract of the investigation meant publicity for a piece of research that reflected well on America's capability in science.

Bache, however, had had some experience with the *Popular Science* editor, who had earlier dealt with the Franklin Institute investigation in a slighting and sarcastic fashion, and Bache so warned Henry. Yet, despite the understanding Henry thought he had with the editor, his abstract was published, as Bache put it, "in a mutilated form, cutting out the pungent part, reflecting in a balderdash way upon brother Jonathan's thin-skinnedness, &c., &c."[35] Bache used his friendship with Wheatstone in an attempt to secure a retraction, but when the editor refused, Bache turned to William Holl, editor of the *Analyst*, which published Henry's abstract in its original form and heavily censured the *Magazine of Popular Science* for its review.[36]

At one level, the episode simply reflected traditional elements of antagonism. To Englishmen, Americans were often brash, credulous, and—as the editor of the *Magazine of Popular Science* claimed—"morbidly sensitive to criticism," while Americans usually resented British smugness and condescension. Had they perceived the matter so, Bache and Henry might well have determined the business unworthy of response, particularly since they enjoyed the immediate support of important people and influential periodicals. However, to Bache, the matter involved his country's reputation in science. As he said to a British colleague (probably John Stevelly in

Belfast) who apparently had written a defense of the experiments for publication in the *Athenaeum*, "So much has been said by [illegible] about the disposition of Americans to make a fuss about what they do, that if the affair were a private one I should be heartily glad to drop it altogether."[37]

The second controversial episode in which Henry was involved occurred at the Liverpool meeting of the British Association for the Advancement of Science in September 1837. The event has already been described in Nathan Reingold's *Science in Nineteenth-Century America*, but the main details are worth reiteration. Before leaving the U.S., Henry had carefully prepared a large map showing all the canal, railway, and steamboat routes in America. As he told John Torrey, he planned to exhibit this map at the Liverpool meeting "if an opportunity presents." Henry had a pretty clear idea of what an American should attempt at such a meeting. Robert Hare had admitted to him that his own efforts at the Bristol meeting of the association in 1836 had been spoiled by an overzealous approach, and Henry had resolved, as he later said to Bache, "to keep myself perfectly cool and not put myself too much in advance."[38]

In the meeting of the mechanical section, before which Henry showed his map, he limited himself to a few remarks on the extent of internal improvements in America and something of its topography. However, just as he was stepping from the stage, Henry said, "A person named De Butts asked me in rather a rough manner to give some authentic information relative to the rate of velocity of the American steamboats."[39] The chairman of the section immediately interceded on behalf of Henry, who nonetheless expressed a willingness to answer at least from his own experience and mentioned some speeds he had observed on Hudson River boats. At that, Dionysius Lardner, author of a popular treatise on the steam engine, "Jumpt up and said the question of Mr. De Butts was an important one but that we did not want popular information on the subject, that he did not believe that a velocity as great as I had stated had ever been obtained."[40]

Lardner's remark caused a great commotion, and he was severely criticized by the chairman for treating a foreign visitor in such a fashion. But Henry, as he said, remained calm and responded, to great applause, "that truth and science should know no country."[41] Yet, although he spoke of internationalism, in Henry's mind the quarrel revealed a wide prejudice against Americans and the "low opinion" in which the English held American science and literature.

As in the *Magazine of Popular Science* fight, the Americans were not without their supporters in Liverpool. Lardner's attack was much deprecated at the meeting, and the *Liverpool Journal* carried a full report of the exchange, including a most flattering tribute to Henry's scientific discoveries.[42] Nevertheless, Henry, Bache, and Gray, too, were much concerned with British opinion and conscious of their nationality. One of Gray's first acts upon landing in England was to be fitted out with a coat cut in the English style, so as to be not quite as identifiable in his origins. Even Bache was a bit intimidated by his first dinner at the Athenaeum Club.[43]

In addition, there were more serious differences. The recent publication of Harriet Martineau's travels in the U.S. proved embarrassing, and Henry was particularly sensitive to her reports of the evils of slavery. On the other hand, both he and Bache were offended by the distinctions drawn in British society between status and intellect. Henry wrote to his brother-in-law, "I have derived more pleasure and profit from several interviews with Mr. Babbage than I could possibly do from kissing the hand of his most greacious and illiterate majesty."[44] Bache, too, compared a desultory conversation with the Earl of Burlington, vice-president of the Royal Society, to the excitement he felt at learning of Charles Wheatstone's research, and he concluded in his diary, "So here rank makes an equality with talent."[45]

The Panic of 1837 seemed to reflect badly on their country, and the Americans were sensitive about that, too. Britons had invested heavily in U.S. internal improvements, and when the collapse was over, many were left holding worthless bonds. Henry, who claimed to have made a study of political economy in an effort to understand the problem, argued that the Panic actually had begun in England. But his explanation for its spread to the United States is interesting in light of his later prescriptions for science. The Panic was produced in the U.S., he said:

by the ruinous system of paper money and the consequent spirit
of gambling and speculation which for two years has infected
everyone. Instead of employing our means on the production
of objects of real value we have been entirely engaged in making
paper fortunes or rather in inflating a bubble which has at
length [burst].[46]

Henry's language in describing the Panic was much like the kind he used to outline the needs of American science. He returned from abroad convinced of the necessity to rid science of the same unsound tendencies that caused the Panic, and of the need for fundamental research of solid value. His analysis of the worth of the British Association for the Advancement of Science followed along the same lines. Its meetings—which mixed "the merest Sciolist and the profound *savant*"—were, to Henry, of little utility. The true worth of the British society, and the objective he later defined for the Smithsonian, was that it supported original investigation.[47]

Asa Gray also returned with some dissatisfactions. In a letter to Hooker sending parts III and IV of the *Flora*, he said, "There are so many errors, so much bad printing, and so many things that we would now do much better that I regret any portion was published before my visit to Europe." Although the trip had provided Gray with unparalleled opportunities for study, with new specimens and useful books, he also came back with a keener sense of his disadvantages than when he started.[48] But if European travel made the Americans more aware of the difficulties under which they labored back home, the experience was also productive in a number of ways, some of which took time to appreciate.

Most immediately, their travels gave Americans an enhanced reputation at home. It was somewhat expected—both in the public press and within the scientific community—that their visits would yield important results. Benjamin Silliman identified the essential concerns of the scientific community. He wanted to have it known in Europe, he said, that "difficult things have been done in this country," and he also hoped that the Americans would return with a rich harvest of the latest scientific news from abroad.[49]

The travelers were aware of these expectations; they carried with them a variety of commissions, from friends and fellow scientists as well as institutions. Elias Loomis, the newly appointed professor of mathematics at Western Reserve Academy in Hudson, Ohio, sent back accounts of his travel to the *Ohio Observer*, and from his experience he argued the need for postgraduate instruction in science.[50]

There is some indirect evidence to suggest the travelers were aware that publishing while abroad lent stature to one's reputation. Joseph Henry claimed to have presented his remarks on American internal improvements at the British Association meeting as much for "my character at home" as from "a desire to show myself at the meeting."[51] John Locke, who from his teaching post in Cincinnati felt remote even from Philadelphia and New York, took such pains to describe in the *Philosophical Magazine* the fruits of his experiments abroad that it is impossible not to conclude he was writing for the folks back home.[52] Finally, when Joseph Henry returned, Thomas Cooper immediately proposed him for an honorary degree from South Carolina College. "He has just returned from a tour of scientific observation in Europe," Cooper wrote Governor Butler, "where his reputation is deservedly established."⁵³

Gray, Henry, and Bache had gone abroad with a sense of mission, and they returned even more convinced of the need to ensure their country's standing among the nations of the world. The willingness of their countrymen to be taken in by specious scientific ideas only heightened the feeling that strong bulwarks were needed. On the grounds that it would encourage just the sort of charlatanism they were trying to stifle, Bache, Henry, and Torrey opposed the formation of an American version of the British Association for the Advancement of Science.⁵⁴ They saw themselves, in a sense, as filters between the U.S. and Europe. In that role, the three scientists worried about proclaimed scientific discoveries that would expose America to ridicule abroad; to mitigate that effect, they wrote reassuring letters to their friends overseas. Gray screened those whom he introduced to his British colleagues, and Joseph Henry seems to have been equally cautious to introduce to Faraday only Americans who would reflect creditably on their country.⁵⁵

Some of the connections the Americans made while abroad ripened into deep friendships. The best example is that which grew between Asa Gray and the Hookers, father and son. That relationship lasted the rest of Gray's life and provided the basis for an important trans-Atlantic cooperation in botany. Bache and Sabine corresponded warmly and at length over the years. They too engaged in cooperative efforts to establish meteorological observations at Coast Survey stations.⁵⁶ Three decades after his tour abroad, Joseph Henry could write to Michael Faraday, with whom he had kept up a correspondence, that he still retained a "vivid impression" of the trip.⁵⁷

Not all the connections were foreign ones; another beneficial result of travel was that Americans met Americans. Joseph Henry, for instance, met John Locke in London and they shopped for instruments together. In Paris, Henry met A. D. Stanley, Lewis Gibbes, and Elias Loomis, all three of whom were touring Europe before taking up their first professorial appointments. "Although I am a bad correspondent," he wrote to Loomis, "I will endeavour not to let it be my fault if the acquaintance we have found in Europe should drop in America."⁵⁸

Perhaps the most important result of their voyages was the renewed spirit the Americans brought home. They returned with a better estimation of their own abilities and a clearer definition of their future objectives. It was particularly crucial to be freed from the sense of intellectual subordination to the Old World. Uncertain of the reception that awaited them on foreign shores, the Americans had begun with an image of Great Britain as a "fairyland" populated by those, as Henry once said, "whom we have long considered almost more than human." However, his own experience abroad, he told Asa Gray, had led Henry to realize they were "but men, inferior perhaps in some respects to ourselves."[59] That understanding made it easier to think of the United States one day assuming an equal place in the international community of science.

The successful inauguration of transatlantic steamship travel in 1838 catches something of the optimism they felt. Steam power suggested expanded views. By cutting the travel time in half, steam navigation promised closer connections between Europe and America and a new era in their relationships. As Joseph Henry reflected on the value of his own experiences abroad, "We will not now be so remote a province of Great Britain in reference to literature and science as we have been."[60]

1. Perry Miller, *The Life of the Mind in America* (New York: Harcourt, Brace, and World, Inc., 1965), pp. 282–85; Rush Welter, *The Mind of America, 1820–1860* (New York: Columbia University Press, 1975), p. 5.

2. Robert E. Spillar, *The American in England During the First Half Century of Independence* (New York: Henry Holt and Co., 1926) is the standard treatment. Also useful is Foster Rhea Dulles, *Americans Abroad: Two Centuries of European Travel* (Ann Arbor: University of Michigan Press, 1964).

3. *Report of the Seventh Meeting of the British Association for the Advancement of Science* (London: J. Murray, 1838). The quote is in a letter from Alexander Dallas Bache to Charles Babbage, London (6 June 1837) Babbage MSS, British Museum.

4. Stanley M. Guralnick, *Science in the Ante-Bellum American College* (Philadelphia: The American Philosophical Society, 1975), p. 72. J. MacLean to Professor Henry, Boston, 25 July 1836, Joseph Henry Papers, Smithsonian Institution Archives (hereafter HPSIA). It is a pleasure to record the considerable assistance I received from Nathan Reingold and his colleagues at the Smithsonian. Their unparalleled resources and unfailing generosity were fundamental to this research.

5. Bache's study was entitled *Report on Education in Europe, to the Trustees of the Girard College for Orphans* (Philadelphia: Lydia R. Bailey, 1839).

6. A. Hunter Dupree, *Asa Gray, 1810–1888* (New York: Atheneum, 1968), p. 74. George Daniels has described the early stages of scientific pro-

fessionalism in "The Process of Professionalization in American Science: The Emergent Period, 1820–1860," which originally appeared in *Isis* 58 (Summer 1967) and has been republished in *Science in America Since 1820*, ed. Nathan Reingold (New York: Science History Publications, 1976), pp. 63–78.

7. Joseph Henry to a friend in Albany, Princeton, 6 February 1837, HPSIA.

8. Alexander Dallas Bache to Joseph Henry, 3 January 1835, HPSIA.

9. Asa Gray to George Bentham, Paris, 3 April 1839, Royal Botanic Gardens Library, Kew.

10. Joseph Henry described William Vaughan as a bachelor of eighty-five and his nephew Petty Vaughan as a bachelor of forty-five (Joseph Henry to Harriet Henry, Paris, 26 May 1837, HPSIA). William's brother John Vaughan was for years the secretary and librarian of the American Philosophical Society and a friendly ambassador for scientific visitors to the U.S.

11. Joseph Henry to Harriet Henry, Glasgow, 27 August 1837, HPSIA.

12. Joseph Henry to Harriet Henry, London, 9 May 1837, HPSIA; Alexander Dallas Bache, Diary, 27 March 1837, Bache papers, Smithsonian Institution Archives. The Vaughans also collected books and periodicals for U.S. institutions, and when the heirs of Nathaniel Bowditch wanted to establish a library to honor their father's name, Benjamin Vaughan solicited the Royal Society for its assistance. See, for example, William Vaughan to Dr. Roget, 6 June 1836, requesting copies of Greenwich observations for the libraries of Bowdoin and Wallaceville Colleges. This and similar letters are in the manuscript collections of the Royal Society.

13. Nathan Reingold, *Science in Nineteenth-Century America: A Documentary History* (New York: Hill and Wang, 1964), p. 68; Joseph Henry to James D. Forbes, Princeton, 7 June 1836, and 19 September 1836, Forbes papers, St. Andrews University Library.

14. Joseph Henry to James D. Forbes, Princeton, 19 September 1836, Forbes papers; Alexander Dallas Bache, Diary, 6 April 1837, Bache papers.

15. "I spent more than a month in Dr. Hooker's family," Torrey wrote to a colleague, "from whom I received every possible kindness." Andrew D. Rogers, *John Torrey: A Story of North American Botany* (Princeton: Princeton University Press, 1942), p. 106. See, for example, Asa Gray to William J. Hooker, Halle, 2 August 1839 and Asa Gray to George Bentham, London, 30 September 1839, Royal Botanic Gardens Library.

16. Joseph Henry to Stephen Alexander, London, 10 April 1837, HPSIA. Alexander Dallas Bache to John Stevelly, 5 April 1837, Bache papers.

17. Alexander Dallas Bache, Diary, 12 March 1837, Bache papers.

18. Joseph Henry to Stephen Alexander, London, 10 April 1837, HPSIA; Alexander Dallas Bache to John Stevelly, 5 April 1837, Bache papers.

19. Joseph Henry, Diary, 22 April 1837, HPSIA.

20. Ibid.

21. Ibid. For further information on the experiment, see Charles Wheatstone, "On the Thermo-Electric Spark," *Philosophical Magazine* X (1837): 414–417; Brian Bowers, *Sir Charles Wheatstone, F.R.S., 1802–1875* (London: HMSO, 1975), pp. 64–65.

22. As quoted in Rogers, *John Torrey*, p. 122.

23. Jane Loring Gray, ed., *The Letters of Asa Gray* (Boston: Houghton Mifflin and Co., 1893), vol. I, p. 21.

24. Asa Gray to William J. Hooker, Paris, 5 March 1839, Royal Botanic Gardens Library.

25. P. Barker Webb to William J. Hooker, Paris, 5 April 1839, Royal Botanic Gardens Library.

26. Rogers, *John Torrey*, p. 105. In a letter that catches the same spirit, Lindley wrote Gray, "Will you come down to see me on Sunday next & eat your mutton with us at 3 o'clock . . . I can give you a bed. You can amuse yourself in my herbarium on Monday, when I will get a few friends to meet you at dinner—all Botanical." John Lindley to Asa Gray, Turnham Green, 28 January 1839, Gray Herbarium Library, Harvard University.

27. Asa Gray to William J. Hooker, London, 11 February 1839, Royal Botanic Gardens Library.

28. William J. Hooker to Asa Gray, Glasgow, 7 February 1839, Gray Herbarium Library.

29. Asa Gray to William J. Hooker, Halle, 2 August 1839, Royal Botanic Gardens Library.

30. Bache, Diary, 20 April 1837, Bache papers; Henry, Diary, 24 April 1837, HPSIA.

31. Joseph Henry to Harriet Henry, Edinburgh, 16 August 1837, HPSIA.

32. Joseph Henry to Harriet Henry, London, 2 August 1837, HPSIA. De Tocqueville's observation that "An American leaves his country with a heart swollen with pride; on arriving in Europe, he at once finds that we are not engrossed by the United States and the great people who inhabit them as he had supposed . . ." is quoted in Dulles, *Americans Abroad*, 5.

33. Joseph Henry, Diary, 4 April 1837, HPSIA.

34. Joseph Henry to Harriet Henry, London, 9 May 1837, HPSIA. The research is described in Bruce Sinclair, *Philadelphia's Philosopher Mechanics: A History of the Franklin Institute 1824–1865* (Baltimore: Johns Hopkins University Press, 1974), pp. 173–84.

35. Alexander Dallas Bache to Joseph Henry, London, 7 June 1837, HPSIA; *Magazine of Popular Science and Journal of the Useful Arts*, II (1836), 114–124 and III (1837), 321–32.

36. "Critical Notices of New Publications," *The Analyst*, VI (1837), 315–28.

37. Alexander Dallas Bache to John Stevelly, 5 April 1837, Bache papers, Smithsonian Institution Archives.

38. Joseph Henry to John Torrey, Princeton, 13 February 1837, Gratz Collection, Historical Society of Pennsylvania; Robert Hare to Joseph Henry, Philadelphia, 13 February 1837, HPSIA; Joseph Henry to Alexander Dallas Bache, Princeton, 9 August 1838, as quoted in Reingold, *Science in Nineteenth-Century America*, p. 85.

39. Joseph Henry to Alexander Dallas Bache, Portsmouth, 1 October 1837, HPSIA. Lardner, the author of popular treatises on steam power, natural philosophy, and mathematics, subsequently lectured on science and its application in several U.S. cities. In one of his lectures while at Philadelphia, he apparently described the Liverpool incident as a mistake, by way of an apology, but Henry was not mollified. Joseph Henry to Jesse Buel, 1838, HPSIA. The Rev. C. De Butts is listed as an annual subscriber in the BAAS *Report*, 1837.

40. Ibid.

41. Ibid.

42. Both the *Liverpool Mail* and *The Albion*, which otherwise carried full reports of the sessions, avoided mention of the quarrel. But the *Journal* gave considerable space to Henry's remarks on internal improvements and the ensuing upset: "In reference to a question as to the efficiency of American steam-boats, the professor stated that they usually performed the voyage from New York to Albany, on the Hudson, at 16 miles an hour. . . . Some irritation, for which we are utterly unable to account, arose in this conversation, in which, however, the professor did not share. It gave occasion to Mr. Russell to pronounce a very glowing eulogium on this distinguished philosopher, which was enthusiastically cheered." *The Liverpool Journal*, 16 September 1837.

43. Dupree, *Asa Gray*, p. 75; Bache, Diary, 8 April 1837, Bache papers.

44. Joseph Henry to Stephen Alexander, London, 10 April 1837, HPSIA. "Miss Martineau's book on America has just appeared and is making something of a sensation here," Henry wrote his wife. "It is very hard on America in reference to the slave question, also our politics religion &c," Joseph Henry to Harriet Henry, London, 2 August 1837, HPSIA.

45. Bache, Diary, 20 April 1837, and see also Bache to his mother, London, 7 April 1837, Bache papers.

46. Joseph Henry to Alexander Dallas Bache, Paris, 19 June 1837, Bache papers. Henry also reported from Paris that the Panic was the source of considerable distress to American travelers, many of whom were

suddenly caught without funds when the failure of U.S. banks ruined their credit. His own personal finances were in order, Henry said, but the $5,000 promised for the purchase of apparatus had been reduced to $500. Joseph Henry to Harriet Henry, Paris, 28 June 1837 and 26 July 1837, HPSIA.

47. Joseph Henry to Charles Coquerel, Portsmouth, 1 October 1837. See Reingold, *Science in Nineteenth-Century America*, pp. 87–88, 91 for other letters between Henry, Bache, and Torrey expressing the same sentiments.

48. Asa Gray to William J. Hooker, 30 May 1840; Asa Gray to George Bentham, New York, 29 December 1839, Royal Botanic Gardens Library.

49. Benjamin Silliman to John Torrey, 17 April 1837, Torrey papers, Library, New York Botanical Garden; 22 September 1837, Gratz Collection, Historical Society of Pennsylvania.

50. Loomis's "Letters from Europe" were published in the *Ohio Observer* from 1836 to 1837.

51. Joseph Henry to Alexander Dallas Bache, Princeton, 9 August 1838. As quoted in Reingold, *Science in Nineteenth-Century America*, p. 85.

52. Locke's article, "On a large and very sensible Thermoscope Galvanometer," was reprinted in the *Journal of the Franklin Institute* XXI (May 1838), 343–45.

53. Thomas Cooper to Gov. P. M. Butler, Columbia, S.C., November 1837, Library, University of South Carolina.

54. An analysis of the founding of the American Association for the Advancement of Science is presented in Sally Gregory Kohlstedt, *The Formation of the American Scientific Community* (Urbana: University of Illinois Press, 1976).

55. I am grateful to Professor A. H. Dupree for suggesting this concept and for pointing out Asa Gray's efforts to filter Darwin's American contacts. For similar concerns on Henry's part, see Joseph Henry to Michael Faraday, Princeton, 9 October 1838, published in *The Selected Correspondence of Michael Faraday*, ed. L. Pearce Williams, 2 vols. (Cambridge: University Press, 1971); Joseph Henry to Charles Babbage, Princeton, 18 May 1838, HPSIA. For other subsequent efforts to protect America's scientific reputation abroad, see John Torrey to Joseph Henry, New York, August 1838, Gray Herbarium Library, Harvard, and Joseph Henry to William Sturgeon, Princeton, 1 November 1838, HPSIA.

56. Two volumes of Asa Gray letters in the Royal Botanic Gardens Library at Kew attest to the close relations between Gray and the Hookers, as do the letters in the Gray Herbarium, which contain the other side of the correspondence. The Sabine Papers at the Royal Society of London describe Sabine's cooperation with Bache in magnetic observations. An indication of the spirit of that correspondence can be suggested by a letter Bache wrote Sabine on 22 May 1860 from Washington intro-

ducing his "particular friend Prof. Peirce of Harvard, Cambridge, U.S. who has charged me only to give him letters to those I love. Take him at once into your magic band and believe me ever affectionately yours." Bache papers.

57. Joseph Henry to Michael Faraday, Washington, 3 November 1866, in Williams, *Selected Correspondence of Michael Faraday*, II, 1032.

58. Joseph Henry to Elias Loomis, London, 2 August 1837, Loomis papers, Beinecke Library, Yale.

59. Joseph Henry to Asa Gray, Princeton, 1 November 1838, Gray Herbarium Library.

60. Joseph Henry to Thomas Thompson, Princeton, 28 September 1838, HPSIA.

ASTRONOMY IN ANTEBELLUM AMERICA

DEBORAH JEAN WARNER
Smithsonian Institution

IN THE EARLY DECADES OF THE NINETEENTH CENTURY, ASTRONOMY in America was a practical science, both literally and figuratively, supported at a steady but very low level. By the time of the Civil War, Americans could be proud of their astronomical institutions and achievements. By the end of the century, Europeans were looking to American observatories as models for their own scientific institutions.[1] What I will suggest in this essay is that the growing popularity of astronomy in America derived largely from the questions astronomers were asking, and the sort of answers they expected to find. These scientific expectations, in turn, had been raised by a series of technological innovations that had occurred in Europe in the previous half century. The turning point was 1834; the vehicle was John Herschel's *Treatise on Astronomy*. To quote Sears Cook Walker, this book signaled "an important era in the history of astronomy," as "it furnished to the general reader information nowhere else to be obtained."[2]

Generally considered a unified subject, astronomy has historically been divided into three distinct branches—practical, theoretical, and physical. Practical astronomers are concerned primarily with measuring the positions of celestial bodies, computing positions for times past or future, and determining local time and terrestrial latitude and longitude. Theoretical astronomers work out the details of celestial orbits on the basis of gravitational forces. Astrophysicists see celestial bodies as real objects and seek knowledge of their physical properties.[3]

American astronomy from the colonial period to the 1830s was essentially practical. Every celestial event—eclipse, comet, or transit of Venus or Mercury—commanded at least a half dozen observers working with good portable instruments. There were, however, no fixed observatories and no astronomers who observed the heavens on a regular basis. The most likely model for an American observatory was the British Royal Observatory at Greenwich, established in 1675. With "the finding out of longitude of places for perfecting Navigation and Astronomy," as their mandate, the Astronomers Royal made repeated observations to establish precisely the positions of the sun, moon, planets, and stars.[4]

With limited resources, Americans could not hope to compete against Greenwich, which was staffed by an astronomer and two assistants and equipped with a large assortment of instruments. But there was no need for America to duplicate the work done at Greenwich. The British *Nautical Almanac* was useful all over the globe. Fundamental catalogs compiled at the Royal Observatory were equally useful on both sides of the Atlantic and, except for unusual celestial occurrences, routine observations made here would add nothing to those made over there. There was no widespread belief, either in America or in England, that longer term or more precise observations would lead to a better understanding of the structure of the universe.[5] That being the case, it is not surprising that Americans ignored John Quincy Adams's 1825 plea for a congressionally funded national observatory.

In the years between 1776 and 1834, there were more than eighty Americans whose astronomical activity was made public in one form or another[6]—above and beyond the countless sailors like the dozen on Bowditch's ship who "could take and work a lunar observation as well, for all practical purposes, as Sir Isaac Newton himself."[7] All these astronomers were amateurs. None earned a living by astronomy, and none would, I think, have identified himself primarily as an astronomer. Nathaniel Bowditch was a mathematician.[8] David Rittenhouse owed his reputation as a savant largely to his precise astronomical instruments and ingenious orreries.[9] Several other observers were instrument makers by profession: William Cranch Bond of Boston was at this time a clock maker,[10] and the first observatory in New York City was erected by the watchmakers, Messrs. T. and B. Demilt.[11]

Many early astronomical practitioners were professional surveyors using celestial observations to establish important locations and boundaries. John Garnett, author of nautical almanacs and charts, was said to be "a good astronomer . . . much accustomed to the use of the telescope."[12] Seth Pease, surveyor of the Mississippi

Territory, was commended as "an excellent astronomer,"[13] and the same would apply to Simeon DeWitt of New York State[14] and Andrew Ellicott, who worked for the federal government.[15]

Teachers who held forth in colleges and academies and on the lecture circuit comprised yet another group of astronomical practitioners; although many of these men taught some astronomy, none billed himself as an astronomy teacher per se.

All American secondary students, girls as well as boys, studied astronomy as a component of natural philosophy, learning broad facts about the astronomical universe, but not how to pursue astronomy as a science.[16] Practical astronomy was confined to calculating the sort of things found in almanacs: times of celestial risings and settings, times and durations of eclipses, altitudes of celestial bodies for various terrestrial latitudes. Theoretical and physical astronomy were represented only by their most general conclusions.

The earliest astronomical texts published in the United States were almanacs, instructions on the use of globes, and astronomical and geographical "catechisms." The more sophisticated books were reprints of English works. William Enfield's *Institutes of Natural Philosophy* (London, 1783), which enjoyed a reputation as the most popular college text until the 1820s, presented mideighteenth-century information enlivened with a few of William Herschel's observations and ideas. Samuel Webber of Harvard, who supervised the first American printing of Enfield's work in 1802, appended a set of solar and lunar tables and explanations for their use; these were derived from John Ewing, Webber's colleague at Philadelphia.

James Ferguson's, *Astronomy Explained Upon Sir Isaac Newton's Principles and Made Easy to Those Who Have Not Studied Mathematics* (London, 1756) remained popular well into the nineteenth century; the first American edition, after the latest English one (which had been somewhat updated by the addition of Herschelian data), was brought out by Robert Patterson, professor of mathematics at the University of Pennsylvania, in 1806.

The most advanced text was Samuel Vince's *The Elements of Astronomy* (2d ed., Cambridge, 1801), reprinted at Philadelphia in 1811. The first American text on a par with the British works was John Gummere's *Elementary Treatise on Astronomy* (Philadelphia, 1822 and later). John Farrar of Harvard made the first break with English tradition in 1827 with his *Elementary Treatise on Astronomy*, an adaptation of Biot's *Traité Elémentaire d'Astronomie Physique*.

Astronomical demonstration apparatus—orreries for the solar

system, charts and globes for the stars—enjoyed quite a vogue. In the late eighteenth century, David Rittenhouse built elegant orreries for the colleges at Philadelphia and Princeton,[17] as did Joseph Pope for Harvard.[18] Functionally similar but less elaborate was the "pendant planetarium" invented by the Rev. Burgiss Allison in 1800,[19] and the "terralunelion" described by Uzziah Burnap in 1822.[20]

James Wilson, America's first commercial globe maker, had pairs of terrestrial and celestial globes on the market by 1815.[21] The first American star chart, drawn by William Croswell and published at Boston in 1810, introduced two constellations peculiar to America: the Flying Squirrel and the Bust of Columbus. Croswell's map, drawn on a mercator projection, is the only such map I've seen. Although an awkward projection for solving purely astronomical problems, the map does yield easy answers to basic questions such as where and when the sun and stars will rise and set at Boston.[22]

Montgomery Robert Bartlett's celestial planispheres, issued in the late 1820s and dedicated to John Quincy Adams, were more astronomically conventional.[23] In 1833, Elijah Burritt came out with his celestial *Atlas* and the *Geography of the Heavens*; the latter, perhaps the most widely known astronomy book in nineteenth century America, had gone through at least eight editions and thirty-three printings by 1876.

Accustomed as we are to certain historical dates—e.g., 1830, when Yale acquired a five-inch achromat, then the largest refracting telescope in the country, and 1836, when Williams College established our first fixed observatory—we tend to overlook the many fine observational instruments here in the early decades of the nineteenth century.[24] For the 1806 solar eclipse, both Andrew Ellicott and Simeon DeWitt used large zenith sectors made by David Rittenhouse, while William Dunbar, at Natchez, Mississippi, used a reflecting telescope of six-foot focus, a 2½-foot focus achromatic refracting telescope, a circle of reflection made by Troughton, an astronomical clock with a gridiron pendulum, and a portable chronometer.[25] In 1834, there were, in and around Philadelphia, four telescopes, "as good as those used in the Royal Observatory of England" for making observations of solar eclipses, lunar occultations, and eclipses of Jupiter's satellites. There [were] besides four achromatics of 3½-feet focal length, three Gregorians of equivalent magnifying power, and several thirty-inch achromatics." In addition, there were four transit instruments, plus chronometers and astronomical clocks in "abundance."[26]

With these instruments Americans observed and timed solar and lunar eclipses, planetary transits, comets, and lunar occultations, primarily to determine terrestrial positions. The solar eclipse of 1806, as William Dunbar observed, was visible both from Europe and North America, which rendered it "a very important phenomenon for settling comparative longitudes."[27] Using observations of the 1834 solar eclipse, Edward Courtenay calculated the longitudes of several places in the United States.[28]

Work of this sort clearly was needed. I find no evidence, however, of Americans charting celestial positions for their own sake. Except for Samuel Williams's observation of the so-called Bailey's Beads during the 1780 solar eclipse,[29] Americans paid little or no attention to the physical features of the sun, moon, planets, or comets. In the realm of theoretical astronomy, America's two contributions were Bowditch's useful English translation and emendation of Laplace's *Mécanique Celeste*, published 1829–39, and a curious nebular hypothesis developed by one Isaac Orr.[30]

In short, although there was a fair amount of astronomical activity in America, none of it was organized, and none appears very interesting. In the words of Denison Olmsted in 1827, "No one of the natural sciences appeared to afford less hope of further discoveries than astronomy."[31]

By 1840 Olmsted had made a radical about-face. In his *Letters on Astronomy, Addressed to a Lady*, for instance, he presented numerous questions about the physical nature of celestial bodies, and he outlined the observations on which various explanations were based. He led his readers to believe that they might well contribute to the new discoveries lurking just around the corner.

Foremost among the factors revolutionizing astronomy at this time was a new telescope technology enabling nineteenth-century astronomers to see fainter objects for longer periods of time and to measure their positions more precisely than could their eighteenth-century predecessors.[32] Moreover, these good instruments were becoming widely available. William Herschel, a Hannoverian immigrant in England, had startled the world with astronomical discoveries made possible by the increased light-gathering power of his large reflectors. Herschel aside, reflecting telescopes did not gain popularity until silvered glass mirrors replaced the solid metal ones in the second half of the nineteenth century. Attention in the meantime focused on refractors. Achromatic lenses had been patented and marketed in the mideighteenth century, but the impossibility of making large homogeneous pieces of flint glass limited refracting telescopes to four or five inches of aperture. That problem was solved in 1805 by a Swiss optician experimenting with different types of stirring rods. By 1824 Fraunhofer in Munich could make a 9½-inch refractor, and by 1862 Alvan Clark in Cambridgeport could make one twice again as large.

With his discovery that monochromatic spectral lines could be used to measure the optical properties of glass, Fraunhofer opened a new era in lens design. Equally important was his development of equatorial mounts sufficiently well-balanced to be driven by clockwork, yet steady enough to support large magnifications. And the organization of the Munich workshop, together with increased prosperity in the United States, enabled Americans in the late 1830s and 1840s to obtain telescopes as large and as fine as those used anywhere in the world.

Other astronomical instruments underwent similar developments. The octant was invented around 1735 but not widely distributed until late in the eighteenth century.[33] The first chronometer sufficiently accurate for the determination of longitude at sea was unveiled in 1764.[34] Meridian circles, used in conjunction with a clock to measure celestial right ascension and declination simultaneously, were introduced around the turn of the century.[35]

With this new technology physical astronomy was born, and practical astronomy moved beyond the solar to the sidereal and eventually the galactic universe. The two most innovative and influential early workers in these fields were William Herschel and Friedrich Bessel, director of the observatory at Königsberg. Herschel called attention to binaries as double star systems motivated by gravity. He also developed a map of the Milky Way galaxy, suggested that some nebulae might be island universes similar to our own, speculated on the nature of the sun, and discovered Uranus, the first telescopic planet.[36]

Bessel showed that good observations systematically reduced could yield star positions accurate to .1 second or maybe even .01 second of arc, and that these in turn led to the resolution of stellar parallax, a determination of proper motions and masses, and a measure of the galactic motion of the solar system.[37] In the late eighteenth century, the scientific payoff of astronomical observations was simply another significant figure; in the nineteenth century, it was nothing less than the structure of the heavens.

In 1832, with knowledge of this astronomical revolution widely scattered and not widely known, George Biddle Airy prepared a "Report on the progress of astronomy during the present century" for the second annual meeting of the British Association for the Advancement of Science.[38] The following year, John Herschel came out with his best-selling *Treatise on Astronomy*, covering much of the same ground but designed to interest as well as inform a much larger audience. Herschel's *Treatise* was first reprinted in America in 1834. The introduction of modern astronomy, which came largely through these two publications, inspired the meteoric rise of astronomical activity in the United States.

In the years after 1834, the American astronomical community experienced a phenomenal growth in personnel and facilities and an expansion into new areas of astronomical inquiry. The two developments were mutually reinforcing. The new institutions facilitated the scientific enterprise; the promise of discovery inspired scientists to pursue certain problems, and patrons to fund their research.

A first indication of America's new enthusiasm for astronomy was the interest shown in German, astronomical advances—an interest consciously developed by a handful of men. Sears Cook Walker was largely responsible for importing the first large German telescope into the United States, the six-inch equatorial refractor made by Merz & Mahler of Munich for the Central High School in Philadelphia.[39] Time and again in their publications Walker, his half brother E. Otis Kendall, and B. A. Gould included information about the latest German methods of reducing observations and analyzing orbits.[40] Kendall's *Atlas of the Heavens* (Philadelphia, 1844) was basically a translation of Joseph Johann von Littrow's *Atlas des gestirnten Himmels* (Stuttgart, 1839). Charles Henry Davis translated and published *Professor Encke's Method of Computing Special Perturbations* for the use of the Nautical Almanac in 1851.

An ever increasing number of Americans began to identify themselves as astronomers. A small but vocal group were professionals, proud of their accomplishments and eager for European recognition.[41] There was also a large group of amateurs, eager to see for themselves the phenomena so thrillingly described by O. M. Mitchel in his popular lectures and his journal, the *Sidereal Messenger*.[42] Some of these amateurs bought telescopes of their own; many more joined local astronomical societies in order to own a share in a large telescope.

The most visible of the new astronomical institutions were the fixed observatories from which absolute as well as relative celestial positions could be determined.[43] The fact that Airy in 1832 knew of none in the United States chagrined American chauvinists.[44] By 1860, America had eight observatories equipped for first-class work and at least twenty others with good quality instruments.[45] The

Cincinnati twelve-inch refractor, mounted in 1845, was then the second largest in the world. The Harvard fifteen-inch refractor, mounted in 1847, ranked, together with its companion at Pulkowa, Russia, as the largest in the world in the antebellum period.

The rise of an American telescope-making industry was both a stimulus and a response to America's new interest in astronomy. First into the field was Amasa Holcomb, whose reflectors were examined by the Franklin Institute as early as 1834. In the 1840s Henry Fitz began making equatorially mounted achromatic refractors that were significantly cheaper than European ones; by 1863, he had sold more than fifty. By the 1850s the Cambridgeport workshop of Alvan Clark & Sons was turning out telescopes comparable to European ones and recognized as such in Europe as well as in the United States.[46] William Würdemann, hired by the Coast Survey in 1834, was but the first of a long line of German immigrants who brought to America the skilled craftsmanship needed to make precision instruments.[47] In 1834 the United States had been dependent on Europe for all good astronomical apparatus; by the Civil War we had gone a long way towards self-sufficiency in this regard.

Many observatories were affiliated with colleges and academies, and they offered instruction in the techniques of observational astronomy, both practical and physical. The trend towards greater observational proficiency can be traced in the popular astronomical texts. John Gummere's *An Elementary Treatise on Astronomy* (2d ed., rev., Philadelphia, 1837) simply lists various astronomical instruments, omitting all mention of telescopes. William Norton's *An Elementary Treatise on Astronomy* (rev. ed., New York, 1845) provides slightly more elaborate discussions of the instruments; in small print, as an afterthought: "An observatory is not completely furnished unless it is supplied with a large telescope for examining the various classes of objects in the heavens. . . ." By contrast, the Elias Loomis work, *An Introduction to Practical Astronomy* (New York, 1855), is a practical manual for observatory use.

While observatory facilities were expanding, the teaching of mathematics was becoming more sophisticated, enabling students to solve problems in celestial mechanics. Modern languages were introduced at this time, enabling Americans to communicate directly with European astronomers. For advanced astronomical training, American students went to Harvard to study theoretical astronomy with Peirce or practical astronomy with the Bonds at the Observatory. They also went to the University of Michigan, the Coast Survey and the Naval Observatory, and to Europe.[48]

As astronomy gained a prominent place in the curriculum, professors were relieved of responsibility for other subjects and enabled to specialize in astronomy. In 1836, Denison Olmsted of Yale took the title professor of natural philosophy and astronomy, thereby becoming the first American formally responsible for astronomical instruction.[49] Stephen Alexander became professor of astronomy at Princeton in 1840.[50] At Harvard, the Perkins Professorship of Astronomy and Mathematics was established in 1842, with Benjamin Peirce its first incumbent; the Phillips Professorship of Astronomy was established in 1858 and given to William Bond.[51]

Some astronomy jobs opened outside the classroom. The Coast Survey, especially as revived under Bache in 1843, the Nautical Almanac, established in 1849, and various government-sponsored expeditions all employed astronomers on a part- or full-time basis. The Naval Observatory hired about a dozen astronomers, mostly Navy men, and used about as many part-time computers.[52] On the other hand, O. M. Mitchel received no money for his work as director of the Cincinnati Observatory.[53] William Bond served without pay as Astronomical Observer to the University for seven years until 1846, when he was hired as Director of the Harvard College Observatory with his son as his paid assistant. The observatory also gave minimal support to a few research assistants.[54] The Dudley Observatory did not pay its director until 1860; while in Albany, Gould and his four assistants had been on the Coast Survey payroll.[55]

Americans continued to report astronomical observations and calculations to newspapers and to general scientific journals and associations. Now, in addition, there was the *Astronomical Journal* (1849–61), which emphasized practical and theoretical astronomy, and the *Transactions* of the American Photographical Society, which served as a forum for spectroscopy and photography, the new tools of an emerging astrophysics.[56] And, as money permitted, observatories published *Annals* of their accumulated observations.

Throughout the nineteenth century, science in general, and astronomy in particular, provided many real benefits in addition to information about the natural world.[57] First were religious benefits; there was a saying that "an undevout astronomer is mad." Political benefits also resulted. Astronomy's historical heroes had defied the authority of church and state in their pursuit of truth, and the science was promoted as conducive to independent thought and a valuable ally of democracy. To quote Maria Mitchell on psychological benefits, astronomy brought "calm to the troubled spirit, and a hope to the desponding; when we are chafed and fretted by small cares, a look at the stars will show us the littleness of our own interests."⁵⁸ The enthusiastic attendance at public lectures on science and the support for local scientific societies suggest the social and cultural benefits expected from these activities. The Baconian notion that science inevitably yielded practical benefits was widely known and often repeated.

Against this background, such factors as economic prosperity, a growing nationalism, and the rise of professionalism strongly influenced the organization of astronomical institutions. Nevertheless, the heavens had been declaring the glory of God for a very long time before Americans discovered their need for fixed observatories. To understand fully why astronomy arose just how and when it did, we must consider internal developments in this particular science.⁵⁹

Practical astronomy was always easy to justify.⁶⁰ As the United States expanded to the west and the south and as railroads tried to keep on schedule, the role of observatories in providing standard time and a fixed point of reference became increasingly important.⁶¹ By midcentury Americans had made two significant improvements in the techniques for determining latitude and longitude. Latitude was found by measuring the meridian zenith distance between pairs of stars with a zenith telescope. This method had been initiated by Capt. Andrew Talcott of the Army Corps of Engineers in 1834 and adopted by the Coast Survey a dozen years later. To aid this work, astronomers developed catalogs of stars culminating near the zenith for various terrestrial latitudes.⁶²

Longitude was determined by comparing local time with standard time at a known station. In the "American method" these times were communicated, essentially instantaneously, by means of the telegraph. Telegraphy also was used in conjunction with a chronograph for automatically recording the times of star transits—a decided improvement over the former eye-and-ear method.⁶³

Despite their infamous appreciation of practical benefits, Americans seem to have been far from indifferent to the pursuit of astronomy for its own sake. In his popular lectures, which directly inspired the publicly funded observatories at Cincinnati and at Albany, O. M. Mitchel concentrated on "the structure of the universe, so far as revealed by the mind of man," and pointedly neglected the "uses of science."⁶⁴ With the Cincinnati refractor, O. M. Mitchel measured Struve's double stars south of the equator, resolving several into their components and showing others to be physical as well as optical binaries.⁶⁵

There were other examples of astronomy for its own sake. The citizens of Albany, who probably did not understand how demanding a precise positional measurement could be, raised money for a meridian circle reputed to be the finest ever made.⁶⁶ The Naval Observatory, originating under the subterfuge of the Navy's Depot of Charts and Instruments, acquired a great deal of sophisticated apparatus and sponsored observations far beyond the needs of navigators. The first volume of its accumulated observations contains a catalog of 1,248 star positions, determined "solely in the hope that they may add somewhat to the precision of our knowledge of the right ascensions and magnitudes of some of the smaller stars."⁶⁷ The federally sponsored Naval Astronomical Expedition to Chile in 1849–52 had as its primary purpose the refinement of our knowledge of solar parallax.⁶⁸ Before the Civil War was over, the Harvard astronomers had compiled a great three-part catalog of some 15,000 faint stars situated between the equator and 1° north declination.⁶⁹

All over the country, seeking comets and asteroids and calculating their orbits were popular and well-rewarded sports. Maria Mitchell's discovery of a comet in 1847 brought her a gold medal from the King of Denmark and international recognition.⁷⁰ Two of the six astronomers who shared the 1859 Lalande Prize awarded by the Paris Academy of Sciences were Americans: Horace P. Tuttle for his discovery of three comets, and George Searle for his discovery of the asteroid Pandora.⁷¹ Alvan Clark won the Lalande Prize in 1862 for his discovery of the companion of Sirius, a faint star predicted by Bessel on the basis of Sirius's apparent orbital motion.⁷²

After 1818, by which time Bowditch had essentially completed his work on the *Mécanique Celeste*, Americans did no theoretical astronomy to speak of until 1847, when European prediction and discovery of Neptune stimulated new interest in this field. Sears Cook Walker at the Naval Observatory found earlier sightings of this planet hidden among Lalande's star observations of 1795. He then computed an orbit that satisfied all the observations but differed considerably from that predicted by Leverrier. Benjamin Peirce refined the orbit further by including perturbations caused by the other planets, and declared that Neptune was not the planet to which geometrical analysis had directed the telescope, and that its discovery had been simply a happy accident.⁷³

Americans tackled a variety of problems in celestial mechanics in addition to Neptune. Miers Fisher Longstreth developed a new set of lunar tables that were adopted by the *American Nautical Almanac*. Charles Henry Davis, director of our *Almanac*, proudly reported to Congress that the American predictions for the solar eclipse of July 28, 1851, as seen from Cambridge, Massachusetts, erred by only 20 seconds, whereas the British *Almanac* erred by 85 seconds.[74] To give more examples, George Phillips Bond and Benjamin Peirce independently argued that Saturn's rings cannot be solid and are probably fluid.[75] And in 1849 Bond published a paper "On some applications of the method of mechanical quadratures," which anticipated by three years a similar paper by J. F. Encke, director of the Berlin Observatory and one of Germany's most acclaimed theoretical astronomers.[76] By midcentury, the United States was clearly producing first-rate work in this area.

But it was physical astronomy that most dramatically caught the attention of the American public. And it was in this realm that America made her most important contributions. Requiring observational competence more than mathematical sophistication, physical astronomy could profit from the work of amateurs—a point Airy well understood when he encouraged O. M. Mitchel thus to concentrate the efforts of the Cincinnati Observatory on physical observations with the great refractor.[77]

At Yale, attention turned as early as 1833 to the periodic meteor showers, with the object of determining speed and direction of motion, and also the kind of matter of which they were constituted. Olmsted and his students asked similar questions about the zodiacal light.[78] Two undergraduates built a fourteen-foot focus reflector—not as large as William Herschel's, but still the largest on this side of the Atlantic—and proceeded to scrutinize nebulae, seeking clues to their physical nature.[79] The Yale geologist, James Dwight Dana, puzzled over the apparent volcanoes on the moon.[80]

The origin of the Harvard College Observatory can be traced to the disappointment among Bostonians when Harvard astronomers could not enlighten them about the great comet of 1843.[81] Clearly, by the 1840s astronomers took for granted the computation of orbits which, since Newton, had been the major cometary problem; in its stead they puzzled over the physical nature of comets. In the words of Elias Loomis, comets "seem to require us to admit the existence of matter of a different kind from anything we witness upon the earth."[82] Here was an interesting and challenging scientific problem. Moreover, the wealthy citizens of Boston, whose money built and sustained the Harvard College Observatory, expected a well-staffed and equipped observatory to produce science—not economic benefits and not public entertainment. Their hopes were not disappointed. The Harvard telescope revealed startling and hitherto unknown features in every celestial object towards which it was directed: an eighth satellite of Saturn, the so-called crepe ring surrounding Saturn, the number of stars in and the form of the nebula in the Sword of Orion, the structure of Donati's comet of 1858, and some structure in the great Andromeda nebula, an object John Herschel had said "may be, and not improbably is optically nebulous."[83]

At the same time Americans were making visual observations, they also were doing pioneer work in celestial photography, spectroscopy, and photometry. The first celestial photograph, a daguerreotype of the moon, was taken by John William Draper in 1840.[84] More detailed photos of the moon, as well as photos of Jupiter, taken with the Harvard refractor, were proudly displayed at the London Crystal Palace of 1851. Six years later, using a collodion emulsion and a spring-governor of their own design on the telescope clock drive, the Harvard astronomers succeeded in photographing stars.[85]

Similar results were achieved by Lewis M. Rutherfurd in New York City and Henry Draper at Hastings-on-Hudson, both of whom also pioneered in celestial spectroscopy.[86] During the Civil War, Rutherfurd observed the spectra of the sun, moon, and planets, and classified stars on the basis of spectral characteristics.[87] In the area of photometry, George Phillips Bond compared the intensity of light from the sun, moon, and Jupiter;[88] Alvan Clark compared the sun with other stars.[89]

Despite these many activities and accomplishments, which all Americans might have rallied behind, the antebellum astronomical community was fractured by hostilities and jealousies. To a large extent, the internal factors of the science determined the external relations between the scientists. It was a conflict between practical and physical astronomy and how observatory time and talent should be allocated between the two. Theoretical astronomy, using positions to develop orbits, sided with the practical.

The most outspoken member of the practical camp was Benjamin Apthorp Gould, who in 1848 had returned to the United States sporting the first German Ph.D. in astronomy granted to an American. Aiming to Germanize American science, Gould established the *Astronomical Journal*, patterned on the *Astronomische Nachrichten*, and pursued practical astronomy as he had learned it from Encke, Gauss, and Schumacher. He frequently criticized the British influence on American science and encouraged his compatriots to visit Germany, the true home of the exact sciences.[90] To Gould, the popular physical astronomy, as pursued in England, was antithetical to the progressive, professional, Germanic method.

With this practical-physical dichotomy in mind, let us look at some actual American controversies. The inspiration behind the Dudley Observatory came from a series of public lectures given by O. M. Mitchel and illustrated with telescopic views of comets, clusters, and nebulae. As was his wont, Mitchel described recent discoveries in both practical and physical astronomy and intimated that a new observatory could make significant contributions in both these fields.[91] Imagine the surprise in Albany when Gould, director of their observatory, declared that "the science of astronomy consists in the investigation of those laws which govern the motions of the heavenly bodies" and insisted that "the Natural History of the Heavens, although a cognate and most interesting field of inquiry, is not included in the domain of astronomy proper."[92] Gould's ouster from Dudley certainly resulted from his vocal denigration of physical astronomy as well as from the conflict between public and professional values.

Another major controversy swirled around the Harvard College Observatory. The first two directors, William Bond and his son George Phillips Bond, were ostracized by the Germanophilic wing of the American astronomical community. The chronograph devised by the Bonds, although superior to other models, never found much favor in the eyes of B. A. Gould or S. C. Walker, the men at the Coast Survey responsible for improvements in longitude determination.[93] Benjamin Peirce, jealous of Bond's position, presented a paper on the constitution of the rings of Saturn, without sufficient credit given to the recent work on the same subject by his colleague across the yard.[94] When the National Academy of Sciences was established in 1863, the younger Bond's name was omitted from the list. The hostility towards the Bonds stemmed not from incompetence, but rather from their predilection for physical astronomy and the fact that their ties, both personal and professional, were closer to England than to Germany. In 1849, William became the first American admitted to the Royal Astronomical Society, and in 1865 George Phillips Bond became the first American honored with that society's coveted Gold Medal.

In conclusion, as Americans took up the new astronomy, as they actively sought answers to impractical questions about the size and nature of the universe, they were able to inspire their fellow citizens to support their work. But this support, lavish by earlier standards, was never enough to support all the rising scholarly enthusiasms and expectations. Prior to 1834, when no astronomical questions were being asked, no conflicts arose. The conflicts of the pre-Civil War period reflect the vitality the American astronomical community had achieved in its first quarter century of action.

1. "Denkschrift betreffend die Errichtung einer 'physikalisch-technischer Reichsanstatt' für die experimentelle Förderung der exakten Nahrforschung und der Präzisiontechnik," in *Reichsandstatt-Etat für die Etatsjahr 1887–88*, vol. IV, app. B, pp. 37–65.

2. John Herschel, *A Treatise on Astronomy*. A new edition, with a preface and a series of questions for the examination of students, by S. C. Walker. (Philadelphia: Carey, Lea and Blanchard, 1836), preface.

3. Agnes M. Clarke, *A Popular History of Astronomy During the Nineteenth Century*, 4th ed., rev. and corrected (London: Adam and Charles Black, 1902), introduction. While the names of the various branches often varied, the branches themselves remained distinct.

4. Eric Forbes, *Greenwich Observatory*, vol. 1: *Origins and Early History* (London: Taylor and Francis, 1975), p. 22. In 1775, John Ewing corresponded with the Astronomer Royal, Nevil Maskelyne, about the possibility of establishing an observatory at Philadelphia; Maskelyne's reply is printed in *Proceedings, American Philosophical Society*, 1843–47, 4: 98.

5. Eric Forbes, *Greenwich Observatory*. There seems to have been very little urgency to reduce or publish any Greenwich observations beyond those needed for the *Nautical Almanac*. David Rittenhouse, "An oration, delivered February 24, 1775, before the American Philosophical Society," in William Barton, *Memoirs of the life of David Rittenhouse, LL.D., F.R.S.* (Philadelphia: 1813), pp. 543–77, concerns history, moral benefits, and current problems of astronomy.

6. Based on astronomical books and on observations and theories reported to the American Philosophical Society, the American Academy of Arts and Sciences, the Connecticut Academy of Arts and Sciences, and the Franklin Institute.

7. Nathaniel Ingersoll Bowditch, *Memoir of Nathaniel Bowditch*, 3d ed. (Cambridge: John Wilson and Son, 1884). Originally printed in vol. 4 of Bowditch's translation of the *Mécanique Celeste* (1839), p. 29.

8. Ibid.; also in *D.A.B.* and *D.S.B.*

9. Brooke Hindle, *David Rittenhouse* (Princeton: 1964); also in *D.A.B.* and *D.S.B.*

10. Edward S. Holden, *Memorials of William Cranch Bond and of His Son, George Phillips Bond* (San Francisco: C. A. Murdock and Co. and New York: Lemcke and Buechner, 1897); also in *D.A.B.* and *D.S.B.*

11. James Renwick, "Longitude of New York ascertained by observations . . . ," *American Journal of Science*, 5 (1822): 143–44.

12. Jose Joaquin de Ferrer, "Observations of the eclipse of the sun, June 16th, 1806; made at Kinderhook, in the State of New York," *Transactions, American Philosophical Society*, 1804, 6 (part 2): 264–75.

13. William Dunbar, "Observations of the comet of 1807," *Transactions, American Philosophical Society*, 1804, 6 (part 2): 368–74.

14. "Simeon DeWitt," in *D.A.B.*

15. "Andrew Ellicott," in *D.A.B.*

16. See, for instance, Montgomery Robert Bartlett, *Young Ladies' Astronomy* (Utica: Printed for the author by Colwell and Wilson, 1825).

17. Brooke Hindle, *David Rittenhouse.*

18. I. Bernard Cohen, *Some Early Tools of American Science* (Cambridge: Harvard University Press, 1950), pp. 64–65, 157; Joseph Pope, "Description of an orrery of his construction," *Memoirs, American Academy of Arts and Sciences*, 1804, 2 (part 2): 43–45.

19. Burgiss Allison, "A Description of the Pendant Planetarium," *Transactions, American Philosophical Society*, 1802, 5: 87–89; "Burgiss Allison," in *Appleton's Cyclopaedia of American Biography*, vol. 1, p. 58.

20. Uzziah C. Burnap, *The Youth's Ethereal Director* (Middlebury, Vt.: Printed by J. W. Copeland, 1822).

21. Silvio A. Bedini, *Thinkers and Tinkers* (New York: Charles Scribner's Sons, 1975), pp. 380–82.

22. "A Mercator MAP of the STARRY HEAVENS, comprehending the whole Equinoctial, and terminated by the Polar Circles . . . W. Croswell fecit 1810"; William Croswell, *Description and Explanation of the Mercator Map of the Starry Heavens; with its application to the solution of problems* (Boston: John Eliot, 1810).

23. Montgomery Robert Bartlett, *The Description and Use of Bartlett's Celestial Planispheres* (Utica: Printed by Colwell and Wilson, 1825). The maps themselves went through several editions, including 1824, 1825, 1827, and 1831.

24. Elias Loomis, *Recent Progress of Astronomy; Especially in the United States*, 3d. ed. (New York: Harper and Brothers, 1856); Willis I. Milham, "Early American Observatories," *Popular Astronomy* 45 (1937): 1–28; Whitfield Bell, "Astronomical Observatories of the American Philosophical Society, 1769–1843," *Proceedings, American Philosophical Society*, 1964, 108: 7–14; David Musto, "A Survey of the American Observatory movement, 1800–1850," *Vistas in Astronomy* 9 (1967): 87–92; W. C. Rufus, "Astronomical observatories in the U.S. prior to 1848," *Scientific Monthly* 19 (1924): 120–39.

25. Andrew Ellicott, "Observations of the eclipse of the sun, June 16th, 1806; made at Lancaster," *Transactions, American Philosophical Society*, 1804, 6 (part 2): 253–58; Simeon DeWitt, "Observations on the eclipse of 16 June, 1806," Ibid., 300–2; William Dunbar, "Observations of the eclipse of the sun, June 16th, 1806; made at the Forest, near Natchez," Ibid., 260–64.

26. Sears Cook Walker, "Note on astronomical observations, particularly with reference to a monthly statement of celestial phenomena, to be calculated for Philadelphia and its vicinity," *Journal, Franklin Institute* 13 (1834): 76–78.

27. William Dunbar, "Observations of the eclipse of the sun, June 16th, 1806; made at the Forest, near Natchez," *Transactions, American Philosophical Society*, 1804, 6 (part 2): 260–64.

28. Edward Courtenay, "On the difference of longitude of several places in the United States," *Transactions, American Philosophical Society*, 1837, n.s. 5: 343–46.

29. Samuel Williams, "Observations of a solar eclipse, October 27, 1780; made on the east side of Long Island, in Penobscot-Bay," *Memoirs, American Academy of Arts and Sciences*, 1785, 1: 86–102.

30. Isaac Orr, "An essay on the formation of the universe," *American Journal of Science* 6 (1823): 128–49; Ronald Numbers, "The nebular hypothesis of Isaac Orr," *Journal for the History of Astronomy* 3 (1972): 49–51.

31. Denison Olmsted, *An Oration on the Progressive State of the Present Age* (New Haven: 1827), quoted in Stanley M. Guralnick, "Science and the Ante-Bellum American College," *Memoirs, American Philosophical Society*, 1975, 109: 80.

32. Henry C. King, *The History of the Telescope* (London: Charles Griffin and Co., 1955); J. Robert Waaland, "Fraunhofer and the great Dorpat refractor," *American Journal of Physics*, 33 (1967): 344–50.

33. "Quadrant," *Encyclopaedia Britannica*, 1st American ed. (Philadelphia, 1798), vol. 15, pp. 727–28.

34. Forbes, *Greenwich Observatory*.

35. King, *History of Telescope*, chap. 11.

36. "William Herschel," in *D.N.B.* and *D.S.B.*

37. "Friedrich Wilhelm Bessel," in *D.S.B.*

38. George Biddle Airy, "Report on the progress of astronomy during the present century," *Report, British Association for the Advancement of Science*, vol. 1, 2d. ed., (London: 1835), pp. 125–89.

39. Elias Loomis, *Recent Progress of Astronomy*, p. 214; B. A. Gould, "Sears Cook Walker," *Proceedings, American Association for the Advancement of Science*, 1854, 8: 18–45.

40. This point was made and suggested to me by Marc Rothenberg. Cf. E. Otis Kendall, "On the longitude of several places in the United States, as deduced from observations of the solar eclipse of September 18th, 1838," *Transactions, American Philosophical Society*, 1841, 7: 67–72; Sears Cook Walker and E. Otis Kendall, "On the great comet of 1843, "*Proceedings, American Philosophical Society*, 1843, 3: 67–85; *Astronomical Journal* as edited by B. A. Gould.

41. Marc Rothenberg, "The Educational and Intellectual Background of American Astronomers, 1825–1875," Ph.D. diss., Bryn Mawr College, 1974.

42. O. M. Mitchel, *The Planetary and Stellar Worlds* (New York: 1848) contains the text of his ten lectures.

43. B. A. Gould, "Sears Cook Walker," *Proceedings, American Association for the Advancement of Science*, 1854, 8: 26.

44. O. M. Mitchel in *Sidereal Messenger*, 1 (1846), 2; John Quincy Adams, *An oration delivered before the Cincinnati Astronomical Society* (Cincinnati: 1843).

45. Alexis Caswell, "Address on retiring from the duties of President," *Proceedings, American Association for the Advancement of Science*, 1859, 13: 1–26.

46. Robert P. Multhauf, ed., *Holcomb, Fitz, and Peate: Three 19th Century American Telescope Makers* (Washington, D.C.: Smithsonian, 1962); Deborah Jean Warner, *Alvan Clark & Sons, Artists in Optics* (Washington, D.C.: Smithsonian, 1968).

47. Edward S. Holden, "The beginnings of American Astronomy," *Annual Report, Smithsonian Institution* (Washington, D.C.: Smithsonian, 1898), pp. 101–8, said Würdemann "had a decided influence on observers and instrument makers throughout the United States, as he introduced extreme German methods and models among us, where extreme English methods had previously prevailed."

48. Marc Rothenberg, *Background of American Astronomers*.

49. "Denison Olmsted," in *D.A.B.*

50. "Stephen Alexander," in *D.A.B.*

51. "Benjamin Peirce," in *D.A.B.* and *D.S.B.*; "William Bond," (see n. 10); Bessie Zaban Jones and Lyle Gifford Boyd, *The Harvard College Observatory* (Cambridge: 1971), pp. 46, 96–97.

52. A. Hunter Dupree, *Science in the Federal Government* (New York: Harper Torchbooks, 1964), pp. 100–8; see also the annual reports of these organizations.

53. "Ormsby MacKnight Mitchel," in *D.A.B.*

54. Jones and Boyd, *Harvard College Observatory*, pp. 40, 56, 116–17.

55. B. A. Gould, *Reply to the "Statement of the Trustees" of the Dudley Observatory* (Albany: 1859).

56. Deborah Jean Warner, "The American Photographical Society and the early history of astronomical photography in America," *Photographic Science and Engineering*, 11 (1967): 342–47.

57. Nathan Reingold, ed., *Science in nineteenth century America* (New York: Hill and Wang, 1964), pp. 134–35; David Musto, "American Observatory Movement," pp. 87–92; Carroll W. Pursell, Jr., *Astronomy in America* (Chicago: 1967).

58. Quoted in Eve Merriam, ed., *Growing Up Female in America* (Garden City, N.Y.: Doubleday, 1971), p. 91.

59. Psychohistorians might consider the image of a large nineteenth-century equatorial telescope mounted under a dome, as contrasted with that of a small portable instrument from the eighteenth century.

60. Edward Everett, "The uses of astronomy," *Annals, Dudley Observatory* (1866), vol. 1 (inaugural address).

61. "The accurate determination of the geographical position of a single such place [in this case, the Hudson Observatory of Western Reserve College in Ohio] in a new state, affords a standard of reference by which a large surrounding territory is tolerably well located through the medium of the local surveys." Elias Loomis, *The Recent Progress of Astronomy* (New York: Harper and Brothers, 1856), p. 213. On a more formal basis, a good many observatories—including those of Lewis R. Gibbes, Lewis M. Rutherfurd, and William and Maria Mitchell, as well as most institutional ones—were official Coast Survey stations. For the Harvard College Observatory time service see Ibid., p. 253.

62. W. L. Marshall, comp., *Notes on Talcott's method of determining terrestrial latitude* (1893); Truman Safford, "A catalogue of the declinations of 532 stars culminating near the zenith of the observatory of Harvard College, Cambridge," *Memoirs, American Academy of Arts and Sciences*, 1860, 8: 299–332.

63. Loomis, *Recent Progress of Astronomy*, pp. 304–67.

64. O. M. Mitchell, *The Planetary and Stellar Worlds* (New York, 1848), p. xi.

65. "Death of O. M. Mitchell," *American Journal of Science* 34 (1862): 451–52.

66. Benjamin Apthorp Gould, "On the meridian instruments of the Dudley Observatory," *Proceedings, American Association for the Advancement of Science*, 1856, 10: 113–19.

67. *United States Naval Observatory, Astronomical Observations*, (1846) 1: preface.

68. Loomis, *Recent Progress of Astronomy*, pp. 293–99.

69. *Annals, Harvard College Observatory*, vols. 1, 2, and 6.

70. "Maria Mitchell," in *D.A.B.* and *D.S.B.*

71. *Comptes Rendus, Académie des Sciences*, 48 (1859): 485–87; *American Journal of Science* 28 (1859): 119–20.

72. *Comptes Rendus, Académie des Sciences*, 55 (1862): 936–37.

73. Loomis, *The Recent Progress of Astronomy*, pp. 9–53.

74. C. H. Davis, "Report on the Nautical Almanac," *American Journal of Science* 14 (1852): 317–35 (esp. p. 323); C. H. Davis, "On the solar

eclipse of July 28, 1851," *Proceedings, American Association for the Advancement of Science,* 1851, 6: 93–99.

75. George Phillips Bond, "On the rings of Saturn," *Memoirs, American Academy of Arts and Sciences,* 1851, 5: 113–21; Benjamin Peirce, "The constitution of Saturn's rings," *Proceedings, American Association for the Advancement of Science,* 1851, 5: 18–22.

76. Warren De La Rue, "Address . . . on presenting the Gold Medal of the Society to Professor G. P. Bond," *Monthly Notices, Royal Astronomical Society* 25 (1864–65): 125–37.

77. Letter from George B. Airy in *Sidereal Messenger* (1846) vol. 1.

78. Denison Olmsted, "Observations on the meteors of Nov. 13th, 1833," *American Journal of Science* 25 (1834): 363–411; 26, 132–74; see articles by Denison Olmsted, Alexander Twining, F. A. P. Barnard, and Edward C. Herrick in, among other places, the *American Journal of Science.*

79. Ebenezer Porter Mason, "Observations on nebulae . . ." *Transactions, American Philosophical Society,* 1841, 7: 165–214.

80. James Dwight Dana, "On the volcanoes of the Moon," *American Journal of Science* 2(1846) 335–55.

81. Loomis, *Recent Progress of Astronomy,* pp. 244–56; Jones and Boyd, *Harvard College Observatory,* pp. 48–50.

82. Loomis, "Scientific intelligence—Astronomy." *American Journal of Science* 2 (1846) 438.

83. George Phillips Bond, "An account of the nebula in Andromeda," *Proceedings, American Academy of Arts and Sciences* 3 (1848) 75–86.

84. "John William Draper," in *D.A.B.* and *D.S.B.*

85. Jones and Boyd, *Harvard College Observatory,* pp. 70–87.

86. Deborah Jean Warner, "Lewis M. Rutherfurd: Pioneer astronomical photographer and spectroscopist," *Technology and Culture* 12 (1971): 190–216; "Henry Draper," in *D.A.B.* and *D.S.B.*

87. Lewis M. Rutherfurd, "Astronomical observations with the spectroscope," *American Journal of Science* 35 (1863): 71–77.

88. George Phillips Bond, "On the light of the moon and of the planet Jupiter," and "Comparison of the light of the sun and moon," *Memoirs, American Academy of Arts and Sciences,* 1861, 8: 221–86; 287–98.

89. Alvan Clark, "The sun a small star," *Proceedings, American Academy of Arts and Sciences,* 1863, 8 (part 2): 569–72.

90. D. B. Herrmann, "B. A. Gould and his *Astronomical Journal,*" *Journal for the History of Astronomy* 2 (1971): 98–108.

91. Mitchel's Albany lectures were most likely similar to the ten published as *The Planetary and Stellar Worlds* (New York: 1848).

92. Quoted in Richard Olson, "The Gould controversy at Dudley Observatory: Public and professional values in conflict," *Annals of Science* 27 (1971): 265–76.

93. *Annals, Harvard College Observatory*, vol. 1, pp. 21–28; Sears Cook Walker, "Report on the experience of the Coast Survey in regard to telegraph operations, for determining longitude, & c.," *American Journal of Science* 10 (1850): 151–60.

94. Jones and Boyd, *Harvard College Observatory*, pp. 97–100.

SCIENCE, PUBLIC POLICY, AND POPULAR PRECEPTS: ALEXANDER DALLAS BACHE AND ALFRED BEACH AS SYMBOLIC ADVERSARIES

ROBERT POST
Smithsonian Institution

Democratic republics extend the practice of currying favor with the many and introduce it into all classes at once....
ALEXIS DE TOCQUEVILLE

NATHAN REINGOLD SEES PUBLIC POLICY AND POPULAR ATTITUDES as likely topics for closer attention from historians of American science.[1] They are indeed, although in the current state of knowledge the reciprocities between these two topics are nothing if not elusive, especially since no good measure of popular attitudes existed before the rise of the popular press in the 1830s and 1840s. Even when the vox populi of the Jacksonian era can be heard loud and clear, one must contend with the problems of relating that voice to a governmental function that naturally seeks exemption from democratic constraints. Suggestive of a periodic disjunction between public policy and popular attitudes is the high-handed behavior Ferdinand Hassler of the Coast Survey could get away with.

That in mind, one might even rekindle the historians' debate about the degree of governmental sensitivity to popular will—"real" popular will, that is—in the antebellum years. I would not try to renew this debate; on the contrary, my feeling is that from the time public policy became an ongoing issue, popular attitudes made a great difference. Moreover, I would take the liberty of adding emphasis to William Stanton's observation that "science was expensive in time and money, and controversy as well. As Congressmen generally preferred the comfortable groove of strict construction, *the only force likely to divert them was public opinion....*"[2] Both Congress and the public could be fooled, it is true, but the "surreptitious creation"[3] of the Naval Observatory seems to have been a unique circumstance for its time.

After admitting that popular attitudes are important, there yet remains to trace processes of mobilization and implementation. G. M. Young once remarked, appropos of diplomatic history, that most of it was "little more than the record of what one clerk said to another clerk."[4] Some of us would give a lot to know more of what certain clerks actually said to each other! However laborious the job, popular attitudes *can* be ferreted out, and public policy is, well, public. Most enigmatic are specific interactions: just *how* did Alexander Dallas Bache become "the most important single person in the evolution of the government's policy toward science and technology"?[5] Whom did he cultivate, who were his adversaries, what were his techniques and theirs?

Thanks to several worthy studies, particularly Sally Gregory Kohlstedt's,[6] we know a good deal about how Bache stage-managed both the American Association for the Advancement of Science (AAAS) and the National Academy of Sciences. Yet what accounts for his phenomenal success at loosening public pursestrings for the Coast Survey, especially as contrasted to Charles Wilkes's fitful fortunes in fulfilling the mission of the U.S. Exploring Expedition he had commanded? What exactly did Maria Mitchell mean when she observed that Bache's "enormous appropriations" derived from his skill at "working sundry little wheels, pulleys and levers"?[7] To some measure, the answers are obvious—e.g., he was the right man in the right place—but I also suspect that we have not even thought of all the proper questions as yet. In any event, one is certain to find in the last century, hardly less than ours, ambivalent attitudes about science, "mingling anger and enthusiasm, lavish support and profound mistrust."[8]

The historian's ultimate mission, nonetheless, lies in discerning causal relationships; thus, penultimately, one must frame problems in a way that makes this discernment imperative in the solutions. With respect to the task at hand, I propose to examine two opposite sets of assumptions about public policy, assumptions that were rooted in fundamentally disparate notions of hierarchy with respect to abstract science, applied science, and technology. One protagonist is Bache, who sought public support for all three, but privately ranked the abstract above the empirical. Bache's skillful pragmatism and his knack for camouflaging his own enthusiasms amid utilitarian programs gave a decisive impetus to public policy in the antebellum years. Had Bache not been so adroit, it might have been otherwise, for there was a formidable counterinfluence to the hierarchical and programmatic concepts he epitomized: the weekly periodical *Scientific American*.

Never did Bache and the editor of *Scientific American*, Alfred Ely Beach, clash head-on. Indeed, they rarely even acknowledged each other, and when, on occasion, Beach did mention Bache in print, it was with due respect.[9] Obviously, then, the two were not literal adversaries. Yet, the matchup is based on something more substantial than just an attractive alliteration. Beach stood for something he called "democratic science." Bache believed in an elitist imperative that amounted, in Edward Lurie's phrase, to "intellectual imperialism."[10]

To reemphasize, Bache and Beach did not confront each other directly. I set them in opposition as ideal types in a sociological sense—the one "the acknowledged spokesman for professional science,"[11] the other the journalist most seriously devoted to developing the "popular taste for scientific information."[12] Real adversaries there were, too—Horace Greeley and Joseph Henry, for example—but Henry eschewed the sort of role Bache relished, and, in contrast to the deliberately abrasive Greeley, Beach's *Scientific American* presented a reasonable case built upon thoughtful precepts. It was no paragon of consistency, but then neither was Bache, especially when it came to the rhetoric and the reality of his criteria for professionalism.

Professionals, the rhetoric had it, took professional matters only to peers, never the public. There could be no such thing as "democratic science." To be sure, the critical proscription pertained to soliciting popular support in scientific controversies, not merely "currying favor." In one sense, however, the two were inseparable, for the latter necessarily entailed propounding a peculiar system of scientific values and priorities. Bache and most of his cohorts presumed that only a few people "understood the essence of true science,"[13] and that anybody who had to ask could probably never know. Beach, on the other hand, answered the question, "What is science?" with a simple definition: "a collection of facts and experience accurately arranged and properly understood." It was *not* "some mysterious intellectual subtlety, revealed to the few and denied to the many."[14]

Obviously, *Scientific American* spoke to far more people than were privy to "true science" as Bache understood it. Thus, even though Beach could not command the sort of resources that Bache did by virtue of his official position and private alliances, he enjoyed an enormous advantage in the sheer size of his audience. At midcentury, before the circulation of the *American Journal of Science* had ever topped a thousand, *Scientific American* was selling more than ten thousand copies every week. By the Civil War, circulation was forty thousand. When Beach expressed concern about misconceived priorities in public policy, Bache had no comparable mode of reply. Moreover, there was some question as to whether he ought even to try. As the leading proponent of an ethic whose sine qua non was professional autonomy, Bache might simply have followed the lead of Ferdinand Hassler, who had characteristically disdained even to assuage Congress, let alone the public.

Bache's forte, however, was his ability to court public favor quite deliberately, to insinuate his own system of values while ever seeming to remain above it all. Occasionally, some capable scientist such as chemist Walter Johnson or electrician Charles Page took on *Scientific American* directly, on its own pages. This approach betrayed a deficient sense of "style and tone," a shortcoming that Reingold has perceived in Matthew Fontaine Maury.[15] A Hassler might declare (and maybe he did) that "any President can make a Secretary of the Treasury, but only God Almighty can make a Hassler."[16] Maury would never have said anything like that; yet, somehow or other, he appeared inordinately fond of "aggressive promotion."[17]

Bache, on the other hand, seemed aloof. For example, there was nothing discernible linking him with the AAAS "Committee of Twenty" that produced a *Report on the History and Progress of the American Coast Survey*—123 pages of uninterrupted eulogium that concluded by earnestly soliciting "the continuance of the Executive favor and Legislative support [the survey] has hitherto enjoyed."[18] There was nothing discernible, yet to all intents and purposes this was Bache's own production.

Such testimonials, ostensibly impartial, were obviously a factor in the American Coast Survey's "enormous appropriations." But they were not the only factor and not sufficient in and of themselves. The fact nevertheless remains that Bache commanded a half-million dollar budget by 1855, a sum almost equal to the entire Smithson bequest. How different this was from 1825, with President Adams propounding all sorts of lavish programs while a Coast Survey, resplendent in fancy instruments but unfunded and essentially moribund, was what truly epitomized public policy!

A constitutional issue seemed to bother our second generation of statesmen more than the first, although from the very outset strict constructionists had blocked much of anything beyond some skimpy funding for a field party here and there. The $50,000 appropriated to the Coast Survey at its inception in 1807 was extraordinary. An explicit reference to "science" in the U.S. Constitution linked it directly to "useful arts," and one had to stretch the bounds of common sense considerably to read anything more into this than an authorization to protect property rights in discovery and invention. It was inevitable, then, that determinations of science policy would get mixed up in the rhetoric of legalism, with debates cast in terms that embraced the fundamental political polarities of the new nation.

It is also important to reemphasize that concern about science per se (as distinct from concern about the constitutionality of supporting it) had been the province of a tiny minority—one scarcely larger than the combined membership of the American Philosophical Society and the American Academy of Arts and Science. Notwithstanding occasional evidence to the contrary, such as Jefferson's image as a man of science proving a bit of a political liability, popular attitudes were irrelevant. What did count were attitudes on Capitol Hill. Although personally captivated by Joel Barlow's plan for a national academy and university with both an instructional and a research mission, Jefferson quickly retreated when he sensed how the wind was blowing in Congress.

What Jefferson *was* able to accomplish, in addition to establishing a survey of the coast, was to set up a patent office and see that Congress appropriated funds for expeditions under War Department auspices. The first and most famous of these was the expedition that headed up the Missouri to see about "extending the external commerce of the United States." Jefferson revealed its mission to foreigners as "literary" (meaning scientific) and thereby established a precedent for pursuing science as an integral part of exploration and the quest for empire.

Here obstacles began to appear that were not just political, one being simply the shortage of trained observers among men whose capabilities also extended to leadership and organization. Meriwether Lewis had had nothing more than a crash course at the American Philosophical Society (APS). Zebulon Pike, overwhelmed by what he encountered in the northern provinces of New Spain, at one point lamented that he should have had "the eye of a Linnaeus or a Buffon."[19] Hassler, the one scientist fully competent at field work, was a geodesist, not a naturalist.

That, it turned out, was a marketable skill in the United States in 1807, with Thomas Jefferson in the White House and a dynamic commercial impulse being fanned to life. Hassler might have gone a long way if he had been attuned even slightly to American sensibilities. But first off he overspent his princely appropriation, mostly in the capital of a country the United States was at war with. More years passed, with no payoff in any terms, practical or otherwise.

Hassler still held a strong hand when Congress passed his second appropriation, yet, after finally getting to work in 1816, he never bothered to see that his progress was broadcast. Politicians began paying more and more attention to people like Cheever Felch, chaplin of the Navy, who insisted that, whatever Hassler was doing, it could all be done better and quicker and cheaper. The congressional influence of patrician sorts who had tried to befriend science was dwindling. Shortly, Hassler was unceremoniously dumped, and there ensued a period when nothing got funded except what appeared to have a quite immediate payoff (small surveys mostly).

The likelihood of direct benefits from the Long Expedition had been clear, as was the case with the revivification of the Military Academy at West Point under Silvanus Thayer. William Beaumont at Mackinac was given a bit of assistance. On the other hand, William Lambert, partisan of an American prime meridian and proponent of a national observatory, could usually get a polite hearing from Congress but never any funds. As for John Quincy Adams, he betrayed an incredible shortage of political savvy with his plans for a national observatory and even more profligate ideas such as a department of internal improvements aimed at subsidizing science in general. The reaction to these ideas was not so totally disastrous as sometimes depicted.[20] Nevertheless, it is true that, when Adams left office, "it would have been difficult . . . to propose a more unpopular measure than the national university or a national observatory."[21] Perceptions of science tended to things ethereal, useless, and dangerously centralized.

The Age of Jackson, the embodiment of paradox in so many ways, provides the historian with a succession of intriguing anomalies in science policy. Natural philosophy became more esoteric and genuinely popular at the same time. Jacksonian rhetoric was strict constructionist, yet the Constitution was stretched into very odd shapes indeed. The common man counted for more politically than ever before (and, it was said, could perform any public function that the government had any right to be involved in), while the government supported abstruse and most uncommon pursuits as never before. The U.S. Patent Office was reorganized with the aim of having scientists measure novelty against abstract criteria, yet technology was fostered almost exclusively by people with little or no affinity for science. Finally, there was this incongruity: although the new generation of professional scientists felt it essential to define their sphere in very exclusive terms, the public seemed more and more inclined to think of science as technology, the latter being largely a pursuit of self-taught mechanics or "mere empirics."

For a time, it became rather easy to invoke the name of *science* to secure federal appropriations for developing *inventions*. Morse's $30,000 subsidy in 1843 was followed by one for $15,000 to Samuel Colt and subsequent sums ranging from $5,000 to $20,000 for investigating gas lights, preservatives, fire protection systems, and sundry other utilitarian devices.[22] By any straightforward reading, all this was unconstitutional. Equally worrisome were questions about the competence of Congress to weigh the relative merits of various petitions for assistance. Although this sort of aid pretty much died out in the face of popular concern about political favoritism,[23] such subventions helped break down residual constitutional scruples and also conditioned the public to spending in contiguous realms—or realms that at least appeared contiguous because of that blurred distinction between "science and the useful arts."

Here is what perhaps was most significant: the persistence of a concept of science that seems remarkably undifferentiated. Under the heading, "Appropriations . . . for the Encouragement of Science," Beach's periodical in 1849 lumped together funds for the Patent Office Library, for agricultural research, for the botanic garden, for "the construction at the National Observatory of a magnetic clock," for "copying abstracts from old sea journals" and preparing a nautical almanac, for surveys and charts, "for salaries of special examiners of drugs, medicines, and chemicals," and "for testing the capacity and usefulness of electro magnetic power."[24] Trials of motors, boilers, combustibles, or building stones, involving as they did prominent "scientific men" such as Walter Johnson and Charles Page (not to mention Bache himself), got popular images of science and technology almost hopelessly mixed up.

To be sure, the distinction between science and technology characteristically has seemed far more important to scientists or inventors than to anyone else, except historians who tend to take rhetoric at face value. Reingold, who does not, has emphasized Bache's role in defining science and technology in "the American idiom," which renders distinctions between pure and applied science purely artificial.[25]

I would suggest another peculiarity of the American idiom: the importance of cultivating popular approbation while appearing to remain both modest and detached from the petty exigencies of democracy. This is a subtle matter. Maury, who yielded nothing to Bache in the realm of ambition, owed many of his professional setbacks to a penchant for using "the first person singular too frequent[ly] to be pleasing."[26] The fact remains, though, that until such time as the mechanics of *funding* became incomprehensible, American scientists dependent upon the public purse (which is to say, the majority) had to try and make their pursuits popularly comprehensible—or rather, they had to appear to take a disinterested stance while others spoke for them.

Hassler was the last minion of the old school, deliberately flaunting a disagreeable image. He survived both an attempt to put the Coast Survey back into the Navy (where it had formerly languished for twelve years) and a congressional investigation that found him insufferable if not irresponsible. Yet the whole problem of science in a democratic society was seething at the time of Hassler's death in 1843.

At heart, Bache may have been every bit as much an elitist as Hassler, but he also distinguished rigorously between the real and the ideal. He was, in addition, a master organizer, able to institute major changes while avoiding anything like a repeat of the quarrels attendant upon the organization of the U.S. Exploring Expedition, which once saw Jeremiah Reynolds and Navy Secretary Maylon Dickerson tilting at each other in New York newspapers. The most noteworthy point, though, is not Bache's talent for organization, for he would never have gained the opportunity to exercise it fully had he not had such a shrewd instinct for public relations. There follows an illustrative episode.

Bache had private allies in Washington, Cambridge, Philadelphia, New Haven—just about everywhere, in fact. As for his adversaries, the bulk were concentrated in New York City, a focal point of opposition to federal funding of science. This attitude had roots, of course, in New York's traditional antipathy to federal internal improvements. More important, however, was the role taken by the New York press in upholding the cause of the American inventor. Horace Greeley combined this sentiment with his peculiar sort of anti-intellectualism to launch diatribes against the Smithsonian. Other journalists, generally more civil but no less adamant, continued to press for diverting the Smithson bequest to some purpose beneficial to inventors, a relatively large group, rather than to the nation's small cadre of self-anointed professional scientists. With no funds available even for something so fundamental as printing patent specifications,[27] it was a simple matter to portray America's inventors as mistreated or underprivileged and public priorities as being awry.

The villains were "scientific men," a term never clearly defined but an onus upon anyone apparently insensitive to the subtleties of design and "reduction to practice." The key point is that a "scientific man" was not the same as a "scientific American." The former conceptualized science in a way that was narrow, illiberal, un-American. Hostility ran especially strong during the early 1850s when the Patent Office, under the direction of Thomas Ewbank, measured inventors' applications against extremely stringent and allegedly unrealistic criteria for novelty, and examiners such as George C. Schaeffer and Jonathan Homer Lane rejected 80 or 90 percent of the applications they reviewed.[28] This got a bad press nearly everywhere, but especially in New York. In popular perception these men, whose personal alliances clearly were with Bache, Henry, and their ilk, had perpetrated an act of usurpation: just as Henry had diverted (or perverted) Smithson's money to his own narrow ends, his friends were corrupting the proper role of the Patent Office.

Indignation reached an uproarious level in 1853, when the proportion of applications the Patent Office approved slid to an all-time low. Coincidentally, that was the year a group headed by Theodore Sedgwick, jurist, diplomat, and former confidant of de Tocqueville, got together to stage the "New-York Exhibition of the Industry of all Nations." This was the New York Crystal Palace Exhibition, the first American world's fair.[29] Nineteenth-century world's fairs were primarily celebrations of technological progress, with a strong component of international rivalry.[30] This was the golden age of the independent American inventor-entrepreneur, and scarcely a one of any significance failed to make a showing at the New York exhibition: Morse, Howe, McCormick, Goodyear, Colt, Singer, Corliss, and many more. A few of these men were resourceful publicists; two or three had already passed from history to hagiography. Yet, when it was all over, it was Bache, of all people, who had best succeeded in capitalizing on the exhibition.

Bache, of course, was a master at getting good coverage in the journals and annuals read by his peers. Earlier in 1853, there had even been a weights and measures exhibit at the Patent Office.[31] But to play to the throngs of Gotham was something else. First, the Crystal Palace was the most extravagant pageant ever staged in America—or, indeed, aside from its London prototype of 1851, in the world. Second, in no place else was there more resentment about the idea of the "scientific men" in Washington and Cambridge presuming an elite status and finagling support for what appeared to be pet enthusiasms of an abstract nature.

True, it was already clear to New Yorkers that some of what Bache's Coast Survey did was beneficial to nearly everybody. Its mandated function, after all, was to furnish "with the utmost attainable accuracy, and in a connected and uniform manner, all the geographical, topographical, and hydrographical data that can be made in any way useful to the navigation and defence of the coast,"[32] and one of its first major achievements had been the charting of the Narrows.

Still, not everyone saw eye to eye with Bache as to what was "in any way useful." His personal interests tended to run to theoretical problems, and occasionally he had even stated frankly that the "essential value" of the Coast Survey lay not in its practical operations but in "the scientific parts of the work"[33]—meaning, the abstract parts. Although it was only natural for him to promote science in terms of its congruence with American values by "playing up projects of immediate usefulness,"[34] the idea of having an elaborate display packed up and sent to the Crystal Palace was a move worthy of a man accurately dubbed a "landmark."[35] It is scarcely something old Hassler would have done.

Besides the Coast Survey and its offshoot, the Office of Weights and Measures, exhibits were sent by the Light House Board, an organization closely tied to the Coast Survey; by the U.S. Mint, also tied to the survey through the Office of Weights and Measures; and by the *Nautical Almanac*, the mariner's guide published by the Navy Department and likewise with natural ties to the survey. Lieutenant Maury's Naval Observatory had an exhibit too, but otherwise the entire supporting cast was resolutely Bachephile.

The Light House Board and the *Nautical Almanac* both were seeking the exposure so crucial to new organizations. Moreover, in an apparent effort to help dispel a speculative air that hovered around the enterprise, actual management of the exhibition had been delegated to two naval officers. One of them was Samuel F. Du Pont, a member of the Light House Board along with Bache and Henry; the other was Capt. Charles Henry Davis, chief of the *Nautical Almanac*. Davis was also related by marriage to Benjamin Peirce, the Coast Survey's leading consultant at Harvard and of course an intimate friend and dedicated ally of Bache.[36]

Complementing this web of personal and institutional ties were the measures taken for officially publicizing the Crystal Palace Exhibition. The key was Captain Davis's position. New York was full of competent journalists, many with technological expertise, yet it was not journalists but scientists who were assigned control over the major publications. Benjamin Silliman, Jr., became chief editor of the *Illustrated Record* as well as one of the compilers of *Science and Mechanism*—then as now considered "the most informative work [on the exhibition] that was published in the United

States."[37] Charles R. Goodrich, chief editor of *Science and Mechanism*, had been trained at Yale as a chemist and physician. Neither he nor Silliman were members of the elite junto commanded by Bache, Davis, and Henry, but both were dedicated fellow travelers and unabashed in their admiration of its leader.

In the preface to a compilation of articles from the *Illustrated Record* called *The World of Science, Art, and Industry*, Silliman explained the heavy emphasis on information about "the condition, methods, and instruments of the United States Coast Survey": The answer was simply that this was the one part of the whole exhibition "most honorable to our country," sufficient by itself to allow us to claim rank "among the foremost nations of the world."[38] And here was an exhibition more "rich and various" than any ever before seen "on this side of the Atlantic"![39]

Silliman attended to the publication of great quantities of flattering material about Bache, his organization, and what it had done "to make the name of America honored over the whole world."[40] There was a eulogistic history penned by Davis in which he detailed "practical benefits" but also called attention to a mission "in abstract science" that was "equally useful and distinguished."[41] There was an appeal from Silliman himself for "liberal cultivation of pure science."[42] But the general theme concerned the Coast Survey's splendid aggregation of human resources and especially its international prestige.

When Hassler first set out after equipage, he had had to buy all his instruments in Europe. No more, visitors to the Crystal Palace were told. Even though names such as Troughton & Simms, Brunner, Gambey, and Silbermann still appeared on Coast Survey instruments, by 1853 there were American citizens skilled enough to match or even surpass the best instrument makers of London and Paris. Bache himself had conceived key improvements in the design of the compensating base bar, used for measuring base lines, while William Würdemann, one of the survey's best craftsmen, had built bars of incredible precision.

Bache's most accomplished instrument maker, his close friend Joseph Saxton, had invented a reflecting pyrometer that measured lineal increments as small as .00001 inch. This sort of thing had enabled a standard of accuracy considerably higher than the British had settled for when surveying India. And *that* was the sort of thing which impressed ordinary Yankees enormously. Also on display was a zenith telescope built by Würdemann, along with accoutrements of the method for fixing longitudes that owed so much to the survey's late, lamented Sears C. Walker. His legacy, the "American method," was world renowned, and this too was something that made ordinary Yankees feel proud and confident.[43]

In the realm of oceanography, the exhibit included an ingenious deep-sea thermometer invented by Joseph Saxton and a lead for bringing up bottom samples devised by a Navy officer detailed to the survey. The study of ocean depths, floor contour, currents, and temperatures was a routine survey activity. Although general laws regarding tides were a pet research interest of Bache, the survey's exhibit concentrated on the plotting of local peculiarities by means of Saxton's self-registering gauge.[44] Saxton was chief of the survey's weights and measures operation, too, and despite the presence of French balances and standards, Silliman thought it unlikely that in all the Crystal Palace there was "any thing superior in point of workmanship" to Saxton's.[45] The message was clear: The survey got honors for one-upping the French as well as the English.

Finally, there was the imposing array of charts and maps reproduced by electrotyping, that difficult art whose unchallenged master was the survey's own George Mathiot.[46] Only the most miserable landlubber could fail to appreciate the value of fine navigational aids. And even the most casual visitor to the exhibit, it seems certain, came away impressed by the diverse applications of science that had been conceived or refined by Bache and his staff.

Silliman also put in a good word for Captain Davis's *Nautical Almanac*, on display nearby in the same "noble court."[47] Davis naturally anticipated that favorable publicity would help guarantee a successful launch for the almanac. As for Bache, there is little doubt about what *he* hoped to gain. Taking the proper cue, Davis concluded his essay about the survey with an explicit plea for "liberal appropriations."[48] Moreover, he indicated that all would be for the best if Bache were given a free hand in allocating his funds. Although Bache had published numerous charts and maps dealing with the Pacific and Gulf Coasts as well as the Atlantic, the audience for these materials was, after all, rather small and homogeneous. The publicity he got through the *American Journal of Science*, the AAAS, and the APS had no broad impact, nor did the survey's annual reports. So Bache had set out to inform the general public about his organization, playing heavily on the theme of national pride. Maury had had much the same idea, but Bache garnered the lion's share of publicity, not only in the official publications, but also in the reports of distinguished foreign visitors such as Sir Charles Lyell. The survey, Lyell wrote, was "by far the most important scientific work now in progress in the United States."[49]

Publicity was what the Crystal Palace was all about. The inventors who exhibited were, as Greeley lamented, often rather inept at promoting themselves.[50] There were exceptions such as Colt, Corliss, and McCormick, but very few were as adept as Bache. A major element in his success derived from his maintenance of a disinterested stance; Bache boasted through the mouths of others. There is, nevertheless, no doubt that he regarded the Crystal Palace as a marvelous opportunity to sell a mass audience on the express idea that science was good and that it deserved liberal support—and, implicitly, on the philosophy of professional autonomy he himself had largely shaped.

One cannot gauge the impact of this foray precisely. The presumptive evidence, however, strongly suggests that it was a stroke of genius. During the years immediately following, the Coast Survey enjoyed consistently fair winds. In 1843, Bache had inherited a permanent staff of twenty-three; in 1853 he had forty; by 1855 he had fifty-eight—more growth in two years than in ten. Or, consider his federal appropriations: the $500,000 budget he was able to command by the mid-1850s represented a five-fold increase from the 1840s, and even then his annual appropriations had been considered lavish.

After midcentury, mutual feelings between those who identified with the community of professional scientists and those who called themselves inventors characteristically ranged from indifference to undisguised hostility. Part of this is thought to have stemmed from conflicting notions about whether it was proper to mount the stage and promote oneself publicly. People such as Bache are supposed to have been reticent about doing so, and indeed he was. But he was not averse to orchestrating the encomiums of others. The grand sachem of American science seized this extraordinary occasion no less avidly than the likes of Samuel Colt and George Corliss.

As for the "scientific men" involved in publicizing the Crystal Palace, they did more than just boost science in the federal government. They also took the occasion to broadcast a favorite theme: that abstract science intrinsically outranked "mere empiricism," and that without a continuous scientific inflow the wellsprings of invention would run dry. It was this presumption more than anything that goaded Greeley and others to blast "scientific men." And it was these blasts, in turn, that precipitated laments about it becoming "too much the custom, in quarters where a better spirit might be looked for, to ridicule the claims of science, and deny the obligations it has conferred upon industry...."[51]

One quarter where a "better spirit" might have been sought was the New York editorial offices of *Scientific American*. Alfred Beach, who presided, knew the techniques of popular journalism intimately, for his father had helped invent the genre, and at one time his *Sun* had outsold every other newspaper in the world.[52] Bache and his cohorts knew that Beach held a formidable weapon which, if not deadly, could do serious damage to a fledgling program for professionalizing American science. Bache's spokesman Silliman insisted that "no nation can neglect such sources of improvement as, known or latent, exist in science, and hope to retain its manufacturing or commercial wealth and importance."[53] Beach was ready to turn that whole notion upside down: "In many things the laboratory is far behind the workshop, and many learned men would learn a great deal by condescending to be taught by the intelligent artisan."[54] Or, more specifically, Beach referred to recent achievements in astronomy: "Signal advance in the mechanic arts and the prevalence of mechanical genius gave the first step, and something like metaphysical or analytical observation followed it up."[55] Usually, the "claims of science" were simply irrelevant to tangible progress.

Beach explicitly disputed the precept that a "true" scientist's fundamental commitment could only be to the increase of knowledge rather than its diffusion, or that the pursuit of abstract principles was a higher calling than the "mere application" of those principles, or that invention somehow "flowed down" from basic scientific principles. Implicit in the contention over the ranking of theory and practice was another disagreement, about whether or not science was any longer a pursuit that should or could assume democratic proportions. *Scientific American* became the last influential exponent of democratic science in America, the last to deny that this entailed some obvious contradiction in terms.[56]

Today, virtually all historians of American science would agree it was important to the maturation of American science for Bache and his allies to have succeeded in establishing professional standards on their own terms.[57] Much of what they sought was selfish, not disinterested, just as it is also true that much—perhaps even more—of what Beach and his constituency sought was self-seeking, too. Still, that fact did not render Beach's indictment of "intellectual imperialism" any less persuasive. He stressed his ideas week after week, year after year, to a mass audience. Had Bache not perceived the necessity of countering this indictment in a popular idiom, he might not have achieved his aims nearly so readily.

Scientific American's catholic definition of science and its insistence about knowledge being "democratic" may now appear as merely quixotic, as an effort to hold science back short of the "edge of incomprehensibility." Consider, however, Musson and Robinson's assertion that "the historian is not entitled to dismiss out of hand the notions that people of the past had about themselves. Human activities are affected by the way in which human beings contemplate their own behavior."[58]

One aspect of the conventional wisdom about the changes that affected science in the nineteenth century pertains to what has been called the "deluge of facts," a development that purportedly forced abandonment of the inductive method.[59] Yet *Scientific American* continued to insist that the primary tasks were inductive. It also contended that the notion of incomprehensibility was often exaggerated, if not a deliberate myth, and that the taxonomic sciences had not necessarily become more difficult, they had just become different. I might stop here to note that this is something Asa Gray would have agreed with, and to indicate my own feeling that, elements of physics and chemistry aside, a general comprehension of most sciences remained possible even for people with only modest formal instruction.[60]

In any event, *Scientific American* would neither admit that science had moved beyond the capacity of ordinary people to understand, nor that narrow specialization was requisite to its advance, nor that "scientific men" possessed an expertise that somehow was more venerable than other sorts of expertise. Beach never lost an opportunity to relate a tale of some "learned man's" ineptitude in the realm of common sense, manual skill, or "the practical arts."[61] Design, invention, "reduction to practice"—all these, he insisted, also were science.

THERE WAS, AFTER ALL, NOTHING PURPOSELY whimsical in the name of *Scientific American*. Beach did not choose it, but it seems unlikely he ever seriously pondered changing it.[62] He saw *Scientific American* as the spokesman for a whole scientific community, rather than just a small part. Here, Bache was especially vulnerable, for, as Dupree notes, his definition of science essentially comprehended only "those branches which the surveying and exploring enterprises of the government had stimulated."[63] The science *Scientific American* promoted was unabashedly Baconian in aim ("The legitimate goal of science is the endowment of human life with new inventions and riches") and methodology ("observe facts, institute experiments, and from effects reason to causes").[64] But it never advocated a simple-minded data grubbing, nor did it denigrate the quest for knowledge "abstract or remote from common affairs."[65] It questioned only the propriety of financing this with public funds.

Since few of the "learned men" Beach criticized were nearly so ready to abandon inductive method as their rhetoric sometimes suggested, his people and Bache's people ideally could have cooperated. Sometimes they did. Much more often they were antagonistic. This antagonism stemmed largely from the one difference of opinion that was stubbornly irreconcilable, namely the perceived hierarchy between abstract science, applied science, and technology. Beach knew that inventors rarely drew upon ideas at the forefront of scientific knowledge and so found it difficult to understand the urgency, insofar as "the endowment of human life with new inventions and riches" was concerned, of promoting abstract science as a matter of public policy. Invention improved the quality of everyday life; the pursuit of abstractions edified only those engaged in that pursuit.

Beach never denied that the scientist in his laboratory and the mechanic in his workshop both were doing God's work. But, with the proponents of a professional ethic apparently trying to arrogate all rewards to themselves, every advance in their program became anathema. The Smithson bequest, which could have been diverted to truly useful purposes—a comprehensive illustrated history of technology, for example—instead went to fund esoteric publications of extremely limited interest. The AAAS, which might have emulated some of the more admirable practices of its British counterpart, tended to encourage useless speculation. The National Academy simply entrenched the worst tendencies of the AAAS. An organization such as the Coast Survey was grandly funded, even though only a portion of its activities promoted the general welfare. In contrast, *Scientific American* was genuinely democratic in that its heroes included individuals all across the theorist-empiricist spectrum, from Michael Faraday to Thomas Davenport.

I am, of course, paraphrasing or extrapolating from *Scientific American*'s own rhetoric. But the partial element of truth to all this suggests that Beach was not merely fighting a rearguard action on behalf of amateurism; he was also intent upon trying to preserve some democratic ideal in the realm of public policy. As with most people concerned about such matters as democratic ideals, Beach's thinking was sometimes fuzzy; as with most people who promote concepts like elite prerogatives, Bache often had a firmer grip on political realities.

By the 1870s, science had attained a position that would have seemed incredible fifty years before. "In terms of jobs, funding, and the existence of organizations to speak for their interests," Daniels writes, "American scientists had never had it so good."[66] In 1876, the Coast Survey had a $857,000 budget. At the world's fair held in

Philadelphia that year, every federal science agency mounted an elaborate exhibit.[67] Had Bache not appreciated the importance of popular attitudes, had he resembled his haughty predecessor more closely, science could scarcely have attained the degree of autonomy and support it enjoyed by the centennial. Many things might have turned out differently, perhaps even the Allison Commission investigation.

Yet, acceptance of the elite imperative mentioned earlier was not something accomplished in the face of total opposition from *Scientific American*. Beach admired basic research. Indeed, he could, if he wished, treat it even more effusively than Bache, for he was not vulnerable to the charge of seeking public money for private enthusiasms. The world, Beach declared, offered a vast range of admirable pursuits: "Some men like to pass their lives in roaming over the ocean, and others in tilling the earth; to many is afforded a peculiar gratification in the accomplishment of difficult undertakings or the triumphs of mechanical skill, while a select few find the purest and highest enjoyment in the pursuit of abstract knowledge. . . ."[68]

This statement even implies acceptance of a Bachean scheme of hierarchy, but no reconciliation was possible on the issue of *public* priorities. Scant accommodation was possible between a man who advocated "rewards for principles, instead of applications,"[69] and another who said, "Let abstract science be measured in fame and honor, and applied science by money."[70] There was, in fact, nothing exceptional about discord between the minions of professional science and the proponents of the mechanic arts. "Scorn for inventors shows repeatedly in the literature of science," Kuhn notes, "and hostility to the pretentious, abstract, and wool-gathering scientist is a persistent theme in the literature of technology."[71] Bache, however, made very sure that the public perceived him as constrained, concrete, and industrious; that it understood his organization advanced "principles" and "applications" simultaneously; and, most of all, that the Coast Survey's attainments redounded strongly to our prestige as a nation.

I began by quoting de Tocqueville, a man whom some historians I talk to these days are not very enthusiastic about. The way de Tocqueville saw America does seem to have much in common with the way Henry Adams saw Mont-Saint-Michel and Chartres.[72] Yet, the Frenchman did not miss the mark everywhere. He was not wrong in his perception of the need for "currying favor," although he might have used a less pejorative phrase and he might have misapprehended the actual options. Nevertheless, it is a fact that here was something that men determined to shape public policy had to consider, even men who would automatically have been afforded deference or aristocratic prerogatives in some place other than a "democratic republic."

Sufficient proof has been adduced, I think, to the proposition that popular concepts of general utility did not determine the whole history of science in nineteenth-century America. It is still true, though, that science did fare best when it most convincingly presented itself as broadly beneficial in *some* sense—be it material, psychological, symbolic, or germane to some sort of status rivalry. "Social utility" is an interesting benefit recently stressed.[73] And the key point remains that public policy was determined by a representative body sensitive to constituent pressures. Professor Bache combined unequaled personal resources with a proper feel for the exigencies of the American political system—combined them more effectively than any contemporary. Small wonder, then, that it was he who best succeeded in promoting science "in the American idiom."

1. Nathan Reingold, ed., *Science in America Since 1820* (New York: 1976), p. 6.

2. Stanton, *The Great United States Exploring Expedition of 1838–1842* (Berkeley and Los Angeles: 1975), p. 62.

3. A. Hunter Dupree, *Science in the Federal Government* (New York: 1957), p. 62.

4. Young, *Victorian England: Portrait of an Age* (Garden City: 1954), p. 155.

5. Reingold, "Alexander Dallas Bache: Science and Technology in the American Idiom," *Technology and Culture* 11 (1970): 165.

6. Sally Gregory Kohlstedt, *The Formation of the American Scientific Community: The American Association for the Advancement of Science 1848–1860* (Urbana: 1976).

7. Quoted in Helen Wright, *Sweeper in the Sky: The Life of Maria Mitchell, First Woman Astronomer in America* (New York: 1949), p. 69.

8. Oscar Handlin, "Science and Technology in Popular Culture," in *Science and Culture*, ed. Gerald Holton (Boston: 1965), p. 198. Nowhere is ambivalence more pronounced than in the way the press responded to the Smithsonian, at once proud that Henry was molding an institution of world stature, and distraught at the essentially un-American dimensions of that mold.

9. "He had the talent to apply to practical purposes the most advanced results of science, and to make his practical work contribute to the progress of science, in a manner which has advanced geography in America far beyond what it is anywhere else." Thus concluded *Scientific American*'s obituary of Bache (n.s. 16 [1867]: 150) on a note that had just the merest touch of left-handedness.

10. Edward Lurie, "An Interpretation of Science in the Nineteenth Century: A Study in History and Historiography," *Journal of World History* 8 (1965): 693.

11. Kohlstedt, *The Formation of the American Scientific Community*, p. 162.

12. "Periodical Scientific Publications," *Scientific American*, n.s. 20 (1869): 187 (hereafter cited as *SA*). I should point out that Beach had two partners, Orson D. Munn and Salem H. Wales, and did not set policy singlehandedly, though evidently he was chief editorialist. Cf. "The Rise, Progress and Influence of the 'Scientific American'," *SA* 14 (1859): 257–58, and Frank Luther Mott, *A History of American Magazines 1850–1865* (Cambridge: 1938), pp. 316–24.

13. Kohlstedt, *The Formation of the American Scientific Community*, p. 154. This quote is paraphrased from a letter to Bache from James D. Dana, September 6, 1851, William J. Rhees Collection, Henry E. Huntington Library.

14. "The Nature of Science," *SA* n.s. 10 (1864): 89.

15. Reingold, "Two Views of Maury . . . and a Third," *Isis* 55 (1964): 372.

16. Thomas C. Mendenhall, "The Superintendents of the United States Coast and Geodetic Survey," *Centennial Celebration of the United States Coast and Geodetic Survey* (Washington: 1916), p. 135.

17. Harold L. Burstyn, "Maury, Matthew Fontaine," *Dictionary of Scientific Biography*, ed. C. C. Gillispie (New York: 1974), vol. IX, p. 197.

18. *Report on the History and Progress of the American Coast Survey Up To the Year 1858. By the Committee of Twenty Appointed By the American Association for the Advancement of Science* (Cambridge: 1858). The chairman was F. A. P. Barnard. The AAAS had a sort of standing committee on the Survey, which went into action whenever its annual appropriation appeared headed for trouble in Congress.

19. Quoted from *American State Papers, Misc.*, I, 390–91, in Dupree, *Science in the Federal Government*, p. 28.

20. A congressional committee reported in favor of an alternative proposal to amend the constitution to allow instituting a system of prizes "for promoting agriculture, education, science, and the liberal and useful arts." *Annals of Congress*, 19 Cong., 1st sess., 13 December 1825, p. 802. Proponents of prizes made themselves heard from time to time until the Civil War. One of the more elaborate plans was developed by Thomas Ewbank during his tenure as Commissioner of Patents. In a

sense such a mode of promoting science might have seemed peculiarly attractive to Americans, but the idea never caught on.

21. Dupree, *Science in the Federal Government*, p. 42.

22. "Bounty Paid By Government for Inventions," *SA* 8 (1853) : 221.

23. Direct subsidies, that is, became uncommon, though indirect assistance continued in the form of special patent legislation. See, for example, Robert C. Post, "Stray Sparks from the Induction Coil: The Volta Prize and the Page Patent," *Proceedings of the Institute of Electrical and Electronics Engineers* 64 (1976) : 1279–87.

24. *SA* 4 (1849) : 283. I have considered the background and outcome of the test of "electro magnetic power" in "The Page Locomotive: Federal Sponsorship of Invention in Mid-19th-Century America," *Technology and Culture* 13 (1972) : 140–69.

25. Reingold, "Alexander Dallas Bache," passim.

26. W. C. Bond, 28 November 1846, quoted in Reingold, "Two Views of Maury . . . and a Third," p. 372.

27. Congress resisted this until after the Civil War, despite more or less continuous pleas from interested parties. See, for example, John J. Greenough, *Memorial of J. J. Greenough on Printing Patents* (n.p.: 1862). Greenough was a New York inventor and patent agent. An overwhelming plurality of "ingenious Yankees" came from New York State (not New England), and the great majority of the important 19th-century inventors either were born in New York City or else moved there.

28. See Robert C. Post, " 'Liberalizers' versus 'Scientific Men' in the Antebellum Patent Office," *Technology and Culture* 17 (1976) : 24–54.

29. Cf. Charles Hirschfield, "America on Exhibition: The New York Crystal Palace," *American Quarterly* 9 (1957) : 101–16; Ivan D. Steen, "America's First World's Fair," *New York Historical Society Quarterly* 47 (1963) : 257–87; and Linda Hyman, *Crystal Palace/42 Street/1853–54* (New York: 1974).

30. See Monte A. Calvert, "American Technology at World Fairs 1851–1876" (M.A. thesis, University of Delaware, 1963), and Robert C. Post, "The American Genius," in *The Smithsonian Book of Invention* (New York: 1978), pp. 27–29.

31. *A Record of the First Exhibition of the Metropolitan Mechanics' Institute* (Washington: 1853), p. 13.

32. Benjamin Silliman, Jr. and C. R. Goodrich, eds., *The World of Science, Art, and Industry Illustrated From Examples in the New York Exhibition, 1853–54* (New York: 1854), p. 7.

33. "The Report of the Superintendent of the Coast Survey, showing the progress of that work during the year ending November, 1849," S. Doc. No. 5, 31 Cong., 1st sess., 27 December 1849, p. 2.

34. Dupree, *Science in the Federal Government*, p. 101.

35. Reingold, "Alexander Dallas Bache," p. 163.

36. While there is a full-length biography of Davis by his son and namesake, *Life of Charles Henry Davis, 1807–1877* (Boston: 1899), more astute is the sketch by Lillian B. Miller in *The Lazzaroni: Science and Scientists in Mid-Nineteenth Century America* (Washington: 1972), pp. 43–48.

37. Earle E. Coleman, "The Exhibition in the Palace: A Bibliographic Essay," *Bulletin of the New York Public Library*, 64 (1960): 471.

38. Silliman and Goodrich, *Science, Art, and Industry*, p. xii.

39. Silliman and Goodrich, *Science, Art, and Industry*, p. xi.

40. Silliman and Goodrich, "The United States Coast Survey," *Science, Art, and Industry*, p. 39.

41. Charles Henry Davis, "The United States Coast Survey," in Silliman and Goodrich, *Science, Art, and Industry*, p. 40.

42. Silliman, "Gifts of Science to the Arts," in Silliman and Goodrich, *Science, Art, and Industry*, p. 10.

43. Silliman and Goodrich, "Latitudes and Longitudes—Coast Survey Methods," *Science, Art, and Industry*, pp. 121–22, 131.

44. Silliman and Goodrich, "Saxton's Metallic Deep-Sea Thermometer," pp. 42–43; ibid., "Tides and Tide Gauges," pp. 99–101.

45. Silliman and Goodrich, "Standard Weights and Measures," p. 120.

46. Silliman and Goodrich, "The Electrotype Process," pp. 53–55. Mathiot, like Würdemann, was an immigrant, but the very fact that he had left his homeland in order to pursue his art here tended to buttress the feeling that America's star was on the rise, Europe's fading.

47. Silliman and Goodrich, "The American Ephemeris and Nautical Almanac," pp. 115–16.

48. Davis, "The United States Coast Survey," in Silliman and Goodrich, p. 41.

49. "Topographical and Hydrographical Surveys; Charts and Maps," *New York Industrial Exhibition. Special Report of Sir Charles Lyell* (London: 1854), p. 39.

50. Horace Greeley, *Art and Industry as Represented in the Exhibition at the Crystal Palace* (New York: 1853).

51. Silliman, "Gifts of Science to the Arts," in Silliman and Goodrich, p. 10.

52. Frank M. O'Brien, *The Story of the Sun. New York: 1833–1928* (New York: 1928), p. 107.

53. "Gifts of Science to the Arts," in Silliman and Goodrich, pp. 10–11.

54. "The Practicabilities of Science," *SA* 6 (1851): 397.

55. "Astronomy," *SA* 6 (1850): 115.

56. It is true that after the Civil War *Popular Science Monthly* did attempt to perpetuate elements of this tradition.

57. See, for example, Reingold, "Definitions and Speculations: The Professionalization of Science in America in the Nineteenth Century," in A. Oleson and S. Brown, eds., *The Pursuit of Knowledge in the Early American Republic* (Baltimore: 1976), p. 33.

58. Albert E. Musson and Eric Robinson, *Science and Technology in the Industrial Revolution* (Toronto: 1969), p. 2.

59. George Daniels, *American Science in the Age of Jackson* (New York: 1968), chapter 5.

60. Thomas Kuhn speaks to this point in a way, though it seems to me not in quite the right way. See "The Relations Between History and the History of Science," *Daedalus* 100 (1971): 277–78.

61. "The Practicabilities of Science," *SA* 6 (1851): 397.

62. Beach and Orson Munn took over a few months after it was founded by Rufus Porter. See Jean Lipman, *Rufus Porter, Yankee Pioneer* (New York: 1968), pp. 49–62.

63. Dupree, *Science in the Federal Government*, p. 118.

64. "Extracts from [Wm.] Fairbairn's Address," *SA* n.s. 5 (1861): 211.

65. "The Relation of Science to the Industrial Arts, *SA* n.s. 3 (1860): 201.

66. Daniels, "The Pure Science Ideal and Democratic Culture," *Science* 156 (1967): 1701, 1703.

67. See Robert C. Post, *1876: A Centennial Exhibition* (Washington, D.C.: 1976), pp. 74–99, and H. Craig Miner, "The United States Government Building at the Centennial Exhibition, 1874–77," *Prologue* 4 (1972): 202–18. Miner quotes a Canadian newspaperman who noted, "Uncle Sam promises well, considering that this is his first experiment of catering to the public entertainment." This, of course, was not so.

68. "The Relation of Science to the Industrial Arts, *SA* n.s. 3 (1860): 201.

69. Presidential address to AAAS, quoted in Kohlstedt, *The Formation of the American Scientific Community*, p. 161.

70. "A Great Field for Chemical Inventors," *SA* n.s. 3 (1860): 265.

71. Kuhn, "The Relations Between History and the History of Science," p. 284.

72. Nathan Reingold, "Reflections on 200 Years of Science in the United States,"*Nature* 262 (1 July 1976): 9. See also Reingold chapter in this volume.

73. Donald Zochert, "Science and the Common Man in Antebellum America," *Isis* 65 (1974): 464.

THE AMERICAN SCIENTIST IN HIGHER EDUCATION, 1820-1910

STANLEY M. GURALNICK*
Colorado School of Mines

THE STORY OF THE NINETEENTH-CENTURY AMERICAN SCIENTIST IS largely inseparable from the story of higher education in America, for prior to the twentieth century, an overwhelming number of American scientists were employed as teachers in institutions of higher learning. Colleges and universities, it seems, were ideologically prepared to offer scientists more varied and more flexible professional opportunities than were either government or industry. The growth of the American scientific community was thus heavily dependent upon the character and growth of higher education in the nineteenth century; higher education itself was in great part shaped by the growing scientific community.

Neither of these circumstances has been adequately explored. The role of the American college in fostering scientific endeavor within the country has been generally underestimated, and scientists have never been properly recognized as significant agents of change in higher education before the age of the university. Indeed, the common contention that science and her teachers did not flourish during the age of the college has perhaps been the single most influential factor in retarding our understanding of the nineteenth-century scientist and college alike. Until the place of science in the college and early university is more accurately represented, the role of the scientists employed in those institutions will necessarily remain obscured.

* The author acknowledges research support from the Johnson Fund of the American Philosophical Society.

We are all familiar with the myth of higher education, propagated through the years by Whiggish proponents of an American booster faith. Whether told from the point of view of an appreciative alumnus of a single institution or from that of a less parochial writer, the myth has generally had the same pasteurized parts and homogenized ending: from humble beginnings, small endowments, ill-prepared but dedicated teachers, and inadequate instructional facilities, academic institutions have prospered under the watchful eye of providence, overcoming every kind of pecuniary inconvenience and illiberal cultural force to achieve their twentieth-century greatness. Every change—whether the introduction of electives, the granting of the research Ph.D., or the establishment of more institutions—has allegedly contributed somehow to the greatness of the national achievement in higher learning.

The story of scientists in higher education has suffered from the same lack of historical perspective. First scientists are few and then, after much progress in the nineteenth century, they become many, overcoming obstacles to their professional advancement by routing the regressive forces in education that initially prevented them from establishing great research centers to rival those of the Old World.

Some of the obstacles cited by the myth makers are in fact real. Scientists did suffer, throughout the century, the usual public indifference to the needs of the abstract researcher; they did experience difficulty in concentrating resources within an uncoordinated laissez-faire system of academic sprawl. But others of the supposed hindrances to professional development are altogether imagined, perhaps because of the suspense they lend to an otherwise dull success story. Thus we hear of the opposition that religion must have offered to scientific study in higher education, and of the fights that must have raged between the stodgy proponents of the classics in the liberal arts curriculum and the advocates of modern science.

Historical inaccuracies are further compounded by the fact that twentieth-century scholars, with their personal attachments to universities and graduate education, have generally interpreted the educational past as a grand drama leading inexorably up to the glorious and seemingly foreordained present. Evaluating all past academic developments in the light of their contribution to the character of some contemporary universities, these historians have essentially left the events of the half century from the 1820s to the 1870s to the play of their prejudices and imagination.

A single example from the writing on higher education of a distinguished American social historian illustrates some of the persistent orientations that must be superseded. Writing in 1963 of his own Columbia University, Prof. Richard Hofstadter demonstrates toward his home institution a filiopietistic attachment that blinds him to the fact that (1) the piety of the old college did not hinder the progress of scientific study within its walls and (2) revolutionary change within academics is inadequately dated by the Civil War: "By solemn tradition, the presidential office had gone to clergymen, and it was secularized at the same time as the trusteeships themselves. Columbia, choosing the chemist and naturalist F. A. P. Barnard in 1864, was one of the pioneers."[1]

Hofstadter does not tell us that when Barnard came to Columbia, he had already been president of the University of Mississippi since 1854—a time and place not usually cited for scientific spirit nor easily fit into the revolutionary schema. Furthermore, Hofstadter overlooks the fact that Barnard went to Columbia during the war because he couldn't find a teaching job in the South he loved, not because Columbia had a new vision. And finally, Hofstadter neglects to mention a piece of information that would have obliterated all sense of the dividing line he draws between a prerevolutionary period, religious and antiscientific, and a postrevolutionary period, secular and scientific. The fact is that Barnard the scientist just happened to be an ordained and practicing Episcopal minister, while Columbia just happened to be of Episcopal persuasion.

Fortunately, a new generation of education historians is beginning to correct the errors of the old. New explanations are being offered to questions about who initiated institutional change during the nineteenth century and who benefited from it. And, with the help of the present generation of historians of science, new and important questions are being posed. In particular, inquiry is finally being made into the broader social significance of nineteenth-century academic scientists, independent of the value of their particular intellectual achievements. There appears a distinct possibility that the massive transformation that took place in higher education throughout the nineteenth century in America was attributable to something recognizable as scientific spirit; that being the case, it is essential that we chart that advance as carefully as possible. If the chart is complex, we need not be alarmed; we are under no obligation to see our educational present as the product of rational forces, every one of which must have contributed to academe's steady progress toward a beneficial, democratic end.

In examining the professional life style of the nineteenth-century professor of science, I will not find my context in the traditional tale of academe's rise from the liberal arts college with its overdoses of Latin and Greek, through the scientific school augmented by land grant institutions of more democratic purpose, on to the crowning glory of the graduate school, most clearly visible at Johns Hopkins a century ago. Like many self-serving reconstructions of the past, this is at best a half truth, overlooking many institutional changes and social relationships that would make its distortions obvious.

I would rather inquire into the varieties of experiences found in the careers of the scientists who populated academe, asking what kind of background, status, income, anxiety, and conflict they exhibited, and what unsolved problems they bequeathed to the future through their machinations. I will discuss some minor individuals, activities, and institutions that can tell us as much about the state of science and American culture in the nineteenth century as the more accessible and familiar examples from Johns Hopkins, Harvard, Chicago, and Stanford.

In this spirit, it will not be my concern to locate "firsts"—whether the first Ph.D. in physics, the first plan for an agricultural school, or the first professor to make more as a consultant than as a faculty member—but rather to concentrate upon the impulses that created these firsts and allowed for their spread. My intellectual bias is simply stated: I believe that a full review of science and higher education in this country should conform to what Nathan Reingold once labeled "our intuitive belief in the continuity of human experience."[2] When that continuity is clearer, it will also be clear that the dates commonly used to describe our educational past have been poorly chosen.

I have decided upon 1820 as a starting point—not because there was a revolution on that date or because a science professor had never existed before it, but rather because 1820 marks the appearance in American higher education of the self-conscious cultural nationalism that began in 1776 and accelerated after 1815. In his study of emerging national patterns of scientific activity, John Greene designated 1820 as that time by which "American science had come of age";[3] commitments to the values of science, even if only shared by a few, were necessary before those values could be incorporated into an educational system.

Although the colonial college instituted some important curricular revisions involving science and employed some isolated, though occasionally even excellent, science professors, it is not until the 1820s that we can find unbroken strings of college science professors serving in an equal capacity with those who professed other subjects. More importantly, it is not until the 1820s that we first begin to hear, in some of their many guises, repeated and still unanswered questions about the proper relationship of scientific education to the social goals of American democracy.

Just as 1820 approximated the beginning of a long period of evaluation and curricular change that involved scientists in new institutional arrangements, so 1910 marked its close. By that time, according to Laurence Veysey, ideals in higher education had "become rather statically frozen."[4] Scientists in particular were to see no change in their academic status until after World War II.

WITHIN THIS NINETY-YEAR SPAN, I HAVE divided the history of the science professor into three periods: 1820–1845, 1845–1875, and 1975–1910. The names I have given these periods are, however, more important than the dates, because one of the recurrent problems in the historiography of higher education has been the failure to disciminate among institutions that at any given time may have been in different states of maturation through possibly similar life cycles.[5] The first period, "Science Expands the Curriculum (1820–1845)," is marked by significant and little-noted additions of science courses to required standard curricula, permanence of the science faculty, and division of teaching labor into single fields of science. The second period, "The Curriculum Fails: Experimentation, Electives, and Engineering (1845–1875)," is marked by the sense of futility that emerged from the college's effort to cram every field of knowledge into the same curriculum or into the same head. A variety of alternatives to the single, fixed curriculum arose during this period; the scientists' more persistent articulation of the relationship they felt should exist between science and utility led, especially in the early 1870s, to the establishment of scientific schools, where the theoretical basis of applied science could be taught. The third period, "Higher Learning Hybrids (1875–1910)," is characterized by the grafting of a new ideal of research onto the already well-articulated ideals of intellectual training and vocational (utilitarian) training, in an accommodation that remains substantially unchanged today.

Science Expands the Curriculum (1820–1845)

Prior to the 1820's, the teaching of science was not recognized as a profession in America. Although there were a few excellent science teachers in the colonial colleges, instruction in science in the early nineteenth century was generally intermittent, deficient, and relatively unimportant to collegiate intellectual life. Moonlighting medical doctors and others not on the regular faculty delivered most of the instruction. In fact, two or three of the country's medical schools probably employed more faculty with scientific interests than all other institutions combined.

That there was no profession is not simply attributable to the fact that there was no prescribed training for the college science professor—a development that still lay far in the future. There was also as yet no defined role for the academic scientist, no assumptions about his behavior shared by practitioners and employers alike. Some now famous colleges employed no science professors during a number of years early in the century, while others made no substantial changes in the science curriculum from the 1790s to the 1820s.

Remarkably and unexpectedly, in the wave of cultural nationalism that followed the War of 1812, European mathematics and science began to receive as constant an attention and as much discussion as any other collegiate concern. Young scientists were employed in faculty positions never again to be left vacant, and they quickly became engaged in the kind of intramural comparison, competition, and standardization that marked them as a body of professional men. Within their institutions, they exerted an influence that resulted in repeated revisions of the undergraduate curriculum, until by midcentury the average college student was required to study more science than anyone in 1800 would have dreamt, and more, too, than he would ever be required to study again. Meanwhile, outside the curriculum and the college, these same scientists exerted an influence on scientific activity and its social organization that was no less profound.

The transformation of the college curriculum during the second quarter of the century and the accompanying growth of the academic scientific community have usually slipped our historical memory. There have been signs, however, that the antebellum college curriculum is becoming better understood. Two decades ago, Walter P. Metzger wrote that the curriculum was "by no means . . . as archaic and as rigid as later reformers thought it had been, nor was science as underrepresented as its votaries usually claimed."[6] Since Metzger's comment, the flexibility of the curriculum has often been noted, although the significance of that flexibility for the growth of science has not been fully explored. The first author to document growth in other areas of college study, particularly the social sciences, was Wilson Smith in his *Professors and Public Ethics* (1956). My own work has amplified Smith's by refining Metzger's observation that science was not as underrepresented as its later adherents found it convenient to assume.[7]

The primary reason that historians have failed to recognize the place of science and scientists in higher education of the early nineteenth century is the unexplored but pervasive assumption that science and religious culture were incompatible. Because a great majority of the more than 500 colleges that existed before the Civil War were religiously affiliated, there is a special tendency to misinterpret this period in particular.[8] Both Hofstadter and Wilson Smith have warned us that even the seventeenth-century Puritans thought of Harvard no more as a theological seminary than as an engineering school. College, they believed, was properly designed to offer an education appropriate for all learned Christian citizens.[9] Yet still we seem unwilling to insist that the Puritans' descendants in the nineteenth century were equally tolerant in educational matters; still we resist the suggestion that the denominational affiliations of the nineteenth-century colleges are not good clues to the intellectual patterns they evolved. Typically, historians differentiate among these colleges by citing their religious affiliations— Baptist, Congregationalist, Presbyterian, Methodist, Episcopalian, and even nondenominational. But the fact is that such affiliations tell little about these institutions' aspirations or characters.

A much more useful mode of differentiation may be the grouping of colleges according to how much they were able to satisfy plans for scientific expansion. For example, at Kentucky's Transylvania, the first trans-Appalachian College, the Board of Trustees learned in 1833 that $5,000 spent on scientific equipment would "be more promotive of the welfare of the [Presbyterian] institution than any single step that could be taken," for such monies would help both in retaining the school's single science professor and in enticing another to come across the mountains to join him.[10] How many institutions in the midwest, where college failure was endemic, found themselves with similar needs and no financial means by which to satisfy them? And how many of these institutions failed precisely because they could not accommodate the expensive demands of science and scientists alike?

These questions require study. At present, it is clear only that the fewer than 20 percent of the liberal arts colleges that did survive in the second quarter of the nineteenth century all incorporated science into their curriculum. And significantly, that curriculum was single, unalterable, and prescribed for all degree candidates. Intellectual justification for rigidly fixing the same program for all students was found in the "faculty psychology." According to advocates of that psychology, each function or faculty of the mind found its proper exercise and expression through a particular course: the faculty of expression was supposed to be exercised by language, the moral faculty by philosophy, the faculty of reasoning by mathematics.[11] However, the correspondence between subject and faculty often was not explained in detail; the faculty psychology of curriculum became so after-the-fact a form of reasoning, with assumptions so widely shared, that detail was unnecessary. Thus no one stated explicitly that there was a separate geological faculty; however, once geology had become part of the required curriculum, it was simply assumed that its study exercised some essential mental faculty and so could not be removed.

Given this educational philosophy, the number of required subjects in the curriculum expanded, and so did the faculty employed to teach them. The pattern—which first made itself evident at the colonial colleges, such as Yale, Brown, and Princeton, and at those Northeastern colleges established after 1795, such as Union, Amherst, and Dickinson—was the same at all schools, although different in its rate of progress. Sometime during the 1820s, the one person who had previously taught all mathematics and science, and maybe languages as well, was joined by a second regular faculty member with a specialty in science. The names of the professorships they held varied with their institutions, but generally, if the first professorship had been called "mathematics and natural philosophy," the two substituted in its place might have been "mathematics and astronomy" and "natural philosophy and chemistry."

So began a refinement of teaching specialty and a division of teaching labor that would continue throughout the period in all areas of curriculum, but most notably in the sciences.[12] Thus in the 1830s, as geology and the divisions of natural history became subjects of study, new faculty appointments were made. By the end of the decade, it was common for an institution to have four professorships, usually in mathematics, physics, chemistry, and geology; these constituted at least a third, and in some cases closer to half, of the total number of positions on the college faculty.

Schools that were slow to increase the size of their science faculties were berated for their shortsightedness. At the University of Vermont, for instance, which did not get its fourth scientist until 1854, the professor who still taught both chemistry and natural history complained to the trustees in 1847 that:

Forty years ago, scarcely anything was done in connection with
our American College on the subject of either chemistry or
Natural History—very little even thirty years ago. But within
that time a great change has arisen. Those branches have formed
so high an importance both in their scientific & practical relations
that all the colleges with which I am acquainted, except this one,
have found it needful or expedient to make large appropriations
for them both, & have found a great extension of their influence &
usefulness from so doing.[13]

Important as specialization of the science faculty was to be to the continued prestige of thriving institutions, an observer in the 1840s would have had great difficulty predicting the future course of any particular institution from the degree of specialization its faculty had achieved. When F. A. P. Barnard went to the University of Mississippi in the 1840s, he was the only science professor, teaching chemistry, mathematics, physics, astronomy, and civil engineering in addition to serving in the Oxford Episcopal Church—this at a school which would be planning university functions in only a decade.

In contrast, at Wesleyan (not to become a university in the nineteenth century) there were already three scientists on a faculty of seven by 1840. And at Amherst, never to become a university, the catalog of 1852–53 shows that five out of nine professors taught science, dividing their responsibilities into the fields of natural theology and geology, mathematics and natural philosophy, chemistry and natural history, astronomy and zoology, and analytical and applied chemistry.[14]

Obviously, teaching assignments varied so much from institution to institution that instructional boundaries among such disciplines as chemistry, geology, and biology were not always clear. Because the research interests of professors were frequently as diverse as their teaching assignments, it was often difficult to associate a given individual with a single discipline until late in his life. Thus it is no wonder that a historian of the period 1800–1860 found specialists at the end, but no critical point in between by which to date their appearance.[15]

Clearly, the proliferation of science courses in colleges from the 1820s through the 1840s was a healthy trend for the academic scientific community. It was a less healthy trend for the college curriculum, into which more and more required subjects were being crammed. By the 1840s, the unlimited expansion of the curriculum had led to increasing superficiality in all the courses taught within the college. Students, moreover, complained that too much work in science was required of them.

The college responded by replacing the single required curriculum with a system of electives. Such reform has often been misunderstood by historians, who imagine that electives were established in order to allow science into a classical curriculum that had been stubbornly refusing to admit it. But the facts of the case do not encourage such interpretation. Indeed, the loudest voices of reform were to be heard not at those institutions where little science was taught but at those where science courses had swelled the curriculum preposterously. Thus, when Brown's Francis Wayland spoke in 1850 in support of a system of elective choice, he did so at a school that had four scientists on a seven-man faculty and a curriculum in which 160 out of 479 hours of instruction were devoted to mathematics and science.[16]

Where, we might ask, had sufficient numbers of scientists been found to teach the courses that swelled Brown's curriculum and those of other schools? In the 1820s, after all, there was still no standard training or apprenticeship for a science professor. The answer is that until well into the century, a significant number of scientists came from that group of bright young men who had been preparing for the ministry—a circumstance which confirms the fact that piety and science were, in fact, congenial to each other. Indeed, the movement from the pulpit to a science professorship was a natural one—not because science was mired in magic or metaphysics, but because for centuries Protestantism had encouraged scientific research on the theory that familiarity with the works of God promoted appreciation of His Word. Such was the faith of the seventeenth-century scientific revolutionaries, as understood in America through the works of Francis Bacon, Bishop Paley, and other Anglo-American writers. Those who held this faith never experienced the intensity of intellectual conflict that historians in a later age would impute to them.

A Christian, even a minister, did not have to denounce any part of his religion to join the new scientific community. Indeed, for many, abandoning the ministry for science was a simple expedient suggested by changing social conditions—namely, the expansion of evangelical sects that deemphasized an educated ministry; the general decline of interest in the pulpit by college students; and the lessened prestige of the orthodox ministers now forced to battle Unitarianism.[17]

The case of Edward Hitchcock is illustrative of the changing social conditions that, in the New England of the 1820s, offered undecided young men—amateur scientists and ministers alike—new careers in higher education. At midcentury, Hitchcock would be one of the most famous geologists in America, author of the most widely employed college textbook, and a cofounder of the Association of American Geologists. But as a young minister of twenty-eight in 1821, he could not have forseen his future.

Hitchcock had loved science since he was a boy roaming over Massachusetts' rural Connecticut Valley. As a teenager, he grew intensely interested in astronomy; however, an eye problem, and the realization that a poor man could not support himself as an amateur astronomer, led him to prepare for the ministry—an obvious career for a boy of his background and intellectual ability.

Because he had undergone a conversion that caused him to loathe Unitarianism and to embrace orthodoxy, Hitchcock decided to study at Yale Divinity School. There he met Benjamin Silliman, a Yale College science professor whose career as an inspirer of young academic scientists-to-be was just waxing. Silliman became Hitchcock's close personal friend, teaching the amateur nature lover how to trade plant specimens with botanists, and offering career advice whenever it was sought. Although never suggesting that Hitchcock abandon the ministry—indeed, he would have had nothing to suggest in its place—Silliman endorsed a "collateral & recreative pursuit of science."[18] Thus Hitchcock, who became the Congregationalist minister of Conway, Massachusetts, in 1821, kept up an active correspondence with Silliman, occasionally sending some botanical observations for the latter's young *American Journal of Science*.

In the next few years, Hitchcock became restive in his professional position, partly because of failing health, but principally because he did not find the job of turning the tide of Unitarianism as engaging or rewarding as he had anticipated. He certainly did not think of abandoning the intense religious convictions he was to hold until his dying day when he wrote to Silliman of a desire to "change [his] mode of life entirely." Neither did he anticipate that a career in science was an alternative to his present one when he asked, "But what can I do? in what business engage? how support my family?"[19]

Silliman did not have an answer, but neighboring Amherst College did. The trustees of the young and expanding school knew of Hitchcock's interest in science and offered to hire him as a second science professor. Now Silliman, himself a religious man, could be more enthusiastic about urging Hitchcock to leave his salaried position as minister. "There are many men who will be glad of your present situation," he wrote; "the situation of a Professor, although arduous, has many more alleviations than that of a clergyman."

While Hitchcock hesitated, he received another job offer from

Middlebury College. Which faculty should he join? Silliman would not make that decision for his friend, but he did ask to be informed of which position Hitchcock turned down, for the information would be of use to a young man studying in Silliman's laboratory who had just dropped out of Andover Seminary and was also looking for a career in science.[20]

Hitchcock finally chose to accept the offer from Amherst, a school he served for decades as teacher, scholar, and finally as president. In his dedication to his college, Hitchcock was no doubt a model of diligence even in his own century. However, it is well to note that the antebellum college imposed upon its faculty, as a matter of course, arduous institutional responsibilities, many of which were unrelated to a professor's teaching or independent work. In the early 1820s, for instance, Benjamin Silliman not only taught all of the sciences except mathematics at Yale; he also administered the entire entrance examination to prospective freshmen, including the portions in Latin and Greek.[21]

Silliman was probably also required to participate in disciplining students; until midcentury, when the paternalistic system of college supervision was modified, faculty members were regularly required to act as college policemen, even to live with students in the dormitories. Little wonder that the role of a professor in relation to student discipline was the subject of debates as bitter as those about curricular reform.

Occasionally, a professor was even forced to part from his institution over administrative disagreements. Henry Vethake, for instance, became professor of mathematics and natural philosophy at Dickinson College in 1821 not only because of the "extent of his scientific acquirements," but also because of his ability to command "at the same time, both the love and fear of students."[22] Yet in 1830 he resigned because he disagreed with the trustees about how to handle college business.[23] His replacement, Henry D. Rogers, later to become a famous dynamical geologist at the University of Pennsylvania and elsewhere, also displeased the trustees by "differ[ing] from the other members of the Faculty about the concern of the College and its management." Distressed by his outspokenness, the trustees were "desirous to dispense with his services as [science] Professor."[24]

A few examples of faculty insecurity, however, may be misleading, for the records show that only a few were ever fired. Thomas Cooper, who in 1833 was forced to resign the chair of chemistry at the University of South Carolina because he was an atheist who repeatedly denounced the biblical view of creation, is the lone exception to the rule that college faculty enjoyed freedom of inquiry into any interest, scientific or otherwise.[25] His case prefigured the day when science would conflict with religion and popular culture. But before Darwin, no scientists were forced to resolve such a conflict, for orthodox science and orthodox religion supported one another, especially on the issue of creation. Even Metzger, no friend to past encroachments on academic freedom, has admitted that before Darwin, "academic freedom in investigation was diminished not by the action of forces alien and antipathetic to science, but [only] by the inhibitions present within science itself."[26]

If professors rarely had to waste time in disputes that could have led to dismissal, they did have to expend so much energy in college affairs that it is remarkable that so many of them engaged in extracollegial work which augmented their teaching duties, replicated their own interests and values, and offered an additional salary. Most interesting, because it uncovers a familiar pattern of scientific justification, was the involvement of science professors in state geological surveys.

As best we can determine, the idea for such surveys originated with Denison Olmsted, a Silliman student, who in 1817 was appointed professor of chemistry at the University of North Carolina. Failing to find support for geological exploration from the state legislature, "the greater part of [whose] members had never heard of the name of geology before," he flooded the newspapers in 1818 with anonymous letters suggesting the great value to the state of a description of its mineral and vegetable productions. His plan to popularize his interest worked, and the legislature grudgingly spent $100 for a survey to be conducted under the aegis of the state board of agriculture. Olmsted had triumphed. But the legislature also had its day, for Olmsted soon found himself in a situation to be repeated in every state. He had to produce a report emphasizing "*economical* geology [because] this was the only part that would satisfy those who employed me." To justify himself to his scientific colleagues elsewhere, he bravely reported that he was "still accumulating materials of a more scientific kind of which I hope to make a separate use."[27]

Thirty years later, when the principle of the survey was well established, the scientist's public report was still understood as an occasion for private apology to professional colleagues. For example, Prof. Charles Baker Adams of Middlebury College, who moonlighted as Vermont state geologist, begged Sir Charles Lyell "to make allowance for the plan and execution of a report, prepared at the commencement of a survey, not for the Geologist, but to appease the craving for mineral wealth in a community not one in ten thousand of whom knows even that there is any such thing as structure to the earth." Only after such a preface did Baker feel free to explain what "scientific geology" he had actually done.[28]

A modern observer has argued convincingly that the decline in the necessity for these "outrageous [public] claims of utilitarianism" is a measure of the rise of science as a profession;[29] yet his thesis requires some qualification. If the rhetoric of utility never disappeared in the nineteenth century, it is not only because scientists were required to be masters of deception. It is also because the ultimate utility of scientific work was often a sincere article of their professional faith.

No matter how science professors justified their personal scientific pursuits to themselves and to the public, however, it is still certain that their permanent careers as teachers allowed them many opportunities to foster the scientific interests of the country, to serve the outside community, and to become what we would call professional scientists. Most authors have acclaimed the work of Benjamin Silliman of Yale, who promoted the *American Journal of Science*; of Alexander Dallas Bache, who left the University of Pennsylvania to head the Coast Survey; and of Joseph Henry, who left Princeton to head the Smithsonian. But those who remained primarily professors in the small colleges engaged in a range of extracurricular activity which, if not so extraordinary as that of their better known colleagues, does at least question the usual historical estimate of the liberal arts scientist.

Early in the century, for instance, Bowdoin's Parker Cleaveland almost singlehandedly introduced the science of mineralogy into America while researching an inquiry from a lumbering company about the value of some rocks uncovered in the excavation of a sluice route in Maine. Years later, Cleaveland was to serve his country again in a wholly different role as commissioner of the Survey of the North Eastern Boundary of the country.[30]

F. A. P. Barnard, while teaching astronomy at the University of Alabama, was commissioned as astronomer by both Alabama and Florida to settle a boundary dispute between the two states.[31] And James Renwick, of Columbia College, one of the most sought after engineering consultants in America, was invited by speculators to search for gold in Georgia and by a president of the United States to inspect inventions intended to prevent steam boiler explosions.[32] In fact, the first two federal appropriations for scientific investigation went to two groups of college professors, one of which explored the same problem of steam boiler explosions, while the other examined the possibility of sugar cultivation in the southern states.[33]

Prof. Alexis Caswell of Brown University was requested by a court to apply the latest calculations of pneumatics to settle the claim of a manufacturer who insisted that a certain dam disturbed the operation of his waterwheel;[34] William Gillespie of Union College became the country's expert on road building, and George W. Benedict of the University of Vermont served as the northern New England authority on constructing telegraph lines.

Similarly, Ebenezer Emmons managed to earn his full salary from Williams College at the same time that he collected additional compensation from the neighboring state of New York for work on its geological survey.[35] The list could be extended indefinitely. In fact, in my study of more than fifteen antebellum colleges, I found that every one had several professors who were involved in the great natural history and geological surveys, exploring expeditions, or private scientific and engineering ventures.

That there was a connection between the homes scientists built in their local teaching institutions and the house of national science under construction is unmistakable. In 1840, Edward Hitchcock of Amherst, with other college professors working on the state geological surveys, formed the Association of American Geologists (AAG)—an organization which, within the decade, became the American Association for the Advancement of Science (AAAS).

The AAAS was to hold its first formal meeting on the afternoon of September 20, 1848, not in a government building or a church, but in the Hall of the University of Pennsylvania. Because the overwhelming number of men in attendance at that first meeting taught science, they established as a category of membership "The Collegiate Professors of Natural History, Physics, Chemistry, Mathematics, and Political Economy, and the Theoretical and Applied Sciences generally."[36] It is a fact of great historical significance that by 1848 there were already enough science professors with sufficiently nonparochial interests to form an enduring national association.

Any evaluation of the importance of the college science professor of the 1820–1845 period is meaningless, then, if it focuses on the vocational requirements he had not met or the university he had not built. By promoting the intellectual demands of scientific reasoning and rigor in a program that defined culture and intellectual accomplishment for the educated, he carried on a significant diffusion of knowledge upon which later changes depended.

The Curriculum Fails: Experimentation, Electives, and Engineering (1845–1875)

What I have labeled the middle period, 1845–1875, has been the subject of less scholarly attention than it deserves. Historians usually note that scientific schools made their appearance during these years, and they pay homage to the passage of the Land Grant Act, usually twisting its context to serve some romantic notion of the democracy of the intellect. They also make much of the dissolution of the fixed curriculum, by then jammed with too many subjects to remain the only route to a college degree.

The period is not accurately described by these few of its features alone. Because of the variety of its experimentations with alternative forms of educational expression (especially those related to the applications of science) and because of its often unrealized plans for expansion, the period's significance may ultimately lie in a sympathetic history of its failures. In such a history, the Civil War, so often used to divide the educational century, may appear to be less a cause of change than a disastrous interlude to its implementation.

Whatever our perspective on this middle period, however, we must recognize its great importance for the academic scientific community, which grew in size, prestige, and wealth through these three decades. The increase in the numbers of scientists on college faculties is a principal characteristic of the period. At Harvard, for instance, between 1845 and 1869, the percentage of scientists rose from 33 to 56 per cent,[37] and the achievement of a majority became common elsewhere as well.

Meanwhile, at institutions public and private, denominational and unaffiliated, trustees began to realize that the same scientists who had worked so hard to expand their own disciplines might well serve the entire college in administrative posts as important even as that of the presidency. Thus, geologist Edward Hitchcock became president of Amherst in 1845; astronomer F. A. P. Barnard became president of the University of Mississippi in 1854 and of Columbia ten years later; natural historian Samuel S. Haldeman, of the University of Pennsylvania, was offered the presidency of the University of Texas in 1851;[38] and chemist Charles W. Eliot was appointed president of Harvard in 1869—the most celebrated president, perhaps, that Harvard would ever have.

The rise of the academic scientist's prestige was accompanied by a rise in his compensation. Individual scientists, to be sure, felt themselves underpaid. Astronomer Charles Pickering lamented to John Torrey in 1851 that "the time is not [yet] at hand when scientific services shall be justly appreciated in the pecuniary way."[39] Harvard chemistry professor John White Webster fell into such debt in 1850 that he did away with his creditor and dissolved the remains in an acid bath—a feat for which he soon after became the first American college professor to be tried, convicted, and hung for murder. But historical records show the scientist's plight to have been less desperate than plaintive letters and occasional exotic acts would suggest. Edward Lurie has described the rather sanguine position of two of the period's more famous scientists:

When Joseph Henry left Princeton University prior to taking his post as Secretary of the Smithsonian, he received an offer from the University of Pennsylvania of $3500 a year. Louis Agassiz, while receiving $1500 yearly from Harvard, could count on additional university and private funds that averaged an equal amount yearly over the decade 1846–1856.[40]

College archives consistently demonstrate that scientists were not undervalued by their institutions. Indeed, they were occasionally the envied darlings of the administration or trustees. In 1854, for instance, when Union College hired an analytical chemist, sent him to Europe for training, and offered him a salary $500 higher than that of the other faculty members, there was an outcry—and not from the chemist.[41] Similarly, in 1855, when the faculty at Dartmouth learned that higher salaries were being meted out to the faculty of Dartmouth's Chandler scientific school, they petitioned the trustees for equity—to no avail.[42] The period's scientists may have argued themselves poorly compensated, but within a reasonable frame of reference, the complaints appear ill founded.

Salary, of course, constituted an official form of recognition for the scientist and signaled his growing status within the profession. But by the middle of the century, scientists had developed their own less tangible system of status rewards, for they had come to recognize as a sign of professional standing a man's ability to perform "original research."[43] "Distinction among scientific men," wrote one science professor, "has been with me a most powerful motive of action."[44] Another, anxiously awaiting the publication of his research results, expressed "mortification of being anticipated in part by a paper" from a different scientist.[45]

More importantly, perhaps, young men seeking teaching positions were praised by older scientists not for their piety, but for their potential as researchers. Time and again, scientists promoted the value of research in communications to one another and to college employers as well. Thus when President Wayland of Brown solicited endorsements for the appointment of William A. Norton as professor of natural philosophy and civil engineering in 1850, he learned from James D. Dana at Yale that Norton was "fitted not only for teaching but also for advancing by his researches the science to which he is devoted;" from Professor E. Otis Kendall, himself a research astronomer at the Philadelphia Central High School, that Norton was a "successful instructor & disciplinarian" whose "investigations . . . attest to his ability as a physicist;" and from Professor John F. Frazer of the University of Pennsylvania, who in three full paragraphs of encomium never mentioned teaching, that Norton's "publications indicate[d] originality of mind, and habits of laborious and accurate research, joined with the best sources of information."[46]

Although universities were not yet requiring that their scientists engage in original work, they were certainly aware that the scientific community esteemed the researcher, while holding as a near moral failure the man who defrauded his research potential by producing nothing. Little wonder that President Eliot of Harvard, fearing to offend the scientific world with an endorsement, refused to lend his name to support the American lecturing tour of John Tyndall in 1871. Tyndall, he opined, had a "capacity for research" which should have precluded him from wasting his time giving "exhibitions."[47]

As this example shows, the growing emphasis upon publication and peer recognition inevitably wrought a change in the average science professor's relation to the lay public. In the 1830s, he might have given popular scientific lectures at the local lyceum or at the Lowell lecture series, but in the 1860s he would leave such goodwill work to the politically ambitious elder statesmen of science or to the profit-minded popularizers.

Interestingly, the burden of professionalization fell squarely upon the shoulders of individual scientists, for in the middle decades of the century, there was still no specialized training for the aspiring college professor. A recent study of a group of scientists active in 1846 shows that "four out of five went through no special scientific or technological course aside from a regular college curriculum."[48] Yet, among more than fifty percent of the scientists, schools were said to have been of primary influence in their choice of career—proof that the liberal arts college actively fostered the scientific disciplines and their practitioners.[49] Although after midcentury it became common for the would-be professor to continue his education in Germany or in one of the newly established scientific schools, he would probably not have anticipated this later study at the start of his career.

There must have been a high sense of exhilaration among these young men preparing for careers with no precedents. But there was agony as well. Usually of modest background, the aspiring scientist, merely by attending college, entered a kind of educational aristocracy that was in no way synonymous as yet with social class or wealth. At a time when college attendance was neither routine nor widespread, the college "misfit" was apt to be looked upon with some contempt in his local community.[50]

So it was with psychologist G. Stanley Hall, who has left us a valuable introspective record of his educational odyssey into a professorship. As a boy in rural New England, Hall felt different from other boys who did not share his intellectual interests; thus, in 1862, he concealed from his neighbors the fact that he was taking his entrance exams to Williams College so that he would not have to suffer their jibes. He entered college with classmates who "were [also] in moderate, and [even] very straightened, circumstances," but even in their potential source of companionship, he could not overcome his loneliness. To the end of his life he was to maintain that "the dominantly sad note of [his] life" was the sense of isolation he always felt as a consequence of the fact that his "ambition" was so infrequently shared.[51]

If boys of Hall's own age were insufficiently inspiring, it was not so with the Williams faculty. In the previous decades, Williams scientists had increased in number and influence, and Hall responded to their influence in his choice of a career in higher education. After graduation, he entered Union Seminary in New York—a last accession to a dying convention—and began to daydream about becoming a college professor.[52] The enterprising young scholar managed to receive $1,000 in aid from industrialist Henry Sage to do advanced study in Germany, where he took a selection of courses that would still only suggest a career, not a discipline: theology, physics, physiology, pathology, and philosophy. Later, after receiving the first Ph.D. from Harvard, he would have to create for himself the specific role of psychology professor, the past offering only a congenial scientific culture that Hall now redirected and refined.

Like Hall, many scientists were required to use a degree of imagination in designing the careers they eventually came to pursue. And the invention they expended upon their personal lives they also expended upon the lives of the institutions they served. With characteristic zeal, scientists influenced the restructuring of academic curricula during the middle period in two ways: by planning courses of study outside of the liberal arts rubric (a matter we will consider later) and by participation in the debates that led to the demise of the required curriculum.

That curriculum found its defense in the "faculty psychology," which emphasized the balanced exercise of each mental muscle. But as science alone had uncovered many more muscles than could be fully developed in one person—mathematics, physics, astronomy, geology, chemistry, and natural history, to mention only the more prominent courses in the curriculum—the faculty psychology itself atrophied, and a new rationale for the undergraduate curriculum grew to replace it. Individual choice came to be regarded as a positive value, at least among those who argued that students ought to be able to choose or "elect" the subjects they wanted to study in their pursuit of a college diploma.

The idea of electives, first conceived early in the century, was now an idea whose time had come. In the colleges, electives were promoted by as influential a man as Francis Wayland, who at mid-century criticized the fixed curriculum. In public, they were generally supported as a device to attract to the classroom more students in pursuit of personal intellectual pleasures. Ideologically, support for electives could be found in classical liberalism, which defended freedom from restraint, or in American democratic theory, which defended freedom of choice.

However, academic scientists (and other faculty as well) were not all immediately convinced that sound academic policy could be extrapolated from liberal political theory. In addition, many scientists doubted that an elective system would result in an increased demand for science. Experiences earlier in the century had shown that when new subjects were added to the curriculum, the average student would more eagerly seek to avoid the new offerings in science than those in any other field—seeking, for instance, to dispense with calculus before social science. A few professors were inclined to find this behavior promising. For example, Harvard mathematician Benjamin Peirce unabashedly championed electives in the hope that the poor students would elect to stay out of his classroom.[53] His cynicism, however, was not a motive adopted by many of his colleagues.

The concern that eventually united most academics in defense of electives was rather the fear that courses were growing increasingly superficial as more of them were crammed into the single curriculum that every student was expected to study. F. A. P. Barnard, for instance, opposed electives for many years until, in the late 1860s, he conceded that they provided the only means for counteracting the dilution of courses.[54] Biologist John LeConte at South Carolina was piqued by the "prematurity" of the elective system, which he felt had grown out of a "spirit of radicalism and innovation;" yet he too saw the system as a compromise that would allow science courses to become more substantial, at least for those students who took them.[55]

Doubts about the wisdom of the principle of elective choice persisted long after the principle had been made practice. Many academics feared to lose the overall symmetry of the balanced liberal arts curriculum. One theorist straddled two worlds of thought in insisting that the "elective faculty itself needs to be trained in order to be used to best advantage."[56] And, as Richard Hofstadter has observed, by 1910 there would be complaints that many experiments in electives had "invited curricular chaos."[57] The Barnards and LeContes would not have been surprised.

Our interest, however, lies not in the history of elective systems but in their effect upon the employment of academic scientists within the country's colleges. And that effect is less favorable than conventional histories have urged. If more scientists came to be employed in colleges after the establishment of elective systems, it is not for the reason that unfettered students chose to study more science than ever before. Indeed, quite the opposite is true.

At the University of Indiana, for instance, where the elective system was adopted in 1876, the records for graduating students from 1833 to 1892 show that "almost half the courses [of study] noted prior to 1876 had more science in them than the average for all the years since." As the interpreter of these data observed in 1892, the expectation of some that the elective system would give "undue cultivation of the physical sciences, finds here no confirmation."[58] At the University of Michigan, less systematically studied records suggest no more sanguine a picture: in 1894, 782 students chose to study French; 736, German; 679, English; 662, Latin and Greek; and 648, history. Only 447 elected to study physics or chemistry.[59]

Scientists of the middle period, then, were finding more academic positions in colleges not because they had profited directly from collegiate reform, but simply because the entire system of higher education was expanding, with the growth rate for faculty surpassing even that of the phenomenally increasing student body. Indeed, far from receiving a larger portion of the growing academic pie, scientists occasionally found that they received less, even as their real numbers grew: thus, in prereformed Harvard of 1845–1869, the percentage of scientists on the faculty rose from 33 to 56 percent. After 1869, during Eliot's presidency, the percentage leveled off to 40 percent.[60]

The elective system may have introduced advantages for faculty and students both, and it may well have contributed significantly to the increase in college population that took place during the Gilded Age. But the system did not, as conventional histories still assume, increase the influence of the scientists relative to that of the rest of the college faculty.

In fact, in the middle decades of the nineteenth century, the academic scientist was to wield his most significant influence not so much in the college proper as in the variety of new programs designed to supplement the college in its offerings. For want of a better, all-inclusive designation for these programs, we can classify them under the heading of scientific and engineering schools; even as we do so, we do injustice to their variety. A more accurate designation might have been "programs in vocational preparation," had not Plato and Thorstein Veblen succeeded in investing the word "vocational" with such invidious connotations.

But by whatever name we choose to call them, we must recognize that midcentury excursions in scientific education—whether they led to schools of engineering, mining, or agriculture, to graduate schools, or to failure—were promoted in the belief that students should be able to pursue college-level work outside the confinement of the liberal arts program.

That belief, grounded as it was in some version of the elective principle, was not new. Collegiate experiments of the 1820s and 1830s had created earlier opportunities for students to study science outside the prescribed curriculum. But such experiments had failed for two reasons: first, because the only science courses offered in these early years were identical to those taught in the established curriculum; second, because prospective students were usually not attracted to nondegree programs.

The midcentury experimenters wisely learned from the failures of their predecessors. In creating alternatives to the standard liberal arts curriculum for students desirous of studying only science, they frequently established separate "schools," even if only a single professor initially taught there, and they empowered these schools to award degrees—a step of some significance for students and programs both. For then as now, accountability was a matter of some concern; and the number of degrees awarded was a clear indication of a program's ostensible success. Thus, when Joseph LeConte went to Harvard's Lawrence School to study advanced biology with Agassiz, he was urged to sign up for a degree so that Agassiz could have positive results to show the administration.[61]

New programs were conceived out of a common desire to "qualify students for commercial, manufacturing, mechanical & agricultural pursuits."[62] Although ends so comprehensive were never actually achieved, the rhetoric and the sentiment with which they were announced was pervasive, and there seems to have been no liberal arts college, however unlikely, that did not attempt to respond to that rhetoric. President Edward Hitchcock of Amherst secretly suggested to President Mark Hopkins of Williams that the two schools create professorships of agriculture and other technical subjects—a suggestion that failed because Williams was already working on its own plan for an agricultural department complete with operating farm.[63]

Wesleyan, not content simply to teach civil engineering, opened a normal school for teachers in the 1840s. Princeton tried unsuccessfully to hire a professor to teach special applied courses in 1853,[64] and Union College, frustrated in its plans to hire a professor of agriculture among others, settled for a separate analytical laboratory.[65] The University of Virginia, which many mistakenly believe to have adopted an elective program in the 1820s, actually established a chemistry course separate from that taken by liberal arts students in 1858.[66]

Such geographically dispersed colleges as Amherst, Dickinson, and Brown also supplied separate laboratory instruction to individuals not enrolled in the standard college program. And, of course, plans for the establishment of the better known Ivy League schools of science were initiated at the same time.

The late 1840s and 1850s, then, witnessed an expansion of scientific schools much like the efflorescence of liberal arts colleges a quarter century earlier—an expansion that can be understood both quantitatively, in terms of the number of schools created, or qualitatively, in terms of the variety of new programs established. Moreover, the efforts that these schools and programs made to attract students had a salutary effect on the popular estimate of higher education. The decade of the 1850s thus became the first of only two in the century during which overall college attendance increased faster than the general population—and this at a time when the general population was fast being swelled by immigration.[67]

The Civil War, however, abruptly postponed or ended designs for the establishment of more scientific schools, particularly in the South, where there would follow decades of economic prostration to delay the old dreams. On the eve of the war there were even plans for a single regional University of the South that would have concentrated resources and scientific talent—an ideal never to be realized.

Meanwhile, at the University of Mississippi, soon to be closed completely by the war, a myriad number of plans were aborted. In 1859, F. A. P. Barnard had secured funds for an astronomical and magnetic observatory that was never to be built. And the university itself had commissioned the celebrated Boston optician Alvan Clark to grind a telescopic objective of $18\frac{1}{2}$-inches diameter—an objective larger than that of the Harvard Observatory, indeed, the largest the firm had yet undertaken. The lens, however, was not completed until the spring of 1861, and it was consequently never sent to Oxford, ending up instead in the Dearborn Observatory in Chicago. Barnard had also cajoled the legislature into converting his university-in-name-only into a school of "Universal Instruction"; and "[b]ut for the beginning of the war . . . there is little reason to doubt that special scientific schools of various kinds would have been inaugurated in the course of that year [1861]."[68]

A generation ago, a historian of the South tried to vindicate his region's intellectual stature by enumerating the achievements of its more prominent antebellum scientists.[69] He would have done as well to recount their plans for an educational future that escaped their grasp and our historical consciousness.

If the war was devastating to the educational development of the South, however, it was merely a temporary hindrance to collegiate growth in the rest of the country. Once the war had ended, the building of scientific schools continued, reaching a crescendo in the early 1870s. According to the U.S. commissioner of education, in 1870 there were 17 institutions reporting "schools of science" (sometimes only a single department), which enrolled 1,413 students and employed 144 instructors. In 1873, 70 institutions housed such schools, replete with 8,950 students and 749 instructors.

A five-fold increase in faculty in only three years is impressive; and it would be interesting to know how such faculty were trained and recruited, what they taught, and what kind of friction existed between them and the scientists who had been teaching in the college proper for years.[70] Even more interesting, perhaps, would be to discover if the hiatus of the 1860s and the recovery of the 1870s were simple responses to economic realities or manifestations of some broader social patterns.

Such information has yet to be acquired, for the scientific schools have not been studied with any degree of thoroughness. Historically, we only know that many of these schools failed, just as many liberal arts colleges had failed in the earlier days. And neither the Civil War nor the Depression of 1873 alone provides an explanation for their failure.

At some colleges, special scientific programs were terminated before the war began; while at some universities, utilitarian scientific programs survived the depression only to succumb during the prosperity in the 1880s. We can speculate upon what factors contributed to the success of those that did survive, creating permanent institutional structures as legacies to the possibility of linking instruction in science to a purpose broader than that of merely disseminating intellectual culture.

At some of the better known institutions, enterprising scientists were able to establish durable programs by attracting students to the study of their personal scientific interests rather than to the specific technical concerns of institutional donors. At Harvard's Lawrence School, Louis Agassiz had an appointment as professor of "mining and metallurgy," a title indicative of the utilitarianism that brought the school into existence and secured Lawrence's patronage. The professorship was conventional enough for a technical school, except for the fact that Agassiz rarely taught the subjects of his professorship, attracting students instead by his reputation in biological science.[71]

At Yale, too, the scientists at the Sheffield School were attracting students who gave the school more the character of a postgraduate school in science than that of the engineering department that Yale's financial supporters had intended.[72] The builders of the Columbia School of Mines did at least create an undergraduate program of engineering study, but one that suited their own interests and not those of the industry that had given its name and support to the school. Academic scientists, who had learned to deal with the larger world on the geological surveys, used their prestige and administrative skill to balance personal interests with the sources of wider social support in the enlargement of the sphere of educational enterprise.

In the case of strictly engineering schools, a more mature understanding of the relationship between theory and practice aided survival. The "critical institutional innovation," as one modern historian of technology labeled it, was the realization that "technology would be pursued in the manner of science,"[73] that a theoretical basis existed for the achievement of certain practical ends. That realization eliminated, at least in part, the need for new engineering faculties to devise intellectual justification for technical education, and it reduced as well the inclination of the older science faculty to trivialize applied knowledge.

William Barton Rogers, first president of the Massachusetts Institute of Technology (MIT), predicted that the liberal arts college would never attract to its special courses and scientific schools *"numbers"* of students interested in the theoretical study of the practical arts,[74] because such schools did not know how to eliminate either the condescension of scientific intellectuals toward the applied science or the requirement that technology be justified intellectually. Some older schools belied Rogers's criticism; others did not.

It was not, however, only the liberal arts colleges that were sometimes slow to heed Rogers's warning. At many of the land grant universities, complained the *Nation* in 1866, the practical ends of technology were thought achievable simply by "combin[ing] manual labor with scholastic pursuits."[75] For more than a quarter century after the passage of the Morrill Act, very few students enrolled in programs to study agriculture, perhaps the most obviously practical of the new courses; not even the constant lowering of entrance requirements could attract them.[76]

When only a disappointingly small number of students were attracted to "mechanical and industrial" schools, critics complained that there were too many state universities competing for too few students—an observation reminiscent of the continuing complaint that there were too many colleges as well.[77] Thus, it would be valuable to know what procedures agricultural, mechanical, and industrial universities had to follow to become acceptable. If they included an emphasis upon the theoretical basis of applied knowledge or the use of a liberal arts program to hold technically oriented students, then demand for applied scientific instruction would explain less of the increased popularity of higher education in the later nineteenth century than we might assume.

In fact, there is reason to believe that science professors themselves were responsible for creating for applied science a demand that was only faintly perceived by the population. It is well known that book-learning farmers were scoffed at almost until the end of the century, when a few practical successes from research silenced their ridicule. It is equally well known that trained engineers faced problems in confronting industrial mechanics reared in the craft traditions.

In 1867, a congressman, presumably familiar with the popular will, advised a scientist who expressed interest in expanding higher education in practical chemistry: "[People] are slow to appreciate the importance of scientific knowledge . . . establish [a college of chemistry]—open it—prove it, and [then] it will be patronized."[78] Francis A. Walker, who witnessed the expansion of technical education as U.S. commissioner of education and then as president of MIT, claimed that schools of technology were established not to fill a popular demand but to answer to the foresight of their founders and professors, who provided a supply before a demand had yet been expressed.[79]

As it stands, this evidence may be too impressionistic and inconclusive, but it does suggest the need for a detailed study of nineteenth-century technical education—one that would reveal the relationship of the professors in the scientific schools to the publics they served.

The middle period is known to every historian of American science as one of scientific society (and institution) building that conferred visibility on the scientific enterprise and its great statesmen. Most scientists, however, still made their living in some form of undergraduate instruction. A study of publications before 1880 in chemical research—a field that readily lent itself to industrial applications—reveals that publishing chemists who taught outnumbered those who worked in industry by about eight to one.[80] But, although they were still primarily teachers, American research scientists were now more numerous, more variously entrenched in their institutions, and more influential than they had been in the days of the single prescribed curriculum.

Higher Learning Hybrids (1875–1910)

The period during which the scientist became finally established in academe, 1875–1910, witnessed the rise of multipurpose universities, development of the graduate school, and the division of the faculty into separate departments—changes that have given the essential intellectual and administrative character to our present university. This is, therefore, the period that has held the most fascination for modern scholars, especially those in disciplines which trace to that period their own professional efflorescence. Moreover, it is the period that has most benefited from the work of recent education historians. Nevertheless, the statistics which show that the period was one of impressive growth do not reveal the tensions caused by the conflicting ideologies that professors supported.

Higher education in the age of university building had much greater continuity with the past than proponents of a theory of revolutionary change have been disposed to see. Institutional piety, for instance, persisted as a feature of intellectual life throughout the nineteenth century; its demise can be more convincingly related to rapidly declining church membership in the era of the gospel of wealth than to any new attitude that the Gilded Age universities adopted.

Indeed, a most thorough historian of the university has reminded us that "although state institutions de-emphasized religion somewhat more rapidly than did private colleges, the remarkable thing is how long officially sponsored religion endured at many state universities."[81] Thus, some state universities required chapel attendance for students and faculty alike into the twentieth century. John D. Rockefeller, representing a curious mixture of new wealth and old piety, gave the initial funds to establish the University of Chicago on condition that they be matched by an appropriation from the Baptist Education Society. Finally, in Baltimore, President Gilman of Johns Hopkins asked G. Stanley Hall, who had lost his interest in religion, to join a church in the interests of lessening local opposition to the university.[82]

There is continuity too, in the quality of the salaries academic scientists commanded. Admittedly, as in the earlier periods, individual scientists were eloquent on the subject of their poverty, and occasionally they convinced their contemporaries of their plight. Yet it is unfortunate, as one historian of nineteenth-century American science has warned us, that:

Their begging letters and imploring essays also convinced some twentieth century scholars, who out of sympathy or shortsightedness have examined nineteenth century science solely through the eyes of nineteenth century scientists on the make.[83]

In reality, the science professor of the Gilded Age made a salary that placed him in a better economic position vis-à-vis the skilled worker than does his "starving" counterpart of today. Part of his discontent arose then, as now, from the fact that the scientist's academic salary compared unfavorably with that of other professionals in private practice or industry. Moreover, such discontent could only have been heightened in the last century by the prominence of successful businessmen who had made their fortunes without ever attending college.

There were, however, three sources of comparison within the university itself that probably explain some of the frustration expressed in the scientists' "imploring essays." First, the academic scientist knew his salary to be lower than those of his institution's administrators. In the earlier college, it had been common for academic salaries, even when low, to be uniformly so: the college president received the same salary or one only slightly higher than that of the college's professors. Later, when the president was paid significantly more than the faculty—presumably so that he could offer an image of institutional solvency to the businessmen he courted—scientists, as well as professors from other disciplines, were often disgruntled.

Secondly, as a natural consequence of the law of supply and demand, the scientist on the arts faculty commanded a lower salary than those in the scientific and agricultural schools: on the same day in 1889 that the University of Kentucky offered a new professor of agricultural chemistry $1,500 a year, it offered geneticist Thomas Hunt Morgan $1,200 to be professor of biology.[84]

Finally, as distinctions among full, associate, and assistant professors began to materialize in the last few decades of the century, assistant professors found themselves often poorly paid in comparison to their higher-ranked colleagues. The tutors who had carried much of the instructional load in days past had also been poorly compensated, but they were not exactly the counterparts of the new Ph.D.'s. The nineteenth-century tutor, usually biding his time while waiting for a pulpit or some other work, rarely intended to make a career of college professing.

Thus, although complaints about the poor salaries awarded academic scientists in this last period were inflated, they were in some sense true—just as they had been both true and false in earlier periods. The only real change is to be found in the fact that a larger and more prominent scientific community made a louder and stronger complaint.

Research, too, does not represent so great a break with the past as we have sometimes assumed. We have already seen that the research ethic had been developed by science faculty some fifty years before the rise of the graduate school, becoming characteristic of the academic scientific community by midcentury. If some schools, however, remained indifferent to that ethic before the age of the university, we shall see that many continued in their indifference even after the university had become a well-established American institution.

Respect for research was not, after all, a purely home-grown product half a century old. Outspoken justification for making "original contributions" to scholarship was at least somewhat of an import from abroad, transplanted to the New World by the flood tide of American academics-to-be who went to Germany for graduate study in the 1870s and especially the 1880s, before our own universities took over the production of Ph.D.'s. Listening to Friedrich Paulsen and other German academics give speeches on the concept of disinterested research, the youthful American scholars failed to appreciate the fact that the Germans had spent the greater part of the century forging that concept into a shield from political involvement or persecution. No such shield was needed in America; consequently, when the defense was brought home, it seemed to amount to a break with the democratic cultural theory of the American past.

Colleges had accepted science on the intellectual ground that its study was essential to the production of the well-educated man who would serve as a moral asset to his entire community. They had even adopted from the general culture a respect for the utilitarian justification for science. But the idea of treating science or any other kind of knowledge as a *Ding an sich* appeared too elitist to many. Not for another generation, after 1910, would proponents of scientific research realize the necessity of effecting a new reconciliation between science education and the national culture and purpose.[85]

For all that it was practiced and defended by the nineteenth-century advocates, then, research never did succeed in monopolizing the nineteenth-century academic marketplace. Only at Johns Hopkins, Chicago, and Clark, all of which began without prior collegiate tradition, was opposition to the research ethic insignificant. Elsewhere, scholarship was held in some contempt. When the Clark psychologist-to-be G. Stanley Hall returned from study in Germany in the 1870s, the president of a large midwestern state university told him that his higher learning was not needed there because it would "unsettle" students' minds.[86]

At the University of Arkansas in 1889, the president voiced concern about "subversive university-minded faculty members" who were trying to introduce high standards of scholarship.[87] Another university president "defeated any overt move to scientific or scholarly inquiry" on the part of his faculty—so at least reports Thorstein Veblen, who consistently complained that too great an attachment to scholarly pursuits regularly hindered the academic advancement of the faculty researcher.[88] Finally, according to a recent study by Robert A. McCaughey, even Harvard's renowned "university builder" and chemist Charles Eliot took office in 1869 with every intention of continuing to make faculty appointments that reflected old friendship ties rather than professional performance or potential.[89]

Both McCaughey and Veblen attribute the gradual institutional acceptance, and indeed advocacy, of research to competition among universities. A famous scientist, according to Veblen, was "inconvenient" to a university but "unavoidable" if the school wanted to establish its rank.[90] Undoubtedly he was right, but it remains to be asked why powers in higher education would have been reluctant to encourage faculty research without such external stimulus. Was it the fact that research did not comport with the idea of teaching and otherwise caring for students—an idea that had somehow come to define American higher education? Or did the reputation and mobility that publication conferred upon a faculty member make him harder to constrain within the limits imposed by a growing campus administrative structure or by a paternalistic president? The answer to such questions is less clear than the evidence that the research goal was only slowly accepted in higher education, especially at the state universities, whose scientific eminence in the twentieth century we too speedily project backwards.

The distribution of earned doctorates in science during the nineteenth century is part of such evidence. From 1863, when J. Willard Gibbs received the first American Ph.D. in science, until 1910, 2,513 doctorates in science were awarded. More than 80 percent of these were awarded by just eight universities, all privately endowed, and all but one in the East: Johns Hopkins, Chicago, Yale, Columbia, Harvard, Cornell, Pennsylvania, and Clark.[91] Moreover, of the approximately 3,000 men (and few women) who made their living as professors of science in 1903, only 838 were employed in institutions bearing at least the title, if not also the function of university.[92]

Clearly, then, not many were yet involved in the graduate training that regularly accompanied research at those schools which fostered it. Indeed, at Illinois's Lombard University, one of the universities by name only, biologist David Starr Jordan, later president of Stanford, still taught "classes in natural science, political economy, evidences of Christianity, German, Spanish, and literature, and pitched for the baseball team," as late as the 1870s.[93]

In the face of such evidence, we might choose to be as cynical as G. Stanley Hall about the mass of science professors early in the twentieth century, concluding that most were still involved in a routine no less dull than that pursued at the most boring old college, that many were even incompetent, and that only a few were known off their own campuses.[94]

A more sympathetic interpreter, however, might observe that some schools, such as many of the small New England colleges, seem to have made a conscious decision not to attract researchers, despite the knowledge that their science faculties would decline in influence accordingly. The concern at these schools for the culture, literary and scientific, that had to be transmitted through an undergraduate curriculum has probably contributed to the belief that they had been unsympathetic to science before the days of the researcher. That belief, of course, is mistaken. Williams College, for instance, had been solicitous of science since the 1820s, had erected one of the first permanent college observatories in the 1830s, and by 1840 had four professors on a faculty of seven who taught science. Yet in the 1870s, when Johns Hopkins chemist-to-be Ira Remsen taught there and asked for a small room to use for his private research laboratory, he was told: "You will please keep in mind that this is a college and not a technical school."[95]

What had happened to change the orientation of New England undergraduate colleges that had offered scientists their first homes decades earlier? Why had they allowed faculty to carry out scientific research for a half century before finding in that activity a threat to their existence? The changing relationship of higher education to American social structure offers some explanation. Apparently, during the Gilded Age, the eastern college—always intellectually elitist—finally became associated with the upper social classes and their aspirants, thus transforming itself into an enclave of the rich.[96] Science, meanwhile, had become increasingly democratized. Now associated with a vocationalism beneath the station of the country's business managers and their progeny, the science that had ridden into the university on a wave of utilitarianism would have to suffer an ebbing in the college. Furthermore, the large institution's difficult task of forming both research universities and polytechnic institutes out of single amorphous scientific schools was of no concern to the small college, which now looked upon either end as poison to its ideal of the educated man.

Two studies of the undergraduate origins of Ph.D. scientists who made reputations in research almost quantify the college's gradual loss of interest in science during the last decades of the nineteenth century and the first of the twentieth. The first study, conducted in 1906, covered scientists who had attended college in the nineteenth century; the other, conducted in 1927, covered many who had attended college in the twentieth. Even though the total number of prominent scientists had grown from 1906 to 1927, the number produced at Princeton had dropped from twenty-three to two; at

Amherst, from twenty-six to four; at Wesleyan, from sixteen to five; at Williams, from fourteen to five; at Dartmouth, from ten to five; and at Oberlin, from ten to two. This pattern is too regular to be discounted: when new wealth sent its sons to college, it clearly did not plan to make science professors of them.[97]

At Dartmouth, in the half century 1821–1870, 86 percent of the graduates had become ministers, lawyers, doctors, or teachers and professors; in the first decade of the twentieth century, the figure had dropped to 50 percent, the other half of the graduates entering business.[98] A more socially homogenous student body was now more immune to the influence of science educators. The leadership that the old college had played in expanding college curriculum and producing scientists was gone, moving west, staying in tune with middle-class social aspirations.

Many students, of course, still aspired to some kind of academic life; those who did certainly experienced a greater career opportunity than had existed for college scientists-to-be earlier in the century. The scientist who planned to teach in a college had a very clear idea of his future, even if it was a future of declining influence. Those who desired one of the newer careers in a research university might not have been able to anticipate the exact focus of their research—whether in one of the older scientific disciplines or in one of the emerging sciences such as psychology or genetics. But even if their research areas were still uncertain, students at least had institutional models upon which to pattern an academic life. One graduate student even admitted to reading the obituary notices in *Science* to anticipate what university openings he might apply for—a novel, if ghoulish, device that had not been possible before the expansion of the market for young researchers.[99]

Also novel to higher education were the conditions of employment that the promising doctorate holders negotiated wih universities congenial to the new research ethos. They typically demanded "time for original work" and insisted that promotion be based on scientific distinction, not on seniority, loyalty, or any other subjective criterion.[100] At a time when the intimate relationship between teaching and research was not well articulated, the teachers generally attempted to keep contact with students, especially undergraduates, to a minimum. Henry Rowland, for instance, could almost boast that he regularly neglected students for research.[101]

In addition, these teachers often allowed themselves a generous measure of eccentricity in their behavior. Thus, Rowland's older colleague at Hopkins, the English mathematician J. J. Sylvestre—

more brilliant and more eccentric than the American academics of earlier generations—would sometimes leave a meal to work out a new idea. On one occasion, he actually reported that he had to sit up during the night with his feet bathed in warm water to prevent all the blood from rushing to his head to meet an idea too quickly.[102]

Such demands, such arrogance, and such eccentricity may have been unavoidable in a university faculty, but even the best of schools were to some extent put off by them. Thus even the universities that would tolerate research genius in the 1890s were not yet willing to establish the research institutes and research chairs that still remained on the academic agenda of a few hopeful scientists.[103]

The behavior of research scientists not only altered their relationships with students and administration but also affected the intercourse they had with one another: narrowing focuses for research and enlarged faculties lessened communication among scientists. Earlier in the century, scientists had often lamented the difficulty of keeping up with the progress of knowledge in more than a few fields. Now, although they enjoyed the stimulation of some men in their own departments, they saw little of anybody else. And if no one appreciated the factionalization of faculty that the fragmentation of knowledge had brought, no one was able to arrest it. Thus although President Gilman in the opening years of Johns Hopkins had hoped that professors, especially younger investigators, would prepare presentations of their work for the entire faculty, his dream had to be abandoned in the face of reality.[104]

The death of old dreams, however, gave rise to new ones. If specialization cut down on intramural communication, it also provided opportunities for university researchers to gain extracurricular sources of income. The new independence and self-confidence that reputation conferred even allowed a few to become wealthy. Henry Rowland, one of the leading contributors of essays on the impoverishment of professors, also became one of the richest. Having discovered, in his theoretical studies of electricity, a principle that was valuable for the design of electric dynamos, Rowland was hired as a consultant to the Niagara Cataract Construction Company, which was engaged in harnessing the energy of the falls. When the company paid Rowland only $500 for his services, he filed suit in U.S. Circuit Court for $30,000, or compensation at the rate of $150 a day (not a bad salary for the depression year of 1893). The company obviously had miscalculated the modesty of a university physicist. After a four-day trial, the jury awarded Rowland $9,000, evidently in recognition of his immodest assertion that he was the best physicist in America.[105] It was certainly a new kind of courtroom drama to watch a professor extol his own scientific qualifications publicly.

Yet if a courtroom jury found it easy to decide Rowland's claim for compensation, no jury, inside the university or without, found it possible to resolve the academic questions that Rowland's activity entailed. What was the nature of the university's claim on the scientists it employed? Was it the university's responsibility to provide researchers with the time and laboratory space to conduct sponsored research? Did the pursuit of industrial research by university faculty express a new institutional purpose that had not yet been made clear, or was that pursuit merely the logical outcome of the ideal of utility?

At Harvard, as elsewhere, a few faculty members sought clarity in the face of institutional ambivalence. Thus chemist Wolcott Gibbs would not do industrial consulting; Josiah Parsons Cooke, another chemist, was even more insistent that a professor must refuse to accept compensation from industry. A professor's primary responsibility, argued Cooke, was to the life of the mind, a life to which the "commercial spirit is fatal."[106]

Harvard itself was not able to concur in such idealism. In 1881, the university accepted money from the American Bell Telephone Company for a new physics laboratory, promising in exchange "that professors could use university laboratories in work for private companies."[107] Thus did the social order of industrial America touch academic scientists once more. The influence it took from some scientists in the undergraduate curriculum, industry gave back to others in the form of the possibility of using faculty positions to attract commercial profit.

Irked by these seemingly unrelated conditions—the failure of some universities to pay due respect to research scientists and the folly of others that worshipped their research indiscriminately—Thorstein Veblen raised a loud complaint against American higher education in the early twentieth century. Thinking the sole purpose of a university to be the pursuit of disinterested research, Veblen lamented the victimization of higher learning by wealth-seeking trustees and administrators—the "captains of erudition." Veblen felt that even scholars and scientists had compromised the research ideal by giving in to the business values of private gain and production. The university, he argued, had become involved in too many illegitimate enterprises. In particular, it had undertaken to serve utilitarian ends by offering its students a technical education that would better have been confined to purely technical institutions.[108]

Admittedly, Veblen's criticisms were acrimonious, but his understanding of the relationship of higher education to American culture was subtle. Intuitively he realized that Americans would attempt to use any institution, especially the university, to achieve all ends. And with society's ends so varied, science educators could no longer identify a single cultural ideal that also expressed their own purpose.

Ironically, perhaps, Veblen never understood that research itself could become part of the vocationalism he deplored, that research could lose the purity of purpose, motive, and method that its coterie of academic boosters have often claimed for it. Indeed, if we were to take Veblen's business metaphysics seriously, we would have to admit that research-oriented academic scientists, even more than professors of strictly applied subjects, had been the salesmen who had created demands for their products where none had existed. Their programs, after all, never had an imperative as clear as those of the vocational and engineering schools. Yet by 1910, scientific research, at least in many institutions, achieved recognition as an acceptable academic end, clearly in advance of a significant call for researchers by industry or government.

The success of these scientist-salesmen can be measured imperfectly in numbers. By 1910, thousands of scientists were employed in higher education, more than a hundredfold increase over their number in 1820. Together they comprised three-fourths of all American scientists.[109] If most men who practiced science still taught science, it is a tribute to their salesmanship that they found so many institutional homes in which to shelter their plans for current activity and future expansion.

The reason for their success lies in the very academic eclecticism that Veblen deplored. For the pluralistic rhetoric of higher education offered three models that professors could manipulate in their formulas for justification: science in the service of intellectual culture (however that culture would be redefined in the changing undergraduate curriculum); science in the service of utility, technology, and practice (even if the proper practical method was only realized in theory); and science in the service of the infinitely inquiring human spirit.

In the three parts of the century discussed before, each use of science education grew to maturity and offered new opportunities to faculty scientists. The proper proportions for a mixture of these components of academic purpose have never been settled. Consequently, after the science-laden liberal arts course spawned new alternatives in the 1840s, the scientist in academe has not been limited to only one role.

It would be misleading, however, to leave the impression that the roles of academic scientists were only the results of their own entrepreneurial efforts. Many times in the last century the supply of academic scientists was less than the demand; that reality must have had origins other than those of their own instigation—demands that originated in commerce, in agriculture, in urban technologies, and in the increased literacy of the population. What is clear, however (perhaps clear enough to be a truism) is the fact that scientists were constantly involved in the transformation of nineteenth-century academic life and, with it, undoubtedly national culture as well.

1. Richard Hofstadter, "The Revolution in Higher Education," in *Paths of American Thought*, ed. Arthur M. Schlesinger, Jr. and Morton White (Boston: Houghton Mifflin Sentry Edition, 1970), pp. 269–90. Quotation on pp. 278–79. The problem of filiopietism in writing about higher education has been recognized before; it is interesting to notice the lamentation of psychologist G. Stanley Hall a half-century ago: "True history in the field of higher education was perhaps never so hard to write as in this country, pervaded as it is with insidious biases for competing institutions." *Life and Confessions of a Psychologist* (New York: D. Appleton and Co., 1923), p. 246.

2. Nathan Reingold, "American Indifference to Basic Research: A Reappraisal," in George H. Daniels, ed., *Nineteenth-Century American Science: A Reappraisal* (Evanston: Northwestern University Press, 1972) pp. 38–62. This quotation appeared in the original oral version.

3. John C. Greene, "American Science Comes of Age, 1780–1820," *Journal of American History* LV (1968): 22–41.

4. Laurence Veysey, "Stability and Experiment in the American Undergraduate Curriculum," in *Content and Context: Essays on College Education* (New York: McGraw Hill, 1973), p. 8.

5. With so many institutions to choose as examples, it is easy to be misled by convenient and accessible unrepresentative examples.

6. Walter P. Metzger, *Academic Freedom in the Age of the University* (New York: Columbia University Press, 1955), p. 10.

7. Stanley M. Guralnick, "Sources of Misconception on the Role of Science in the Nineteenth-Century American College," *Isis*, 65 (1974): 352–65; *Science and the Ante-Bellum American College* (Philadelphia: American Philosophical Society Memoirs, 1975).

8. Donald G. Tewsbury, *The Founding of American Colleges and Universities before the Civil War* (New York: Columbia University Teachers College Press, 1932) is still the standard treatment of college founding. For unsympathetic interpretation of denominational ties see

Richard Hofstadter, *Academic Freedom in the Age of the College* (New York: Columbia University Press, 1955), p. 211, passim. The last few years have witnessed a number of articles in the *History of Education Quarterly* that stand as the best challenge so far to the traditional historiography including the erroneous assumption about religious culture: James Axtell, "The Death of the Liberal Arts College," 11 (1971): 339–52; Hugh Hawkins, "The University Builders Observe the Colleges," 11 (1971): 352–62; David B. Potts, "American Colleges in the Nineteenth Century: From Localism to Denominationalism," 11 (1971): 381–87; Natalie A. Naylor, "The Ante-Bellum College Movement: A Reappraisal of Tewksbury's Founding of American Colleges and Universities," 13 (1973): 261–74; and Jurgen Herbst, "American College History: Re-Examination Underway," 14 (1974): 259–66.

A discussion of college founding emphasizing new community "booster spirit" is found in Daniel J. Boorstin, *The Americans: The National Experience* (New York: Vintage Books, 1965), pp. 152–61.

9. Richard Hofstadter and Wilson Smith, *American Higher Education: A Documentary History*, 2 vols. (Chicago: University of Chicago Press, 1961), I, p. 2.

10. B. O. Peers to Board of Trustees, 1833, College Papers, Transylvania College Library.

11. A full discussion of faculty psychology is found in Walter B. Kolesnik, *Mental Discipline in Modern Education* (Madison: University of Wisconsin Press, 1958).

12. Frederick A. P. Barnard observed that tutors were first assigned subjects rather than classes at Yale in 1828, close to the time of Yale's defense of the prescribed curriculum—a development related to specialization. John Fulton, *Memoirs of Frederick A. P. Barnard* (New York: Macmillan and Co., 1896), p. 64.

13. George W. Benedict to the Corporation of the University of Vermont, 4 August 1847, University of Vermont Archives. The corporation appointed three eminent scientists to appraise Benedict's remarks and soon made appropriate adjustments in staff and supplies.

14. Fulton, *Memoirs of Barnard*, p. 198; *A Catalogue of the Officers and Students of the Wesleyan University for the Academical Year 1840–41* (Middletown: William D. Starr, 1840), p. 6; *Catalogue of the Officers and Students of Amherst College, for the Academical Year 1852–53* (Amherst: J. S. and C. Adams, 1852), p. 5.

15. Donald Beaver, "The American Scientific Community, 1800–1860: A Statistical-Historical Study" (doctoral diss., Yale University, 1966), p. 151.

16. *A Catalogue of the Officers and Students of Brown University, 1850–1851* (Providence: John F. Moore, 1850), p. 15.

17. Metzger, *Academic Freedom*, has a list of clergymen who also taught science, p. 15n.

18. Silliman to Hitchcock, 5 December 1822, President Hitchcock Collection, Amherst College Library.
19. Hitchcock to Silliman, 1825, ibid.
20. Silliman to Hitchcock, 14 April 1825 and 6 August 1825, ibid.
21. Fulton, *Barnard*, p. 32.
22. *Harrisburg Chronicle*, 23 October 1821, copy in Dickinson College Library.
23. Henry Vethake, *A Repy to "A Narrative of the Proceedings of the Board of Trustees of Dickinson College, from 1821–1830,"* (n.p.), printed copy in Dickinson College Library.
24. Minutes of the Trustees of Dickinson College, 10 July 1831.
25. It is interesting to note that the atypical experience of Cooper at South Carolina took place at a public, not a denominational, institution.
26. Metzger, *Academic Freedom*, p. 18.
27. Denison Olmsted to Parker Cleaveland, 11 December 1827, Film 1106, American Philosophical Society (original in Bowdoin College Library).
28. Charles Baker Adams to Sir Charles Lyell, 5 December 1845, History of Science Film 4, American Philosophical Society (original in University of Edinburgh Library).
29. George H. Daniels, "The Process of Professionalization in American Science: The Emergent Period, 1820–1860," *Isis* 58 (1967): 161.
30. Leonard Woods, *Address on the Life and Character of Parker Cleaveland*, 2d. ed. (Brunswick, 1860).
31. Fulton, *Barnard*, pp. 102–3.
32. John Adams Dix to James Renwick, 11 July 1833 and Martin Van Buren to Renwick, 19 July 1838, both in Renwick family papers, Columbia University Library.
33. *Manual on the Cultivation of the Sugar Cane and the Fabrication and Refinement of Sugar* (Washington: F. P. Blair, 1833). Principal authors were Charles U. Shepard, then working in Silliman's laboratory at Yale; Silliman; and Silliman's son-in-law Oliver P. Hubbard, who was professor of science at Dartmouth College.
34. John Kingsbury to Geo. Keeley, 20 December 1830, Brown University Archives.
35. Trustees' Minutes, 17 August 1836, Williams College Library.
36. *Proceedings of the American Association for the Advancement Of Science, First Meeting, Philadelphia, September 1848*, (Philadelphia: 1849), pp. 9, 144–56.

37. Robert A. McCaughey, "The Transformation of American Academic Life: Harvard University 1821–1892," *Perspectives in American History*, 8 (1974): 267n.

38. Samuel Stehman Haldeman to John Fries Frazer, 22 December 1851, Frazer papers, American Philosophical Society.

39. Charles Pickering to John Torrey, 4 April 1851, Film 628, American Philosophical Society.

40. Edward Lurie, "An Interpretation of Science in the Nineteenth Century: A Study in History and Historiography," *Journal of World History* 8 (1965): 689.

41. Jonathan Pearson, Diary, 24 July 1854, Union College Archives.

42. Minutes of the Trustees of Dartmouth College, 24 July 1855, Dartmouth College Archives.

43. Examples of this phrase are ubiquitous by midcentury. Printed samples may be found in the more than score of letters submitted to the University of Virginia in support of a prospective natural scientist: *Testimonials of the Qualifications of Richard S. McCulloh* (Washington: J. and G. S. Gideon, n.d.; letters dated 1840–43).

44. Edward Hitchcock, Private Notes, December 1843, President Hitchcock Collection, Amherst College Library.

45. Joseph Henry to Benjamin Silliman, 10 December 1830, Silliman family collection, Yale University Library.

46. James D. Dana to Francis Wayland, 25 July 1850; E. Otis Kendall to Wayland, 12 July 1850; John F. Frazer to Wayland, 23 July 1850; all in Brown University Archives.

47. Charles Wm. Eliot to J. P. Lesley, 16 August 1871, Lesley papers, American Philosophical Society.

48. Robert V. Bruce, "A Statistical Profile of American Scientists, 1846–1876," in *Nineteenth-Century American Science: A Reappraisal*, ed. George H. Daniels, (Evanston: Northwestern University Press, 1972), p. 88.

49. Clark A. Elliot, "The American Scientist in Antebellum Society: A Quantitative View," *Social Studies of Science* 5 (1975): 102, found in his statistical analysis of scientists of all degree of reputation and accomplishment "that the colleges were a surprisingly important factor in promoting the study of science in antebellum America."

50. For this attitude in the sixties see William Jewett Tucker, *My Generation: An Autobiographical Interpretation* (Boston: Houghton Mifflin, 1919), p. 34 passim.

51. G. Stanley Hall, *Life and Confessions of a Psychologist* (New York: D. Appleton and Co., 1923), pp. 156, 159, 594.

52. Ibid., pp. 173, 184.

53. McCaughey, *Perspectives in American History*, p. 261.
54. Fulton, *Barnard*, pp. 381–82, passim.
55. John LeConte to William Sharswood, 22 September 1867, Sharswood papers, American Philosophical Society.
56. Seth Low, "Higher Education in the United States," *Educational Review* 5 (1893): 5.
57. Hofstadter, "The Revolution in Higher Education," p. 289.
58. Richard G. Boone, "Results under an Elective System," *Educational Review* 3 (1892): 60–61, 65.
59. Thomas Bertrand Bronson, "Some Interesting Statistics," *Educational Review* 9 (1895): 87. Statistics are for arts students and do not include engineering schools.
60. McCaughey, *Perspectives in American History*, p. 267n.
61. William Dallan Arnes, ed., *The Autobiography of Joseph LeConte* (New York: D. Appleton and Co., 1903), p. 141.
62. Minutes of the Trustees of Dartmouth College, 23 July 1844.
63. Edward Hitchcock, confidential letter to Mark Hopkins, 12 July 1848 and Mark Hopkins to Hon. Jos. White, 3 February 1853, both in Williams College Library. Mark Hopkins to Edward Hitchcock, 25 July 1848, President Hitchcock collection, Amherst College Library. Amherst had been thinking of establishing its own professorship of agricultural chemistry since 1843 and in 1852 added a plan for a "Scientific Department designed for graduates [of colleges]"; Minutes of the Trustees of Amherst College, 8 August 1843, 6 August 1844, 10 August 1847, and 11 October 1852, Amherst College Library.
64. Minutes of the Trustees of Princeton College, 20 December 1853.
65. Minutes of the Trustees of Union College, 25 July 1849 and 21 July 1855.
66. Minutes of the Rector and Visitors, 12 March 1858, University of Virginia Archives. At Virginia the prescribed arts curriculum led to a masters degree—the source of confusion about its curriculum. This program was not a graduate course, did not allow elective choice, and enrolled virtually all the students at the university who were not candidates for a professional degree.
67. Arthur Comey, "Growth of the Colleges of the United States," *Educational Review* 3 (1892): 128; 130.
68. Fulton, *Barnard*, pp. 236, 244–45.
69. Thomas C. Johnson, *Scientific Interests of the Old South* (New York: Appleton-Century, 1936).
70. *Report of the Commissioner of Education for the Year 1874*, (Washington: Government Printing Office, 1875), p. LXVIII.

71. Howard S. Miller, *Dollars for Research: Science and Its Patrons in Nineteenth Century America* (Seattle: University of Washington Press, 1970), p. 80.

72. Russell H. Chittenden, *History of the Sheffield Scientific School of Yale University* (New Haven, Yale University Press, 1928), I, p. 115.

73. Edwin Layton, "Mirror-Image Twins: The Communities of Science and Technology," in *Nineteenth-Century American Science: A Reappraisal*, ed. George H. Daniels (Evanston: Northwestern University Press, 1972), p. 217.

74. William Barton Rogers to Henry Darwin Rogers, 29 March 1848, University of Virginia Archives.

75. "Educational" (editorial) *Nation* 3 (18 October 1866) : 305.

76. Frederick Rudolph, *The American College and University: A History* (New York: Vintage Books, 1962), pp. 258–60.

77. "Educational," p. 366.

78. J. Ross Browne to William Sharswood, 7 November 1867, Sharswood papers, American Philosophical Society.

79. Francis A. Walker, "The Place of Schools of Technology in American Education," *Educational Review* 2 (1891) : 211.

80. Robert Seigfried, "A Study of Chemical Research Publications from the United States before 1880" (doctoral diss., University of Wisconsin, 1952), pp. 173–81.

81. Laurence R. Veysey, *The Emergence of the American University* (Chicago: University of Chicago Press, 1965), p. 112.

82. Hall, *Life of a Psychologist*, p. 245.

83. Howard S. Miller, "The Political Economy of Science," in *Nineteenth-Century American Science: A Reappraisal*. ed. George H. Daniels (Evanston: Northwestern University Press, 1972), pp. 95–112. See also p. vii. The quotation given was in the original oral version.

84. Minutes of the Trustees, 5 July 1889, University of Kentucky Archives.

85. Ronald C. Tobey, *The American Ideology of National Science 1919–1930* (Pittsburgh: University of Pittsburgh Press, 1971).

86. Hall, *Life of a Psychologist*, p. 196.

87. Rudolph, *American College and University*, p. 345.

88. Thorstein Veblen, *The Higher Learning in America* (New York: Hill and Wang, 1957), pp. 119, 164.

89. McCaughey, *Perspectives in American History*, p. 270. In his analysis of Eliot's communications with science faculty, Hugh Hawkins concludes that through the 1880s, Eliot thought research only "suitable for summer vacations when it would not undermine teaching." See

Between Harvard and America: The Educational Leadership of Charles W. Eliot (New York: Oxford University Press, 1972), p. 64.

90. Veblen, *Higher Learning in America*, p. 128.
91. J. McKeen Cattell, "The Origin and Distribution of Scientific Men," *Science* 66 (1927): 514.
92. Cattell, "The American Society of Naturalists: Homo Scientificus Americanus," *Science* 17 (1903): 565.
93. Hofstadter, "The Revolution in Higher Education," p. 284.
94. Hall, *Life of a Psychologist*, p. 14.
95. Quoted in Rudolph, *American College and University*, p. 271.
96. On the social transformations in the late nineteenth-century college, see George E. Peterson, *The New England College in the Age of the University* (Amherst: Amherst College Press, 1964). On student backgrounds in particular see David F. Allmendinger, *Paupers and Scholars: The Transformation of Student Life in Nineteenth-Century New England* (New York: St. Martin's Press, 1975).
97. Both studies contrasted in J. McKeen Catell, "The Origin and Distribution of Scientific Men," p. 515.
98. William Jewett Tucker, *My Generation: An Autobiographical Interpretation* (Boston: Houghton Mifflin, 1919), p. 354.
99. McCaughey, *Perspectives in American History*, documented on p. 306.
100. McCaughey, *Perspectives in American History*, quoted on p. 297.
101. Hall, *Life of a Psychologist*, p. 237.
102. Hall, *Life of a Psychologist*, p. 236.
103. Veysey, "American Undergraduate Curriculum," p. 17.
104. Hall, *Life of a Psychologist*, p. 241.
105. Hall, *Life of a Psychologist*, p. 237, for a view of the Rowland affair; *Baltimore News*, 20 January 1894 (on American Philosophical Society Film 117), for the figures; Nathan Reingold, ed., *Science in Nineteenth Century America: A Documentary History* (New York: Hill and Wang, 1964), pp. 323–28, for excerpts of Rowland's address as first president of the American Physical Society, 1899.
106. Hawkins, *Between Harvard and America*, quoted on p. 214.
107. Hawkins, *Between Harvard and America*, pp. 214–15.
108. Veblen, *Higher Learning in America*, pp. viii, 11, 21, 196.
109. Cattell, "Origin and Distribution of Scientific Men," p. 515. This percentage would decline rapidly in succeeding decades as new patterns of employment outside of academe developed.

RATIONALIZATION AND REALITY IN SHAPING AMERICAN AGRICULTURAL RESEARCH, 1875-1914 [*]

CHARLES E. ROSENBERG
The University of Pennsylvania

THE 1960s AND 1970s HAVE PRODUCED AN INCREASINGLY ACERBIC critique of American agricultural research and, in particular, its social consequences. Doubts as to the quality of agricultural research have a much longer history. Not all critics agree, of course, but the cumulative burden of their charges is clear enough. First, the quality of research—both in the U.S. Department of Agriculture (USDA) and the state universities and experiment stations—has been spotty and in a sense provincial. Second, American agricultural research has been marked by the lack of a guiding social vision, by an inability to predict, evaluate, and contend with the social and economic consequences of innovation.[1]

Such criticisms pose grave questions in regard to the place of applied science in American institutional and political life. These questions are by no means ephemeral or superficial, for those aspects of agricultural research that seem problematical today were present a century ago in nascent form. Indeed, they were implicit in the social realities that allowed the vigorous growth of agricultural research to take place. The following pages will attempt to demonstrate this continuity through a discussion of the agricultural experiment station scientist in the last quarter of the nineteenth and opening years of the twentieth century.[2] I shall emphasize the ironic gap between the conscious social goals of pioneer agricultural scientists and administrators and the consequences both of their policies and of the arguments that rationalized these policies.

[*] This chapter first appeared in a slightly different form in *Social Studies of Science* 7 (November 1977): 401–22. Reprinted with permission of that journal.

One fundamental contradiction facing these scientists grew out of the ambiguous status of agriculture in American society, an ambiguity paralleled by that surrounding applied research in the world of science. Honored in the pantheon of formal social values, the farmer was in reality often viewed as a hayseed, a figure of fun and bland contempt. In the hierarchy of scientific achievement, applied science possessed a similarly problematical quality: enshrined in the formulas of conventional rhetoric, the applied scientist was nevertheless part of a scientific world in which abstractness generally correlated with status. The utility so appealing to mid-nineteenth century Americans constituted a dubious virtue in the world of pure science. In their appeals for public support, moreover, would-be entrepreneurs of agricultural science had no choice but to affirm a necessary interdependence between science pure and science applied. Yet here too lay an ironic gap between ingenuous expectation and the unfolding of a more complex reality. For scientists and administrators could not foresee the possibility of conflict between the demands of pure and applied science—inconsistent demands that were to shape the careers of individuals and institutions.

A final and perhaps most poignant irony centered on the gap between the vision of a good society these scientists shared with most of their generational peers and the social consequences of their professional acts. Almost all these scientists seem to have nurtured a real faith in the family farm and the necessary social virtues of its yeoman proprietors. Yet, as we will see, circumstances dictated that the small agricultural producer would not be the most prominent beneficiary of experiment station research and development. Administrators and scientists almost invariably had to work with larger farmers, with the more highly capitalized producers of fertilizers and seeds, of livestock and horticultural varieties. Stated crudely, experiment station scientists and administrators never considered the possibility that insofar as their work proved successful, it would mean helping to enrich the rich, impoverishing and ultimately forcing many ordinary farmers from the land. Nor did the experiment station workers foresee an ultimate contradiction between their advocacy of an increasingly sophisticated, government-supported research effort and the gradual demise of a self-sufficient decentralized world.

These late-nineteenth-century Americans were neither unintelligent nor cynical; the inconsistency between their conscious design and the consequences of their acts arose from a more fundamental consistency: that between their social assumptions, their professional needs, and the nature of the cultural context in which these needs had to be elaborated.

American agricultural scientists have always justified their role in terms both revealing and compromising. The formulas are ubiquitous and almost timeless. With the scientific community generally, they have shared the belief that pure science is a necessity if applied science is to flourish; mundane applications grow inevitably, and ultimately only, out of increased understanding of fundamental principles.[3] Innovation cannot be predicted, and therefore scientist-formulated research must be supported generously and nonintrusively. Perhaps most significantly, the agricultural scientists' ideological stance rested on an unquestioned faith in the transcendent virture of productivity. To increase the productivity of the soil— to make two blades of grass flourish where one had before—was to act in an unambiguously moral fashion. In political terms, the promise of increased productivity was the socially visible component of the agricultural scientist's habitual dissolution of the distinction between pure and applied science, the tangible return to lay supporters.

CAREER OPTIONS FOR AMERICAN SCIENTISTS were severely limited in the 1870s, and those committed to serving America's agricultural community faced a particularly bleak future. The colleges established under the Morrill Act of 1862 had seemingly failed American agriculture; bereft of students or appropriate curriculum, they displayed little prospect of accumulating either.

Science offered at least the possibility of economic application and thus public support. To the more rhetorically inclined, the application of science to agriculture promised to raise the American farmer's moral and intellectual status at the same time it improved his economic position. The successful infusion of scientific procedures and ideas might, it seemed, make the ordinary farmer a man of learning, no longer an object of casual scorn but a professional like those lawyers or physicians who too frequently regarded themselves as his superior. Aspirations such as these meshed well with the professional ambitions of scientists and administrators who orchestrated demands for a national system of agricultural experiment stations. The first of these novel institutions were founded by individual states in the 1870s. Further agitation culminated in passage of the Hatch Act of 1887, which provided each state with an annual endowment of $15,000 for the support of a station.

The scientists and administrators associated with these institutions necessarily had to deal with a demanding and impatient constituency—one little capable of distinguishing between the activities of a model farm and those of a laboratory-oriented experiment station, between the testing of fertilizer content and genuine chemical research.[4] Confronted with such ambiguous circumstances, scientists and administrators were forced into an implicit quid pro quo with influential laymen—that is, the promise of concrete economic return in exchange for social support and a degree of autonomy. Thus the appeal of rhetorical tributes to the inevitably positive interactions between science pure and science applied. Results of seemingly little practical value might, scientists reiterated, prove of the greatest practical relevance. "It is wisdom's way," as a pioneer student of plant disease put it, "to hold the position that all truth, no matter how abstruse, is for the improvement of the masses." Yet the difficulty, as chemist W. O. Atwater warned farmers, was:

that the seemingly simplest and most pressing problems reach down to the profoundest depths of abstract law; that often the things which appear theoretical are at the bottom the most essential, that not infrequently the practical interests of the farmer require the theoretical problems to be considered first, for the same reason that the foundation of a house and not the walls is first to be built.

But, Atwater reassured in 1889, this reality need not "necessarily involve the postponement of important discoveries to a remote future, for experience shows that when men in any line have attacked problems in a thorough and scientific manner, good results have speedily followed."[5] The importunities of America's farmers would not, and should not, go unanswered. For without scientific guidance, Atwater and his like-thinking peers urged, most American farmers could not hope to survive. The scientist-entrepreneurs who administered and spoke for the experiment stations represented their institutions as playing a potentially central role in a dismayingly changeable economy.

In some ways, this analysis of political and economic realities was acute, even if self-serving. Like most of their thoughtful peers, American agricultural scientists were convinced in the last quarter of the nineteenth century that they lived in a world of crisis. Social health implied the maintenance of a stable yeoman class, yet an ever more competitive world marketplace made the lot of such proprietors increasingly difficult. The American farmer had to understand that newer modes of transportation:

have made the supply of an agricultural product almost anywhere in the world an appreciable factor in supplying the demand

for that product almost anywhere else in the world. If he be
thoughtful he must recognize the fact that in the future he must
work on a narrower margin of possible profit than in the past.

Modern warfare would not be conducted with sword and cannon, but with the weapons of trade and economic competition.[6]

Only the most efficient could hope to survive; a Darwinian calculus dictated that marginal farmers, those unable to incorporate the teachings of science and logic of efficiency in their operations, must leave the land to more capable hands.[7] Increasingly sharp price competition in staple crops implied both the need for specialized production and marketing and the ordered rationalization of that production. Science seemed the most plausible and responsible means of attaining these economic ends and thus keeping America's farmers solvent, moral, and immune from the pressures of an increasingly baleful urban life.

The connection between social stability and an economically viable agriculture soon became a cliche in experiment station circles. A Missouri scientist argued in 1904 that:

The agriculture of the future is destined to become the most
important factor in the development of the American
nation. Agriculture in any age if conducted by free land owners
contributes to independence of thought, integrity of character
and fearlessness of action. . . . The man in America who owns
160 acres of land, a good home, flocks and herds, is in a position
to fight for the truth without fear or favor.[8]

The image of the city, on the other hand, constituted a negative emotional counterpart of the farm's moral and civic virtues. The city bred anarchism and violence, served as the battleground for violent confrontation between labor and capital. "It is in these great cities," as a prominent agricultural politician and experiment station advocate put it in 1883, "that rings, strikes, frauds, trades-unions, socialism, centralization and political corruption, crime and immorality are born, fostered, and best flourish. In agricultural districts . . . the tendency is in the opposite direction; the people are there, as a rule, devoted to peace, justice, and good order."[9]

Such social and economic views—that is, the emphasis on crisis, on the need for maintaining the family farm yet upgrading its productive capacity—played a prominent role in lobbying for the Hatch Act in the 1880s and was reiterated again and again in the following decades. Not only would science have to make two blades grow where one had before, it must also dictate what variety of blades should be grown, how they should be cultivated, and where they might best be marketed. Thus a legitimating ethos of stewardship, a term often invoked by experiment station publicists, informed the work of scientists and administrators. Through selfless intellectual leadership they might, it seemed, play a significant role in maintaining a uniquely valuable social order, while at the same time advancing their own careers.

Late-nineteenth-century scientists and administrators were drawn from families oriented toward learning and achievement; they harbored little sympathy for Grangers or later for Populists, the Farmer's Alliance, or other groups dedicated to improving the farmer's market position through political action. Not surprisingly, they rejected Bryan in 1896 and overwhelmingly endorsed McKinley Republicanism; typical was the authority in veterinary medicine who could not "swallow Bryan's anarchism," and planned to "support the greatest of Americans—Major McKinley." Similarly, a South Carolina professor of agriculture could describe Ben Tillman and his agrarian supporters as driven by the rage of the sans-culotte. Even earlier than the 1890s, the natural ties between academic supporters of agriculture and the wealthier and better educated farmers were apparent; in 1877, for example, the superintendent of the Louisiana State University could explain that his institution depended on the larger sugar and cotton planters "with their more liberal ideas" to serve as a "break-water against the narrow and constraining ideas of the little hill farmers."[10]

The ironies implicit in these scientists' and administrators' understanding of economic matters and the potential role of science in resolving them, so apparent today, were hardly evident a century ago. The small farmers who, in the categories of formal rhetoric, were to be the beneficiaries of such scientific largesse ultimately benefited far less than larger producers. Experiment station and agricultural college scientists naturally cooperated most easily with larger farmers, with the more highly capitalized producers of fertilizers and seeds, of livestock and horticultural varieties, and—most importantly—with the leaders of specialized producers' associations. States with little in the way of a well-differentiated organizational structure offered a discouraging prospect to would-be scientific entrepreneurs; there were no preexisting organizational levers to pull, no means of appealing to an interested and influential leadership structure.[11]

From their inception, the experiment stations worked closely with agriculturally related business and producers' associations. Insofar as they were tied to regulatory work with fertilizer manufacturers, seedsmen, or breeders, the stations worked with larger producers and promoters of purebred lines. Inevitably, a sincere desire to improve productivity made station staff members publicists for these improved practices. Demands for the effective dissemination of existing knowledge—a moral as well as political imperative in this generation—implied cooperation with banks and agriculturally related business, especially railroads. Scientists and administrators had contracted a firm alliance with farm publishers as early as the middle of the nineteenth century; scientists furnished news and features, while editors and publishers provided political support in state legislatures and farm organizations.

This cordial coexistence with business implied both immediate and ultimate problems. Dependence on influential laymen often meant an active role for such men in the planning and even execution of research, a constraint that many working scientists found increasingly distasteful. Continued success in attracting business support only increased a sense of obligation and mutual identity of interest, as well as the habit of justifying requests for support in concrete economic terms. Thus the most powerful and convinced of lay supporters was, by virtue of that commitment, potentially the most egregious and intrusive of allies. Only after decades have we succeeded, as one administrator plaintively explained in 1901, in gaining the farmers' confidence "in the face of much opposition from dense ignorance, prejudice, and the extraordinary suspicion of motives." But, he concluded, "it has been accomplished by . . . getting down to their level, and the more legitimate and better work of the Station has suffered in consequence."[12]

Some young academics turned such compromising relationships to their mundane advantage, becoming evangelists of productivity, promising more than could reasonably be expected from even the most prudent farming, and thus advancing their careers proportionately to the whetted enthusiasm of their audience. Personal aggressiveness could be an important, in some cases indispensable, asset. "I know that modesty is a good quality," one older professor wrote to a more youthful aspirant, and that self-assertiveness usually goes:

. . . with a shallow, cheap man. At the same time, if with your splendid training and good sound sense you would combine some of this aggressiveness and boldness and self-assertiveness, self-advertising, self-exploitation, you would be appreciated at more nearly what you are worth than you are at the present time.[13]

Thus the necessary contacts between professional staff members and potentially influential supporters implied recurring problems of institutional discipline.

Shaping American Agriculture

Other problems surfaced as well. Community of interest implied conflict of interest. Station directors, for example, were repeatedly urged to endorse agricultural products or to insert commercial advertisements in station bulletins, while staff members might on occasion invest in products they had endorsed in their professional capacity.[14] Agricultural business also provided an unwelcome tribute of a rather different sort: competition for able young men at salaries far higher than those available to junior scientists and professors.

BY THE LATE 1890s, A GOOD MANY STATION leaders had become convinced that these circumstances demanded change. Even if they were committed to applied science, well-trained scientists and ambitious administrators still sought increased research support and greater autonomy for the implementation of administrative policies. But such demands always had to be formulated in terms of an equally categorical commitment to science pure and science applied; they could neither make ultimate distinctions between the two nor anticipate the development of possible conflict between them. Applied science must in time languish, they contended, if pure science were neglected. The logic of institutional supplication always pointed toward the same rhetorical stance: investment in pure science was prudent, its practical returns inevitable. There was no escaping this formulation, for it was sufficiently vague to mask the conflict implicit in the differing perceptions and needs of laboratory scientists, administrators, and influential laymen.

This rigidity of vision was particularly marked among the leaders of experiment station work at the end of the century, men who had matured into that leadership during the formative years of the 1870s and 1880s. Hoping to foster mutually supportive relations with influential farmers and farm organizations, they had come to regard their constituency's demands and the compromises implied by such demands as legitimate as well as practically unavoidable. Leaders of this generation, men as well known to their contemporaries as Liberty Hyde Bailey of Cornell and Samuel William Johnson of Connecticut, are easily distinguishable from a minority of more laboratory-oriented scientists who sought to impose increasingly academic standards in the conduct of experiment station research.[15]

Established leaders reacted in two ways. As administrators and (necessarily) political intermediaries, they defended their policy of service and accommodation. On the other hand, they were careful to affirm their unshaken allegiance to the world of scientific research. The question was one of timing. Bailey, for example, stated his position with characteristic clarity in 1893:

It is necessary that the immediate dollars and cents value of the investigation should usually be kept directly in sight by the experimenter until such time, at least, as the stations shall find themselves universally established in the popular mind. But the time cannot be far distant when some of the more abstruse and far-reaching problems of unapplied science must demand investigation from the experimenters, for these principles lie at the bottom of all permanent progress, and their solution is, in the long run, the most practical line of research. . . .

"The time will come," as another experienced director put it four years later," when we shall be liberated for this work of investigation, but it will only come if we improve our present opportunity to show the farmer what we are trying to accomplish."[16]

Such men of worldly wisdom could make stern taskmasters for young scientists anxious to escape the lecture platform, the routine of answering letters and analyzing fertilizers. Men such as Bailey and Johnson regarded the discipline-oriented—and in many cases German-inspired—demands of some of their younger colleagues as premature. These experienced administrators rationalized their position in what might be described as evolutionary terms: present realities were a necessary transitional stage, creating an institutional base for the support of more abstract work in future decades. The experiment station's constituency had to be "educated" gradually, and if genuine conflict between the desires of staff scientists and influential laymen should develop, the agricultural community's immediately perceived needs must be attended to first. As the influential and outspoken E. W. Hilgard put it, his California experiment station could hardly adopt Mr. Vanderbilt's policy of damning the public, especially in a state with novel climatic and geographic features. There the dismissal of popular demands would be unwise, if not immoral. Said Hilgard: "I do not believe that a station so situated ought to make it their [sic] business to pursue recondite studies in vegetable physiology or animal chemistry, unless they have first satisfied this legitimate demand, on the strength of which the passage of the Hatch Bill became possible."[17]

It must be emphasized that men such as Hilgard were not simply time-serving careerists, willing to trim their administrative sails in the direction of plausible accommodation. They doubted neither the legitimacy of their client constituency's demands nor the social utility of the experiment station. E. H. Jenkins, for example, though a vigorous advocate of improving experiment station research, still contended that he would immediately drop research he had planned or even begun if it should conflict with the demands of a particular farm organization for cooperative work. "The value of any station," Jenkins explained, "is the usefulness of its work to the everyday farmer. George Eliot has said that the object and aim of any rightly constituted man is use; it is for use he exists. The same may be said of the experiment station."[18]

Yet, at the same time, Jenkins never doubted that most investigation designed to answer immediate questions was simply "makeshift," a means of attracting support for more scientific work, "which is in the end more practical." Most of his administrative contemporaries were far harsher than Jenkins in demanding that staff scientists dedicate their energies to the demands of the public that supported them.[19]

As already implied, policy conflict represented in part a generational difference. Not only were younger men more likely to have received advanced training, they were also less likely to feel a nationalistically tinged commitment to agriculture. Yet not all men of the 1870s and 1880s were content with a modus vivendi based on accommodation. (Nor, it must be understood, were all youthful scientists at the turn of the century discontented with their professional lot). An elite of well-trained and intellectually ambitious experiment station scientists had, since the 1880s, been dissatisfied with the compromises demanded by the agricultural constituencies that paid their salaries.

Perhaps the most articulate of such workers was agricultural chemist W. H. Jordan. Since the 1880s, Jordan had been sensitive to pressures from a short-sighted and sometimes short-tempered constituency. He had been aware, as well, of the painfully inadequate supply of well-trained scientists.[20] Nevertheless, he worked tenaciously to channel increasing portions of experiment station budgets toward genuine research and away from the routine field trials and extension work that so often served in the place of research. The greatest threat to America's infant experiment stations, Jordan warned in 1888, lay in the pressure for immediate results. "In order to avoid this," he contended, "every station needs to be imbued with the spirit of scientific investigation, and to desire to know the truth more than to obtain a scientific reputation with the station's constituency." Jordan contended again and again that American work lacked rigor; we must, he urged, look to German standards and German results.[21] There was no such thing as national truths in the investigation of natural phenomena.

Despite increasing success—or at least growth—Jordan continued to warn of the pitfalls faced by America's agricultural scientists. "Have we full scientific liberty?" he asked in 1906. "Is it desirable to combine within the activities of the same individual the work of research and exploitation?"[22] Teaching demands also intruded increasingly on the scientist's precious research hours. "What kind of a man do you want for an investigator?" Jordan asked in 1904. The answer was clear enough: "a man absorbed in the things he is doing and who shall not be turned aside and wearied by having to drill a class or to do anything else but hunt his subject and the truth."[23]

W. H. Jordan represented an emphasis clearly alternative to those policies followed by his more compromising peers. But like them, he maintained a commitment to the agricultural enterprise itself; he doubted neither the importance of promoting agricultural productivity nor the necessary role of science in attaining that goal. He too emphasized the equal legitimacy and necessary interdependence of pure and applied science. However, Jordan, like a self-conscious minority of his colleagues, contended that a necessary balance had been destroyed by the unrelenting importunities of community demands.

But even Jordan's particular compromise proved uncongenial to a number of even more disaffected agricultural scientists. To these investigators, agricultural research was nothing more than a way of earning a living until more appropriate posts could be found. For at least some scientists before World War I, an experiment station position meant long years at work that their personal aspirations and their discipline's standards categorized as routine and even demeaning. "I am sick of it," a North Carolina scientist complained in 1890, "and want to go where there is some degree of civilization —some appreciation of scientific labor and something to make life worth living.[24] But positions in which research constituted a primary responsibility were few, and carping at repetitive analytical work was a luxury confined to private letters and occasional professional meetings.

BY THE FIRST DECADE OF THE TWENTIETH CENTURY, the fundamental pattern of experiment station research had been well established. Problems had emerged, as well as tentative solutions. Additional support for research, the guaranteeing of institutional autonomy, the sharper differentiation of diffusion from innovation were all goals either achieved or well in sight by the outbreak of the First World War. But in modifying unsatisfying realities, these solutions tended to reaffirm old rigidities.

The Adams Act of 1906 addressed itself to the problem of research support. This measure doubled each station's level of federal support from $15,000 to $30,000 and—at the urging of the scientists and administrators responsible for lobbying it through Congress—

limited the expenditure of these additional funds to "original investigation."[25] But despite the optimism surrounding the passage of this measure, none of the problems it sought to address found immediate resolution. Those investigators who had found the research environment before 1906 confining still faced annoying hindrances to their work in most states.[26] Scientists still had to provide bulletins and reports on order—so many each year and with subject matter immediately relevant to the state's farmers.

Specialized terminology had to be carefully avoided in official publications, while in some instances authors were denied even the formal attribution of authorship. More important, ambitious investigators might be denied the privilege of publishing in their discipline's achievement-defining journals and required to publish instead in a station's official bulletins (which were ordinarily ignored by academic biologists). At least one investigator contended that papers "of the utmost fundamental value might, if published in the regular series of station bulletins, be actually ridiculed and bring a station into disrepute with certain classes of their constituents."[27] In a good many states, experiment stations still bore an onerous burden of regulation and administration.

Perhaps most time consuming for station men were the demands of what today would be termed extension work: lecturing and the demonstration of new or improved techniques of culture or varieties to individuals and groups. "Some of us," as one anxious scientist put it, "have felt as though the extension work would swallow up the other, as a sort of side show blanketing a circus."[28] Moreover, success either in innovation or the dissemination of new techniques only increased lay demands for such services. As late as 1922, a California scientist expressed the dilemma clearly:

> As a means of supplying the information the Extension workers and farmers are demanding, fundamental research is absolutely essential. But the mixing of research and extension activity inevitably superficializes the research; the mind of the research worker becomes focused on the application rather than on the principle.

Real research required not only intense concentration, he concluded, but understanding of the basic sciences, "which are constantly undergoing profound advances in method."[29] There were only so many hours in the day. Most directors, moreover, remained anxious to hire people with an agricultural background and an ability to communicate with farm groups at the expense of other, more academic, criteria.

Insofar as the stations could provide economically relevant innovation and gain the support of influential laymen, they seemed successful indeed. But such success could do little to redefine the compromising nature of the scientist's contextual realities. To succeed was to succeed in convincing farmers that agricultural science had something to offer them and thus intensify their demands for advice and information. Given the peculiar conjunction of political power and the ideological centrality enjoyed by agriculture, scientists found themselves little able to moderate unwelcome demands. They were necessarily dependent upon the political sagacity and possible scientific commitment of deans and directors.

Especially in those stations located on university campuses, success also implied adjustment to parallel university structures. A desire to upgrade experiment station standards, for example, implied new relationships with parallel university departments—agricultural chemistry with chemistry, breeding or animal husbandry with biology, and so forth. Should the agricultural colleges create their own advanced-degree granting programs and if so, how specialized should their requirements be?

In any case, experiment station and agricultural college scientists had to deal with the ill-disguised disdain of their "purer" basic science counterparts. Nevertheless, those experiment stations located away from university settings regretted their social and intellectual isolation (and regretted in some ways as well a political isolation that left their stations unprotected from "the petty nagging of the partisan spoilsmen and the legal shyster" who, as one long-time administrator put it, often constituted local boards of control).[30]

The agricultural colleges and experiment stations responded with a number of institutional modes corresponding to these complex and in some ways conflicting realities. Most significantly, the agricultural sciences were, with growth in scale and levels of support, able to evolve a series of disciplinary options conforming in a rough way to the structure of these contextual realities. New fields emerged with their own peculiar internal ethos, specialized knowledge, and formal requirements; these new skills and roles conformed in a rough way to the manifold tasks and responsibilities that confronted the pioneer experiment stations and their leaders at the end of the nineteenth century.

It had become clear that there was no agricultural science as such. Attempts to organize associations and journals under this rubric had all failed. An earnest Society for the Promotion of Agricultural Science did linger on into the twentieth century, but a journal dedicated to *Agricultural Science* (founded in 1887) had failed for lack of support by the mid-1890s.[31]

Agriculture was an academic category with a decreasingly precise relevance in a world of increasingly specialized scientific aspirations. The necessarily omnicompetent "professor of agriculture" characteristic of agricultural colleges and experiment stations in the 1870s and 1880s had settled into his constituent functional parts: professorships of agronomy, horticulture, biochemistry, animal husbandry.[32] Soon professorships of rural sociology and agricultural economics would be added.

With the establishment of the cooperative extension service in 1914, specialists in the dissemination of knowledge came to occupy another segment of the burdensome tasks assumed by the professor of agriculture earlier in the century. Just as that nineteenth-century academic category called moral philosophy sorted itself out into economics, politics, ethics, philosophy, sociology, and psychology, so the professorship of agriculture disintegrated into a host of scholars identifying themselves with a variety of roles and disciplines allied only by their relationship to the economic needs of agriculture.

Perhaps the most novel aspect of this fragmentation was that bureaucratic blend of education and proselytizing that came to be called extension. If the intimate and in some ways compromising interaction of experiment stations and agricultural colleges with their nourishing social substrate caused difficulties—in some states and in particular fields almost insurmountable ones—this same link provided in the extension network a uniquely effective apparatus for the diffusion of applied knowledge.[33] In terms of institutional innovation, extension was perhaps the most significant legacy to develop out of that ambiguous social context we have sought to describe.

In the twentieth century, constituency pressure came to be applied increasingly to the now visible and self-conscious applied disciplines. Horticulturists, for example, for long years had to withstand demands that they provide acres of immaculately pruned and thriving varieties; animal husbandry experts had to prove that they could be successful stock raisers and exhibitors.[34] On the other hand, scientists with little personal commitment to the world of agriculture or little interest in applied problems had, as the twentieth century progressed, a far richer universe of institutional options. When, for example, the Johns Hopkins School of Hygiene was established with Rockefeller support in 1917, two of its early luminaries were experiment station men, Wisconsin's E. V. McCollum and Maine's Raymond Pearl.[35] Neither had originally sought experiment station work.

As the twentieth century progressed, "basic science"-oriented departments at the more prestigious agricultural colleges and experiment stations effectively insulated themselves against the demands of a lay constituency. An intermediate layer of applied scientists and educational specialists now dealt with the needs of laymen, supervised demonstrations, answered letters, and regulated agricultural products.

An equilibrium had been reached by the 1920s; increasing productivity allowed scientists to avoid confronting the difficult choices implicit in their adherence to the gospel of science pure and science applied. The more research-oriented found increasing freedom in which to define their projects and funds to support their work. Applied scientists created new professional disciplines with securely internalized identities and well-defined institutional niches. Influential laymen garnered appropriate equities in the form of trained researchers and teachers and, ultimately, increased productivity. By midcentury, not two, but three and in some cases even four, blades grew where one had before.[36] Both pure and applied science at last had a secure institutional context. If the conflicts implicit in this ingenuous formulation could not be avoided ultimately, neither in practice need they be generally confronted.

This evolution brought unlooked-for consequences. Productivity had come to be defined in exclusively material terms and, as we have contended, its social implications were assumed to be necessarily benevolent. That this productivity and the scientific and educational activities of the experiment stations would fail to reinvigorate the family farm was simply inconceivable to the scientists and administrators who staffed America's experiment stations in their first quarter century.

Eugene Davenport, for example, dean of the College of Agriculture at Illinois, was, among his administrative contemporaries, perhaps the most eager in his courting of large-scale agricultural business and most entrepreneurial in his administrative policies. Yet he could write to President-elect Wilson in 1912, warning that the state universities and experiment stations must remain free of corporate, capitalistic influence; it was their function to instruct the average citizens of the state. "Our National hope is that our farmers may not yet become peasants and that in the meantime their unequal burden be lightened. With *our farmers peasantized, our democracy would be for the present lost.*"[37] The ultimate defenders of this "farmer citizenship" were the state universities and experiment stations. The next generation of American farmers must be enlightened, Davenport contended. "If not," he warned, "it will be a serious thing for the whole country, for it will be impossible for the Anglo-Saxon race to continue its existence with old-time methods of farming."[38]

There is no reason to doubt Davenport's sincerity, nor the irony implicit in the gap between his conscious intent and the consequences of his administrative actions. Like a well-crafted automaton, the agricultural research establishment moved ponderously into the twentieth century, rigid in ideology and wedded to a habitually compromising interaction with its client constituency, yet diverse and well-funded enough to support the ambitions and ideals of most scientists and administrators. Unless mistaken for growth, change could not come easily.

1. See, for example, Nicholas Wade, "Agriculture: NAS Panel Charges Inept Management, Poor Research," *Science* 179 (5 January 1973): 45–47; Wade, "Agriculture: Critics Find Basic Research Stunted and Wilting," *Science* 180 (27 April 1973): 390–93; Wade, "Agriculture: Signs of Dead Wood in Forestry and Environmental Research," *Science* 180 (4 May 1973): 474–77; Wade, "Agriculture: Social Sciences Oppressed and Poverty Stricken," *Science* (18 May 1973): 719–22; Wade, "Agriculture: Research Planning Paralyzed by Pork-Barrel Politics," *Science* (1 June 1973): 932–37; André Mayer and Jean Mayer, "Agriculture, the Island Empire," *Daedalus* 103 (Summer 1974): 83–96.

2. There is no definitive modern study of the agricultural experiment station in America. The most useful general history of agricultural research in the United States is still A. C. True, *A History of Agricultural Experimentation and Research in the United States, 1607–1925* (Washington, D.C.: USDA misc. pub. 251, 1937). More recent and useful, though concerned largely with formal policy debate, is H. C. Knoblauch, E. M. Law, and W. P. Mayer, *State Agricultural Experiment Stations: A History of Research Policy and Procedure* (Washington, D.C.: USDA misc. pub. 904, 1962).

3. Late nineteenth-century scientists had to contend with an endemic suspicion of the abstract and academic. They were forced to remind themselves again and again, that, in the words of one chemist, "the public will in the end judge him to be the 'practical man' who can present his order at the bank of Nature and have it cashed in crops, when the order bears the endorsement of a fine spun theory or not." Editorial, *Agricultural Science* 4 (March 1892): 144.

4. There is no detailed study of the lobbying that culminated in passage of the Hatch Act, but see Charles E. Rosenberg, "Science, Technology, and Economic Growth: The Case of the Experiment Station Scientist, 1875–1914," *Agricultural History* 44 (1971): 1–20 and the relevant portions of the studies by True and Knoblauch et al. cited in note 2.

5. Byron D. Halsted, "The Station Bulletin and What Should Go Into It," *Agricultural Science* 4 (June 1890): 164; W. O. Atwater, *Experiment*

Station Bulletin No. 2 part I, June 1889, (Washington, D.C.: USDA, 1889): p. 17; Atwater, remarks, *Proceedings at the Third Annual Convention of the Association of American Agricultural Colleges and Experiment Stations, 1889* (Washington, D.C.: USDA misc. bull. 2, 1890), p. 99. Atwater was not only a leader in American nutrition research but a prominent lobbyist for the establishment of experiment stations on the German model and first incumbent of the directorship of the USDA's Office of Experiment Stations. See Charles E. Rosenberg, "Wilbur Olin Atwater," in *Dictionary of Scientific Biography*, ed. Charles Coulston Gillispie (New York: 1970), I: 325–26.

6. The quotation, typical of many scores of such reflections, is by George Morrow, "President's Address," *Proceedings of the Eighth Annual Convention of the AAAC & ES, 1894* (Washington, D.C.: USDA—OES, bull. 24 1894), p. 26.

7. As one director put it, "The hundreds of thousands of farmers who are making but a bare living owe their poverty, not to low prices, but to wasteful methods of production." Charles Thorne to H. W. Collingwood, 2 April 1907, director's letterbooks, Ohio Agricultural Experiment Station, Wooster, Ohio. Despite agreement that farmers suffered because of generally difficult economic conditions, it was always assumed that the most alert could escape bankruptcy. As one farmer put it, "bad, shiftless management" caused farmers to fail. "The wide awake, industrious man does not complain." F. H. Williams to Peter Collier, 16 September 1890, Geneva Experiment Station papers, Cornell University Collection of Regional History.

8. F. B. Mumford to Miss M. A. Turner, 9 May 1904, reel 35, College of Agriculture papers, University of Missouri Archives. Mumford went on to assure his inquirer that: "The farmer of the future will be an educated man. He will possess power and influence by reason of this fact. The arts and culture of the ages will contribute to this happiness. Scientific discoveries and their application will contribute to his economic advancement."

9. Henry E. Alvord, *The Farmer and His Family* (Hartford: 1883), p. 16. Cf. William H. Brewer, *The Brighter Side of New England Agriculture* (Manchester, N.H.: 1890), p. 20; Brewer, *The Farm and Farmer, the Basis of National Strength* (Boston: 1890), pp. 18–19; L. H. Kerrick, *Agriculture the Master Science* (Bloomington, Ill.: 1901), p. 1; W. D. Hoard to H. D. Hoard, 5 May 1886, letterbooks, vol. IV, W. D. Hoard papers, State Historical Society of Wisconsin.

10. The three examples cited are drawn, respectively, from Leonard Pearson to H. J. Waters, 13 October 1896, reel 34, College of Agriculture papers, University of Missouri Archives; J. M. McBryde to C. W. Dabney, 28 November 1890, C. W. Dabney papers, Southern Historical Collection, University of North Carolina; D. F. Boyd to E. W. Hilgard, 4 January 1877, folder 87, box 5, E. W. Hilgard papers, Bancroft Library, University of California, Berkeley. Even the chemist's mind-deadening routine analyses of soils and fertilizers was often seen as a political necessity since "it is those generally of influence who desire such work done . . ." Report of M. A. Scovell, Director, Experiment Station, 1887,

21–30 June 1887, James Patterson papers, University of Kentucky Archives.

11. One would-be scientific entrepreneur elucidated the situation clearly in attempting to explain his frustration at conditions in North Carolina: "They say plainly they don't know what to expect and don't know what they want and above all there is no organization and no way that I can see to get at the farmers as at the north." Milton Whitney to W. O. Atwater, 10 June 1886, reel 2, W. O. Atwater papers, E. F. Smith Library, University of Pennsylvania (originals deposited at Wesleyan University, Middletown, Conn.)

12. H. C. White to A. C. True, 4 April 1901, Georgia file, records of the Office of Experiment Stations, National Archives, RG 164.

13. H. J. Waters to T. J. Mairs, 21 January 1907, reel 37, College of Agriculture papers, University of Missouri Archives. Agriculture, as one prominent administrator warned in 1906, "is a field so undeveloped that a man with a ready tongue and a fertile mind can say almost anything, which will likely be a plague to the institution later." E. Davenport to George A. Harter, 12 June 1906, "Personal" letterpress books, dean's papers, College of Agriculture, University of Illinois, Champaign.

14. Such ethical issues represented a continuing problem for those Office of Experiment Station administrators charged with overseeing expenditure of federal funds. See, for example, A. C. True, "Advertising in Experiment Station Publications," *Proceedings of the Eleventh Annual Convention of the AAAC & ES, 1897* (Washington, D.C.: USDA—OES bull. 49, 1898), pp. 41–43.

15. For a contemporary contrast of the older and younger generations, see A. C. True to J. Howe Dumond, 13 June 1902, Massachusetts file, records of the Office of Experiment Stations, NA, RG 164. Johnson and Bailey actually represented a middle-of-the-road position, one far more sympathetic to the world of academic science than that assumed by most of their administrative contemporaries. For a more typical "old-guard" position, see I. P. Roberts, "Science Versus Art," *Proceedings of the Eleventh Annual Convention of the AAAC & ES, 1897* (Washington, D.C.: USDA—OES bull. 49, 1898), pp. 69–70.

16. Bailey, "Is the Experiment Station Movement a Success? The Work in Horticulture," *Agricultural Science* 7 (1893): 448; remarks of Charles Thorne, *Proceedings of the Eleventh Annual Convention of the AAAC & ES, 1897* (Washington, D.C.: USDA—OES bull. 49, 1898), p. 25. For a similar statement by the influential S. W. Johnson, see "Annual Address by the President," *Proceedings of the Tenth Annual Convention of the AAAC & ES, 1895–96* (Washington, D.C.: USDA—OES bull. 41, 1896), p. 45.

17. Hilgard to A. C. True, 10 June 1896, California file, records of the Office of Experiment Stations, NA, RG 164. See also in the same place, Hilgard to True, 4 March 1901 and Hilgard to B. I. Wheeler, carbon, 31 January 1902.

18. E. H. Jenkins, "The Cooperation of Stations with Farmer's Organizations in Experiment Work," *Proceedings of the Eighth Annual Convention of the AAAC & ES 1894* (Washington, D.C.: USDA—OES, bull. 24, 1894), p. 51.

19. Ibid.

20. "If a station is organized," Jordan warned in 1887 in response to a request for advice in the possible central planning of experiment station research, "with a working force so unfitted for investigation as to be at a loss to know what to go to work on, a force that must be supplied with a piano and a score, . . . then that station had better let its funds remain in the U.S. Treasury," Jordan to W. O. Atwater, 28 November 1887, reel 14, Atwater papers.

21. Jordan, "The Necessity of Caution in Agricultural Research," *Agricultural Science* 2 (1888): 261; *Proceedings of the Twentieth Annual Convention of the AAAC & ES, 1906* (Washington, D.C.: USDA—OES bull. 184, 1907), p. 63; Jordan's remarks in the course of a debate on "How much teaching, if any, is it desirable that a station worker should do?" in *Proceedings of the Eighteenth Annual Convention of the AAAC & ES, 1904* (Washington, D.C.: USDA—OES, bull. 153, 1905), p. 131. Jordan had sought to shape more adequate research conditions since the very beginning of his career. Cf. Jordan to S. M. Babcock, 24 May 1882, box 2, S. M. Babcock papers, State Historical Society of Wisconsin.

22. Ibid.

23. Ibid.

24. Gerald McCarthy to W. O. Atwater, 2 November 1890, North Carolina file, records of the Office of Experiment Stations, NA, RG 164. The papers of contemporary scientists contain many private letters detailing the dissatisfaction of scientists with the conditions at agricultural colleges and experiment stations.

25. For a detailed account of this measure and the agitation for its passage, see Charles E. Rosenberg, "The Adams Act: Politics and the Cause of Scientific Research," *Agricultural History* 38 (1964): 1–10.

26. A good many would-be scientists were, of course, satisfied with conditions before 1906 and little comforted by the prospect of having to conduct "original investigations." They were content enough with a routine of lecturing, demonstrations, and the appropriately varied repetition of plausible experiments. But the dilemma of that minority of well-trained and research-oriented scientists employed at agricultural institutions continued.

   As an example, let me cite the case of A. F. Blakeslee, holder of a Harvard doctorate and employed at the University of Connecticut (Storrs). When Blakeslee sought in 1912 to study heredity by inducing mutations in molds and fungi by means of the X-ray and the introduction of foreign substances in their substrate, he found little encouragement. "With 22 hours of work at present in the classroom," he complained, "and a heavy burden of committee and administrative work there is no chance to run cultures." He failed as well to persuade the director to pay for an assistant out of Adams funds. Blakeslee to W. G. Farlow, 3 January 1912, carbon, Blakeslee papers, American Philosophical Society.

27. H. J. Webber, "A Plan of Publication for Experiment Station Investigations," *Science* 26 (October 18, 1907): 510.

28. J. B. Lindsey to A. C. True, 10 December 1914, True personal file, records of the Office of Experiment Stations, NA, RG 164.

29. W. P. Kelley to E. W. Allen, 18 January 1922, records of the Office of Experiment Stations, RG 164.

30. The phrase is extracted from Charles Thorne's manuscript, "History of the Ohio Agricultural Experiment Station," chap. VIII, p. 8, Ohio Agricultural Experiment Station, Wooster.

31. The content and history of *Agricultural Science* provides an excellent guide to the hopes and sentiments of its founders. The first issue appeared in January 1887 and the magazine expired from lack of support in 1894. It contains, for example, editorial comments bewailing lack of contributions from those many scientists employed in agricultural institutions who preferred, nevertheless, to publish in their discipline's journals, plus those teachers of agriculture (proto-agronomists) who often preferred to publish in farm papers. See *Agricultural Science* 3 (1189): 5–6. E. W. Allen's papers in the records of the Office of Experiment Stations (personal file box 1054, NA, RG 164) contains his correspondence as secretary of the Society for the Promotion of Agricultural Science, 1910–14, and the papers include a number of letters indicating the increasing marginality of the society to the concerns of many members.

32. There is no good account of this development, but for a useful introduction, see Margaret W. Rossiter, "The Agricultural Sciences in the United States, 1860–1920," in *The Organization of Knowledge in Modern America 1860–1920* ed. Alexandra Oleson, (forthcoming from Johns Hopkins University Press in 1979).

33. For the general history of the extension movement, see A. C. True, *A History of Agricultural Extension Work in the U.S., 1785–1923* (Washington, D.C.: USDA misc. pub. 15, 1928); Roy V. Scott, *The Reluctant Farmer: The Rise of Agricultural Extension to 1914* (Urbana, Ill.: 1970).

34. Not all administrators objected to such demands, for success provided a strong public relations argument for the institution generally. Thus, the ambitious dean of the University of Wisconsin's College of Agriculture could boast in 1904 that, "The Minnesota Station had the finest fat steer in the whole great International Show in Chicago. . . . Last year it was Nebraska that won. For two years now the colleges have beaten all of the private breeders and feeders of the country. Ten years ago these men were sneering at the colleges saying that a professor

was all right to talk, but he could not 'do.' Now they are beginning to talk about not allowing the colleges to compete because the farmer and feeder have 'no show' in competition with the professors." W. A. Henry to H. C. Adams, 27 December 1904, Henry file, University of Wisconsin Archives.

35. E. V. McCollum to E. B. Hart, 4 May 1907, biochemistry files, University of Wisconsin; Pearl to C. B. Davenport, 5 October 1903, 6 January 1904, C. B. Davenport papers, American Philosophical Society; E. V. McCollum, *From Kansas Farm Boy to Scientist: Autobiography of E. V. McCollum* (Lawrence, Kansas: 1964).

36. Agricultural economists and historians have in recent studies doubted the actual level of increased productivity before the 1920s and 1930s. For better or worse, late nineteenth and early twentieth-century Americans lacked the modern perspective and econometric sophistication and based decisions to continue and increase support for agricultural research on the optimistic and carefully nurtured perception that science had indeed increased productivity substantially before 1914. In areas such as dairying and horticulture, there was a good deal of evidence to encourage farmers and officers of producers' organizations to support colleges and stations. The structure of power and influence was such that aids to productivity, which seem in retrospect small and inconsistent, were nevertheless disproportionately effective in encouraging support for continued research.

37. E. Davenport to Shelby M. Cullom, 18 September 1912, file drawer 6, folder on "USDA Lever Bill Legislation, 1912–1913," dean's files, College of Agriculture, University of Illinois; E. Davenport to Woodrow Wilson, 23 December 1912, folder on "USDA, Department of Agriculture, 1912–1913." Agriculture, Davenport had contended earlier, "is not a speculative industry. . . . It is a solid business to be engaged in by families and by nations, rather than by individuals climbing over each other to become rich in a few years." One could not earn millions, Davenport conceded," but the only stock about the business that is watered is its cattle, its horses and its hogs." Davenport to C. L. Camp, 16 January 1906, Davenport letterpress books, dean's papers, College of Agriculture, University of Illinois.

38. Ibid.

FROM THE GRAND CANYON TO THE MARIANAS TRENCH: THE EARTH SCIENCES AFTER DARWIN

STEVE PYNE
North Rim, Arizona

IN 1858, THE STEAMBOAT *Explorer* PADDLED UP THE COLORADO River to inaugurate the scientific exploration of the Grand Canyon, the most spectacular continental gorge known. A little over a hundred years later, in 1960, a very different sort of boat, a bathyscape named the *Trieste*, plummeted to the bottom of the Marianas Trench near Guam, the deepest gouge known in the oceans and, at 11,033 meters below sea level, nearly seven times deeper than the Grand Canyon. The Ives Expedition, which sponsored the *Explorer*, belonged to an age initiated somewhat inadvertently by the international effort to survey the transit of Venus; the *Trieste* expedition belonged to the modern age initiated by the International Geophysical Year (IGY). The two surveys may be taken as symbolic: their geographic and theoretical context epitomize the earth sciences for their respective centuries.

The Ives Expedition to the Grand Canyon was a representative sample of a wave of global, geographic exploration begun with the voyages of Captain Cook and terminated with the conquest of the poles. That wave of discovery corresponds in time with the Industrial Revolution; together these two phenomena created an intellectual revolution in terms of both natural history (that is, the earth and life sciences) and the physical sciences. The descent of the *Trieste* into the Marianas Trench however, belongs with another wave of intellectual and industrial revolutions inaugurated largely after World War II: the exploration of the ocean basins and earth interior and satellite surveillance of the planet—all within the context of a technological revolution.

Each expedition contributed to a particular theoretical synthesis of the earth sciences. The nineteenth-century synthesis described the earth in terms of the theory of evolution and the laws of thermodynamics; its fundamental metaphors were those of the organism and the heat engine. The twentieth-century synthesis analyzed the earth in terms of systems theory and plate tectonics; its metaphors were those of the computer and the spaceship.[1]

The purpose of this paper is to analyze that transformation in the earth sciences, especially as it was manifest in America. Between the geologic studies of the Grand Canyon and of the Marianas Trench there is a distance of nearly a hundred years, and more significantly, a phenomenal change in the information base, methodology, conceptual apparatus, and philosophical assumptions of the earth sciences. It is that distance of change I will attempt to measure.

1882: "The Sublimest Thing on Earth"

From its beginning, the interpretation of the Grand Canyon has exhibited two distinct traditions: the views from the river and from the rim. The Ives Expedition of 1858 inaugurated both. Its commander, Lt. Joseph C. Ives, wrote the river view—an adventure narrative. Its highwater mark was the thrilling descent at Diamond Creek into the gorge of the Grand Canyon. Meanwhile, the expedition's chief naturalist, John Newberry, interpreted the region from the rim with some brilliant geologizing. The Ives Report became an international sensation when it was published in 1861, and Newberry's speculations marked a major horizon in the stratigraphy of American exploration geology.

The perspectives of rim and river found their classic expressions, however, with the explorations that followed. In 1875, John Wesley Powell recast the river genre into its dominant form with *The Exploration of the Colorado River of the West and its Tributaries*, an account of his epic 1869 voyage down the Colorado River. In 1882, Clarence Dutton updated Newberry with a wonderful synthesis of Canyon geology and aesthetics as viewed from the rim, *The Tertiary History of the Grand Canyon District*. The two books have remained celebrated landmarks in American nature writing ever since.

The *Tertiary History* was also a milestone in American geology, and its composition was strikingly indicative of its theme. The text proceeded as a narrative, a series of "imaginary journeys" to different scenes in the Canyon region. Each journey culminated in a climactic panorama that illustrated a geologic and aesthetic lesson. These episodes were in turn integrated by the majestic journey to Point Sublime. There the inspired Dutton consolidated his lessons in two rapturous chapters: one depicted the scenic evolution of a day, summarizing the aesthetics of the region; the other dramatically scanned the geologic evolution of the Canyon, epitomizing the scientific lessons of the landscape.

At Point Sublime, the panorama of history was as complex as the landforms about it and as vast as the horizons that framed it. Dutton disliked subjects that, as he put it, "had neither heads nor tails," and at the Grand Canyon he resolved that problem by casting his material into a history that had a beginning, middle, and end—in short, a history written like a novel.

The ready reception of the *Tertiary History* and its frank evolutionism are hardly unrelated. What Dutton had done was to give nineteenth-century American geology a glamorous emblem of its unifying principles and primary data. Of all the discoveries that had dazzled natural history, this place was the most stunning; of all the regions that revealed the elementary theme of geology, the vast longevity of the earth, and its systematic development through time, this was the most incredible. Dutton's book on the Grand Canyon gave an unrivaled cross-section of evolutionary earth history. For our purposes, it was also an unequaled cross-section into the earth sciences.

THE CLASSICAL EARTH SCIENCES HAD developed from the stimuli of an exponential growth, both of industrial power and of information about nature. As a result, two traditions had emerged: one grounded in natural philosophy and modernized by the intellectual and technical achievements of the Industrial Revolution; the other grounded in natural history, swollen by the cascade of discoveries in the earth and life sciences. The first tradition led to geophysics and organized its data according to the principles of mechanics and thermodynamics; the second and much greater tradition led to geology proper, organized according to the principles of (that is, by analogy to) embryology and evolution. In a few spectacular instances, the two traditions clashed, but to a much greater extent they shared broad philosophical assumptions, while differing in methodologies, and fully complemented each other.

Classical geology had appeared in response to an intellectual crisis: the enormous expansion of knowledge about the geography

and history of the earth. As new continents were discovered, old limits of geography were expanded. As new landscapes of geologic time were unearthed, they rendered meaningless the old limits of history—the seven days it took the Old Testament to create the earth and the six thousand years it subsequently took to shape the earth into its modern form. This discovery was the greatest single theme in classical earth science. As George Scrope lectured: "The leading idea which is present in all our researches and which accompanies every fresh observation, the sound of which to the ear of the student of Nature seems continually echoed from every part of her works, is
TIME!—TIME!—TIME!"[2] The analogy of geologic time to astronomical space was commonplace, and the chief debate throughout the century (although, like an avatar, it appeared in many forms) was the age of the earth.

At the core of all this was the fossil. Its explanation was not only a major problem but a major solution to the organization of the earth and life sciences and to the question of the earth's age. By fossil correlations, one could integrate the landscape geographically; by fossil successions one could integrate it historically. It was natural enough that the organism be taken as a metaphor for the organization of natural history. In the early development of geology—say with James Dwight Dana—this meant an analogy with embryology and, after Darwin, an analogy with evolution. As a result, paleontology became the mathematics of geology and stratigraphy, the first specifically geologic subject ordered by fossils, became its mechanics.

The thrust of American geology after Darwin can be interpreted as a search for fossils—not only the remains of organisms by which to refine the paths of biological evolution, but also geological equivalents to fossils that could be organized historically on the model of stratigraphy. Rivers, mountains, and landforms, for example, all became fossil-equivalents—"bits of earth history," as Powell labelled them—that could be classified taxonomically, as though they were new phyla of ancient organisms, and then arranged in evolutionary sequence. Every phenomenon, so it seemed, had its life cycle. Like chemists filling up the periodic chart with new elements, geologists filled up the blank maps and geologic columns with their fossils. It was precisely such a landscape that Dutton had staged at the Grand Canyon. And if his text was composed somewhat like a novel, that is because both literature and earth science were predominantly historical. Indeed, the historical mode of explanation was typical of nearly all intellectual fields, from economics to anthropology, throughout the century.

The evolutionary models used were Neo-Lamarckian, that is, progressive. In this respect, too, Dutton intersected the age's syndrome of thought. For ethical, religious, and scientific reasons, few American geologists were willing to concede a chance universe. They might dispute Clarence King's bias for catastrophic events, but few would protest his caricature of Darwinian man as the "greatest fighting machine the dice box of the ages has thrown."[3] For American geologists, it was apparent that the earth progressed systematically through time, and that, as evidenced by the fossil record, the analogy to an organism was apt. Dana, for example, worked out "an analogy between the progress of the earth and that of a germ." He elaborated:

In this, there is nothing fanciful; for there is a general law, as is now known, at the basis of all development, which is strikingly exhibited even in the earth's physical progress. The law, as has been recognized, is simply this: Unity evolving multiplicity of parts through successive individualizations from the more fundamental onward.[4]

Meanwhile, fresh from his Western exploits, John Strong Newberry spelled out the value of fossils: they were "labels written by the Creator on all the fossiliferous rocks, and . . . no one can be a Geologist who has not learned their language." In an 1867 address to the American Association for the Advancement of Science (AAAS), Newberry showed that, although fossils were the alphabet of natural history, progressive evolution supplied their syntax and made them the language of nature.[5] Dana went even further, eventually proclaiming that "instead of saying that fossils are of use to determine rocks, we should rather say that the rocks are of use for the display of the succession of fossils." Along with Dana, and in a move that helped win acceptance of evolutionary theory, Newberry insisted that this historical design was so substantial that "the boldest and more irreverent of modern philosophers will strive in vain to dethrone the Great Creator from the rule of the universe, or from His place in the hearts and minds of men." Thus it was necessary for the earth system to evolve to ensure the orderly evolution of organisms, for without progressive biological evolution, culminating in man, there could be no design in nature. The alternatives were unthinkable.[6]

Where classical geology developed by elaborating on the theory of evolution, classical geophysics emerged by expounding on the theory of a contracting earth. By assuming that the earth was being progressively compressed, geophysicists could account for a host of phenomena rather as geologists could by finding fossil equivalents.

Mountains formed by crustal shortening; the source of volcanic heat; the extrusion of volcanic rocks, gases, and water; the geothermal gradient; and the relative rigidity of the earth could all be interrelated mathematically by assuming a progressive compression of the planet. The chief concerns of geophysics were thermodynamic—the distribution of energy, as measured by temperature and pressure. In short, every geophysical phenomenon was subordinated to the universal tendency of the earth to contract progressively, just as every geologic phenomenon was subordinated to the universal tendency of the earth to evolve progressively. It was the ambition of the best minds of the earth sciences to reconcile these two universal histories into one.[7]

It is appropriate, then, that the first textbook in classical geophysics, Rev. Osmond Fisher's *Physics of the Earth's Crust*, should be published simultaneously with Dutton's *Tertiary History*; that the leading American evolutionists—Dana, Dutton, and LeConte—should be the American authorities Fisher most often referred to; and that it was Dutton who favorably reviewed the book for an American audience. The reason for this intellectual harmony is simple. The geophysical portrait of the earth that emerged from Fisher's geophysics was readily bound to the evolutionary portrait presented in Dutton's geology by assumptions that are really versions of Spencer's evolutionism. A fixed amount of matter (the earth), acting under an irreversible force, had progressively differentiated and reintegrated itself. In biology, this evolutionism had led to a multiplicity of species from a common parent; in geophysics, to a variety of earth structures and rocks from an undifferentiated mass of earth material.

That both groups should, after a fashion, share the organic metaphor is not so incongruous as first appearances suggest. After all, the first law of thermodynamics had developed out of the interface of biology and physics, and Spencer, who sought to unify the concepts of energy and evolution, had begun his career in physics and engineering. That Dutton and Fisher were united on geologic assumptions is no more surprising than that Helmholtz and Mayer were on biological ones.[8]

The Reverend Fisher rightly insisted that he had treated his material "in a manner that can hardly offend the most strict disciple of the uniformitarian school." But of course the conclusions reached by classical geology and geophysics, especially in the rancorous pontifications of Kelvin, did not always square. The debate ran deeper than the superficial competition between a science grounded in mathematics and mechanics and one grounded in stratigraphy and evolution. The assumptions of both groups were unproven, and there were numerous mathematical cul de sacs in both. Efforts to describe the continents in terms of solid geometry contributed as little to the dynamics of the crust as measuring bones did to the dynamics of evolution. Moreover, both sides were capable of stating their estimates of the earth's age in quantitative terms.

Nevertheless, classical geology and classical geophysics were stalemated on a number of intellectual fronts. There were other points of discord. Geologists insisted that the earth was plastic below its crust; geophysicists, though divided on the question, generally held that the earth was as rigid as steel. To provide a progressive chronology for earth history, geology looked to the remains of organic evolution, denudation rates, and sedimentation rates; geophysics looked to calculations based on tidal retardation or secular refrigeration. To furnish the tangential stresses that led to crustal shortening, each group invoked a different version of a contracting earth: geology, using the planetismal hypothesis after 1900, looked to gravitational collapse; geophysics, using the nebular hypothesis, looked to thermal cooling. Actually, most members of both camps probably followed the example of Fisher in accepting contraction and begging the question of its source. Classical geophysics, in general, only allowed theories that were derivable from the equations of a contracting earth; classical geology, for the most part, accepted only hypotheses inherent in the formulas of progressive evolution.

It was natural that these differences in models and techniques should be magnified when applied to the central question of the classical earth sciences: the age of the earth. After all, the numerical differences between the ages arrived at by geophysical techniques were not as far from certain geophysical estimates as were other geophysical estimates. Kelvin's figures were as distant from estimates arrived at by other geophysical means, such as tidal retardation, as they were from figures derived from geological assumptions.

BY THE EARLY TWENTIETH CENTURY, THE consolidation of classical earth science was nearly complete. If they differed on specific issues, geophysics and geology had agreed on a number of critical assumptions. For example, both agreed on the nature of geologic history, while disagreeing on its length. For both, the description of earth history was the guiding concern of the science, and both conceived that history as the chronicle of a universal, irreversible process.

The intellectual edifice of earth science was enormously satisfying to Americans in particular. By 1910, Eduard Suess, an Austrian whose encyclopedic *The Face of the Earth* had synthesized the new geographic and geologic discoveries of the past century, had been translated into English and the contents of his tome digested. Two Americans, meanwhile, had expanded classical geology into the limits of its potential interrogation of the earth: the origin of the planet and the sculpture of its crustal veneer—that is, into the cosmology of Thomas Chrowder Chamberlin and the geomorphology of William Morris Davis. Together, the Americans gave body and physiognomy to Suess's *Face*.

Chamberlin, collaborating with the astronomer F. R. Moulton, worked out a brilliant compromise between the two traditions of the classical earth sciences. The planetismal hypothesis managed to retain both a contracting earth—by substituting gravitational for thermal energy—and organic evolution, by removing the critique on geologic time and atmospheric processes thundered out by Kelvin. But Chamberlin achieved this result by giving geophysics a peculiar, but for the time predictable, twist. The dynamics of the earth's planetary motion, he argued, were fossils—vestiges of its planetary evolution as fully as brachiopods were for the shallow seas of the Permian. Chamberlin's exposition in *The Origin of the Earth* had a clear Darwinian echo even to the title. And like Dana's egg, this view passed the earth through stages of youth to adulthood.

At the same time, Davis consolidated the landform studies of Newberry, Powell, Dutton, and G. K. Gilbert into what was proudly proclaimed as an American school of geology, geomorphology. Davis envisioned landscapes as passing through an evolutionary sequence of forms, in stages that he termed young, mature, and old—collectively, a life cycle. These stages could be petrified as relic landforms and hence, as fossils, were useful for compiling surface histories. Like the heat engine cycles underwriting the representation of the second law of thermodynamics, Davis's geographic cycle involved an irreversible process in a closed system. The energy of the landscape was expended in the work of erosion; through time, the entropy of the system increased. If left undisturbed, an uplifted block of land would erode into a peneplain, a state of maximum entropy and grisly testimony to the irresistible force of geologic time. Eventually this ideal cycle was relativized, so that it could be applied, with appropriate transformation equations, to practically every climate and environment.

If they squared off on a number of methodological fronts and numerical estimates, the classical earth sciences also shared a number of assumptions. When confronted with the prospect of continental drift, both geophysics and geology held the proposal up to ridicule. Both agreed that the mechanisms for drift were laughably inadequate. Bailey Willis captured this spirit when he scornfully observed that "in general the attitude of advocates of the theory is that since continents did drift there must have been some competent mechanism of some kind."[9]

Geologists pointed out that continents could not drift because, as had been shown from Dana to Chamberlin, the continents and ocean basins were permanent features of the planet. Geophysics noted loftily that continents could not move because, as Kelvin and G. H. Darwin had shown, the earth was too rigid. And both groups abhorred the spectre of "drift," an apparently random motion that the theory implied. Classical geophysics was no more ready for a chance universe than was classical geology. It was the prospect of chance mutations, after all, that had incensed Kelvin about Darwinian evolution. An outraged Rollin Chamberlin summarized the whole position: "If we are to believe Wegener's hypothesis, we must forget everything which has been learned in the last 70 years and start all over again." That, of course, was precisely the point.[10]

Drift theory was a point the classical earth sciences were not ready to accept, or even properly outfitted to test. When continental drift was first publicly debated in 1926, the idea ironically coincided with a period of stagnation in the sciences that even the discovery of radioactivity could not much stir. After all, radioactive decay promised to make absolute the relative stratigraphic chronologies; like evolutionary or thermodynamic processes, it furnished an irreversible history. And if radioactivity meant that the upper mantle could behave plastically, this was, after all, the position geologists had argued for decades. The stagnation of thought that developed at this time was especially apparent in America.

It may be appropriate to return for a moment to Clarence Dutton perched on the promontory at the Grand Canyon that he named Point Sublime. The Canyon, exclaimed Dutton, was "the sublimest thing on earth." In terms of its scientific meaning, no less than its aesthetic majesty, most American geologists would have agreed with him. Nowhere else could American geology have displayed a greater symbol of what it had learned and experienced. So it is particularly ironic that it was Dutton who put forth a concept that challenged classical geology and classical geophysics equally. In fact, the *Tertiary History* is poised between several devastating critiques on the contractional hypothesis. In 1889, Dutton epigrammatically stated his case: the contractional hypothesis was "quantitatively insufficient and qualitatively inapplicable. It is an explanation which explains nothing which we want to explain." He quickly put that criticism into a form that made him famous. He conceived that the earth's crust existed in nearly perfect gravitational equilibrium. He named that concept "isostasy."[11]

1928: The Interlude

Potentially at least, the concept of isostatic adjustments delivered a devastating blow to both classical geophysics and classical geology. By showing how the earth compensated for the transfer of mass on its surface, the concept attacked the geophysical assumption that the earth was rigid and unyielding. Yet, by showing that the earth's larger features are in isostatic balance, the idea destroyed tenets dear to classical geology. For one, isostasy made the emergence and substance of land bridges (and hence of evolution) extremely implausible, and for another, it damaged the major theories of orogeny based on either a thermal or gravitational contraction. Isostasy was a conservative mechanism, and despite efforts by Dutton and others, it could not be made an orogenic one. Nevertheless, its discovery became, like the Michelson-Morley experiment for physics, a primary consideration in any larger theory of the earth. By 1928, the problems it posed were acute.

The year 1928 may be taken as the divide between the classical and modern earth sciences in America. In that year came publication of the final books from the two last giants of the heroic age: Chamberlin and Gilbert. Also occurring was the theoretical consolidation of America's most vigorous laboratory tradition, the petrological investigations of physical chemistry. In 1928, building on the conceptual field and laboratory work of Cross and Iddings, and outfitted with the powerful analytic tools that were the legacy of Willard Gibbs, N. L. Bowen published *The Evolution of the Igneous Rocks*.

The title of that work was not fortuitous: Bowen attempted to explain the immense diversity of igneous rocks by differentiating them from a common parent magma—in effect tracing their descent with modification through time. The selective force was fractional crystallization. The analogy to organic evolution was not only deliberate but, from a modern perspective, somewhat surprising. So saturated were the earth sciences by the evolutionary model that even in the realm of physical chemistry nothing was more natural than to explain the many species of igneous rocks by petrogenesis, and to organize that schema by analogy to evolution.

Although the data for doing so were tenuous, Bowen cautiously tried to correlate the parent magma with the fundamental rock created by a contracting earth. Combined with the geographic cycle of Davis and the planetismal hypothesis of Chamberlin, the American school had a comprehensive and thoroughly unified theory of the earth. The evolutionary model unified the earth sciences from its microcosms (minerals) to its macrocosm (the solar system).

The year 1928 saw as well the published proceedings of the first American (and indeed international) symposium on the subject of continental drift. The symposium might have explored the startling suggestions of Alfred Wegener that the continents had drifted apart to answer some of the riddles posed first by isostasy and later by a host of observations that were making a scandal of the classical earth sciences. The symposium did not do this; on the contrary, American earth scientists basically closed ranks, apparently satisfied with their already imposing achievements.[12]

Within a year after the symposium proceedings were published, Bailey Willis reaffirmed the American legacy with a masterly summary of its propositions in an article titled "Continental Genesis." Not surprisingly Willis, who perhaps became the chief exponent of the American school, published a memorial on Chamberlin at the same time.[13] Meanwhile, Harold Jeffreys established himself as heir to the British tradition of geophysics by reconfirming the basic tenets of Darwin and Kelvin in a revised edition to his monumental text, *The Earth*. As Willis had for classical geology, Jeffreys turned the powerful mathematical and conceptual apparatus of classical geophysics scornfully onto the prospect of large-scale continental displacement. What should have been a divide in 1928 consequently became a plateau that spanned more than 30 years.[14]

The achievements of American geology were real enough. It had been asked to integrate the vast domains of geography and time that had been opened up, and with the aid of fossils and evolutionary theory it had done so. The last continental frontiers were mapped. With respect to the American West, geology had filled in the blank spaces as the westward migration had populated them. The "American school" of geology had a firm institutional base, an impressive catalog of discoveries and theories, a methodology boldly outlined by Gilbert and Chamberlin, a constellation of outstanding teachers, solid journals, and a glamorous heroic age of exploration to celebrate. Furthermore, the school had standardized geologic explanation to mean the construction of a geologic map and the composition of a geologic history. And in the course of all this, American geologists rarely failed to acknowledge the superiority of the American landscape that had made it all possible. The global questions of international geology were being solved largely by the interrogation of the American West.

The year 1928 also witnessed the fiftieth anniversary of the Geoolgical Society of America, the major association of professional geologists. Naturally, the Society wished to commemorate its successes, so it commissioned Herman Fairchild, the only surviving member from its 1888 origin, to write a history. Fairchild took the occasion to indulge in some speculation about the future of geology. He wrote:

> The unknowns of the earth laboratory are comprised in a few groups of broader scope, and most of the problems relating to the planet can not be resolved by strictly geologic evidence and methods. The stratigraphic record of the earth's history since preCambrian time, once the principle theme of geology, now offers only details for future discovery, and the life history on the globe is well known in its principal elements. Like geography, mineralogy, botany, and zoology, stratigraphy and paleontology have already made their great discoveries.
>
> The present interrogations of the world of scientific curiosity are mainly biologic and geophysical. The biologic problems are in the province of the biologists and chemists. The physical questions belong to the physicists and mathematicians.[15]

One doesn't have to read too far between the lines to realize that geology had simply run out of significant questions, and that is exactly the impression one gets by reading the literature. Granted, for external reasons, the thirties and forties were not the most propitious times for geology, but other sciences did not stall as badly as the earth sciences did for largely internal reasons.

By discounting drift theory, American geology had cut itself off from a potential stimulant equal to that biology had acquired with genetics and physics with quantum theory. Geology had defined itself through geographic exploration, and in 1928, with both poles conquered, there would be no new geographic frontiers until the advent of special vehicles equipped to explore outer space and oceanic basins. Until then, Americans were largely indifferent to regions of geological significance—many important to drift theory— not in North America. Instead, the old problems bequeathed by the heroic age were reexamined, often with the old authorities. In some disciplines the tendency verged on scholasticism.

We do not have to think about the effect of all this; it is self-evident. What requires explanation is why the fields that eventually produced modern geology began to expand. There the reasons are complex. In part new techniques developed, mostly after World War II, which made it possible to ask certain questions, to decipher the internal structure of the earth, and especially to survey the ocean basins and to place the planet under satellite surveillance. New data and literally new pictures of nature demanded a new explanation. In part, genetics had gutted the old role of geology as defender of progressive organic evolutionism. The metaphysical and moral energy behind its theoretical superstructure vanished. The chance universe was scientifically acceptable, and geological evolutionism became merely a baroque ritual of ancestor worship.

In this regard, it is significant that the first symposium on drift theory was sponsored by the American Association of Petroleum Geologists (AAPG), men who were responsive to new techniques and who had an economic rather than ethical interest in the possibility of continental displacement as an orogenic agency. After all, the inability of classical geology and geophysics to explain mountain building was a serious embarrassment to the science and a matter of imminent concern to those who worried about the distribution of mineral resources.

In part, the new sciences were revivified by the presence of twentieth-century physics in ways that classical geology was not affected by its traditional ally, biology. In short, despite a strongly conservative background, classical geophysics was prepared to open up after 1928 in ways that classical geology was not. Moreover, the intellectual migration from Europe, which significantly stimulated physics, and indirectly geophysics, was totally irrelevant to geology. The advocates of drift theory at this time were mostly from the southern hemisphere and had little urge to immigrate to America. Given impetus by the example of the new physics and outfitted with instruments designed during the second Industrial Revolution, geophysics was ready to propel the earth sciences into a modern era. Given its moribund state after 1928, any other alternative was unthinkable.

1960: "The Jaw Crusher"

This is not the place to recapitulate the narrative of how continental drift, first proposed in 1912, gradually revolutionized the earth sciences by 1970 in the form of plate tectonics. That has already been done in half a dozen full-length books. But I would like to sketch the broad outline of that narrative with an analogy to the more familiar story of evolutionary theory. What Lamarck was to evolution, Alfred Wegener was to plate tectonics. With his training in astronomy, physics, and meteorology (he wrote a book on the thermodynamics of the atmosphere) and his passion for exploration (he died on a Greenland expedition), Wegener was a perfect transitional figure.

It is entirely appropriate that Wegener deliberately tried to restructure geology and geophysics into a new alliance based on the principle of continental displacement. But his theories were premature. His contention that the continents, made of light rock (sial), plowed across a yielding oceanic floor, made of composed denser rock (siam), was a brilliant hunch. Unfortunately, there was little beyond circumstantial evidence to sustain the view, and Wegener could only conceive of two embarrassingly insufficient forces to power continental motion. They were no more adequate than Lamarck's mechanism of adaptation.

The Origin of Continents and Oceans was revised and republished four times, the first in 1915 and the last in 1929. A few years before this final edition, the book was translated into English. It caused a minor sensation among Anglo-American earth scientists, not unlike the contemporaneous impact of genetics, relativity, and quantum theory. Despite efforts to rationalize classical geophysics and classical geology, they had divided along broadly national as well as methodological lines: Americans excelled in geology, the British in geophysics. It is suggestive that the pioneer in the modern synthesis of the two disciplines should be a German physicist and meteorologist coming from a milieu that also spawned the new physics. However, as with Lamarck's premature theory, continental drift was submerged after a great debate.[16]

The symposium of 1926 was the equivalent of the Cuvier-St. Hilaire debate of 1830. A significant difference between the two events was the corporate nature of the 1926 symposium. Especially as it metamorphosed through time, drift theory would never have a single figure to rally around. It grew out of symposia as quantum theory did from physics congresses. However, the more important similarity between the two debates was the fact that, despite some telling arguments on its behalf, the new theory was effectively crushed. The nabobs of American geology (most of whom were emeritus professors) discredited the subject as unfit for scientific discussion. Eventually, after the theory had settled into a Bohemian existence as disreputable but annoying, Bailey Willis recommended that the theory be permanently liquidated "since further discussion of it merely encumbers the literature and fogs the minds of fellow students." When Wegener died in 1930, the movement supporting drift lost even its figurehead.[17]

Revival began slowly after World War II. Although its origins are multiple, modern geology was clearly catalyzed by the International Geophysical Year in 1958 and may be dramatized by one of the many exploratory probes the IGY spawned. This, of course, is the descent of the *Trieste* into the Marianas Trench. That scientific adventure heralded a new age of exploration as Ives and Powell had a century earlier, for the Marianas Trench posed the central problem of modern geology as the Grand Canyon did for classical geology.

Appropriately enough, this descent into the Pacific occurred simultaneously with the launching of the first weather satellite by the U.S., for if the one effort forced reconsideration of crustal dynamics, the other demanded a new appreciation of the process by which the atmosphere interacted with the crust. And if there is a Darwin figure in the story, he may well be Harry Hess, a Princeton professor of geology. Hess, like Darwin, was puzzled by some submerged volcanic islands he had visited in the Pacific; in grappling with their meaning, Hess eventually produced the ideas that led directly to the modern version of continental drift.

Hess published his ideas the same year the *Trieste* plunged into the Marianas Trench. He proposed simply that oceanic crust was created along ocean rift and ridge systems and destroyed along island arcs and continental flanks in those great trenches like the Marianas—the "jaw crushers," as he termed them. Convection currents in the upper mantle supplied the driving force. The continents rode atop these cells. Hess protested that his speculations were "an essay in geopoetry," but it was the same geopoetry that erupted from Dutton at Point Sublime. The ocean trench, as had the Grand Canyon before it, intersected the syndrome of an age.[18]

The decisive evidence followed a few years later. Paleomagnetic studies of oceanic rock showed unmistakable proof that oceanic crust was spreading symmetrically from ridge systems. In 1967, another symposium in America completely reversed the stance of its predecessor forty years before. Continental displacement was accepted, though in a form quite different from that imagined by Wegener: it was not simply continents but large crustal plates which moved. The continents were a crustal adornment of lighter rock in the plate.

The revised theory became known as plate tectonics. By this time, too, plate theory had found its Huxley—a cosmopolitan

Canadian geophysicist named J. Tuzo Wilson, who brazenly announced that the revolution was at hand. Of course, the myth of Wegener was revived. He was cast as the romantic genius, shunned by his own age, courageously driven by his conviction and vision to a tragic death while dogsledding across Greenland. Naturally, the wisdom of future ages has vindicated him.

After 1967, the conversion of geologic thinking to plate theory was nearly instantaneous. Curiously, the mechanism for driving the plates is still unknown. Convection currents are in vague disrepute, replaced as a mechanism by hot plumes rising to the crust. Wegener lamented that drift theory did not yet have its Newton; in keeping with our analogy to evolution, he should have said that it lacks its Mendel. But the stock criticism against the theory had at last vaporized. Earth scientists no longer denied continental displacement because they had no force to account for it, any more than natural philosophers had claimed that apples cannot fall to the ground because Newton had not explained the cause of gravity.

To put it all in synoptic form, plate theory holds that the earth's crust consists of gigantic plates, each about sixty kilometers thick. On the globe today there are six large plates and about nine smaller ones, many of which are inactive relics of former large plates. The plates, in turn, rest on a plastic-behaving aesthenosphere, a low-velocity seismic zone between the crust and mantle. The thrust of the theory is to study the relative motions of these plates. In general, there are three kinds of plate boundaries: first, zones where crust is generated; second, zones where crust is consumed; third, zones where motion between plates is translational, but crust is neither created nor destroyed.

At present, zones of crustal generation, or spreading centers, are synonymous with oceanic ridge systems. Once initiated, even in continental material, a spreading center will lead to an ocean, with distinctive oceanic rock and geology. Two contemporary examples are the Red Sea and the Gulf of California. The consumption of crust occurs in a variety of ways, depending on the densities of the colliding materials, their velocities, and the angle of collision. The crust may be subducted, that is, one plate is dragged under another, spawning a deep trench, a diving belt of earthquakes (a Benioff zone), and an arc of volcanoes where the cool subducted plate finally melts and rises. This would occur if oceanic plates were involved.

If continental crust were involved, the two plates may simply compress one another, folding and thrusting into a mountain range like the Appalachians. Or if the velocity is high enough, one plate may splinter and override the other, such as happened when India collided with Asia. A zone where translational motion occurs is the San Andreas fault zone in California. In short, almost all significant geologic activity occurs where the plates intersect.

Needless to say, this sketch is highly synoptic, but even in this form it illustrates a radical shift of interest in the earth sciences. In 1867, John Newberry referred to fossils as "labels put on the fossiliferous rocks by the Creator." Today the earth has been characterized as a "veritable junkyard of old plates," buried or compounded in the crust. The transformation from the image of an alphabet to that of a junkyard is parallel to that apparent in Clarence Dutton surveying "the sublimest thing on earth" at the Grand Canyon and Harry Hess referring to trenches like the Marianas as "jaw crushers."

There are, to be sure, fossil plates, and the recognition and historical reconstruction of these plates is still the function of the earth sciences. Yet the pattern of history that results has small connection to the amalgamation of romantic exploration, evolutionism, and "little science" that typified its classical period. In establishing the dynamic relationships within the crust, plate theory commemorates the advent of technological exploration, cybernetics, and "big science."

THAT BROAD TRANSFORMATION CAN BE DELINEATED on four levels. First, there is an exponential increase of information, a change not only of amount but of kind. Second, under the pressure of this information, particular theories of classical geology and geophysics have been scrapped in favor of newer versions. Third, the philosophical context (paradigm, if you will) underwriting those working theories has correspondingly changed. Fourth, the cultural context of the earth sciences has altered so that they perform a different role and ask different questions.

First, information. The stagnation apparent in American geology after 1928 occurred both because the science exhausted its significant questions and because it discovered no new techniques by which to ask alternative questions. There were no new lands to survey, so geologists began rereading the old accounts about former discoveries. It was not until the technological revolution after World War II made possible the geologic exploration of the ocean floors and the satellite inventory of the planet earth that enough new evidence accumulated to overwhelm the old theories. Much of the new data came from new geophysical fields and brought with it a clarion call for greater quantitative refinement as a means of invigorating the science.

Correspondingly, much of the new data was quantitative. Many of the old questions could not be well answered mathematically, as the wide variance in the estimated age of the earth showed. Many of the new questions are susceptible to mathematical analysis, both in the smelting of the information and in its synthesis. This is largely the result of greater sophistication in geophysical concepts, more precise instrumentation, and new mathematical techniques—none of which were available to the classical earth sciences.

Classical geophysics, dealing with the cumulative effects of infinitesimal forces, naturally relied on calculus; modern geophysics, and even geomorphology, rely more on probabilistic techniques. Of course, the digital computer has made possible the calculation of data inaccessible to the geology of Dutton's and Fisher's time. New quantitative questions can be asked because the methods of processing them quantitatively are available. The obverse view, of course, is that a number of quantitative efforts in the past had branched off from the mainstream into eddies. They became cul de sacs, aesthetically mesmerized by the elegance of useless calculation, done it seems out of the belief that one could describe a stream by counting its pebbles.

Second, there is a change in basic working concepts about the earth. Plate theory has repackaged the hypothesis of a shrinking earth on which was erected the assumption of an evolutionary geologic history. Included are beliefs in the permanence of continents and ocean basins, the crustal shortenings and geosynclines that led to mountain building, and the whole set of accretionary theories by which the continents, the atmosphere, and the oceans all grew in tandem. Nor has plate tectonics completely ignored the irreversible changes occurring over geologic time. But it has made them less significant. The age of the earth has been determined, and the age of a system of events is less important a problem than how the system operates. The formerly receding horizon of the earth's past has been made into a triangle and fixed. The whole question of base levels and diastrophism as dating devices, for example, becomes meaningless when crustal plates are stitching a crazy pattern of jigsaw motions and a patchwork quilt of continents—and when one realizes that no ocean basin is older than 200 million years.

Perhaps this last realization, of the comparative youth of the oceans, best summarizes the third broad change, that of philosophical assumptions or paradigms. In the baldest terms, this development has meant the exchange of an evolutionary worldview for one approximating cybernetics or general systems theory. Classical earth science conceived the planet as a closed system. Although Hutton's famous phrase that he saw "no vestige of a beginning, no prospect of an end" was arresting (and one of the few quotable passages in all of Hutton), it condemned earth science to the formlessness of pure uniformity. Yet because the earth system was closed, because its energy reserves were finite, the system must have had a beginning, middle, and end.

It was precisely this history that earth scientists were anxious to date. Every procedure for dating the earth required that it be a closed system in which one measurable process—tidal friction, secular refrigeration, gravitational contraction, the accumulation of oceanic salts, or organic evolution—irreversibly progressed. It was in this sense that Spencer's famous epigram underwrites all of the classical earth sciences. Whether operating out of the context of thermodynamics or Neo-Lamarckian evolution, the resulting vision was the same. It was the interpretation Dutton applied to the Grand Canyon; Davis to the erosion of landmasses; Chamberlin, Fisher, and Kelvin to planetary evolution; Bowen to petrogenesis; and Dana to the North American egg.

This vision is currently being scrapped. Instead, the earth is conceived as an open system through which energy flows along certain pathways and within which matter is recycled. Earth history represents a succession of such systems, each exhibiting a tendency to the steady state. It is certainly true that there are broad, irreversible changes over the course of earth history. The development of the oceans, the atmosphere, and of plant and animal life had enormous geologic consequences, and certainly residual energy derived from the core, mantle, and radioactive crust is decreasing over geologic time. Yet modern earth science finds these questions—which hypnotized classical earth science—less interesting than the mechanics of particular, self-regulating systems.

Whereas the nineteenth century could imagine the earth by analogy to an organism, recent work suggests that more apt metaphors are a spaceship, a computer, or a chemical engineering plant. The emphasis is not on the history of the lithosphere, atmosphere, and hydrosphere, but on their mechanisms of self-regulation. It is in this context that the observed youth of the ocean basins becomes significant. The greater part of the earth is not the residue of past eons of geologic work but a product, comparatively speaking, of the present.[19]

The dating procedures used in plate tectonics theory reflect this orientation. They depend less on irreversible processes than on rhythmic ones, especially the pulse of reversals in the earth's magnetic field. The difference between a stratigraphic column dated by comparing it to the corresponding tree of evolution and a column dated by comparison to paleomagnetic reversals epitomizes the divergence between two conceptions of earth history. The one suggests an implicit progressivism in history; the other, resembling more the alignment of genes on a chromosome than a tree of evolution, suggests an on-off rhythm such as the circuit switches of a computer.

It is no accident that paleomagnetism brought about the confirmation of plate tectonics. Development of a new timepiece was required to complement the essentially steady state mechanisms of the earth systems. In short, the organic metaphor has succumbed to a cybernetic one; the conception of the earth as a growing organism or evolving species is giving way to the conception of the earth as an information-processing system, a huge computer. The acceptance of time's arrow unified the classical earth sciences; the modern version insists that time's arrow have a feedback loop.

The elimination of historicism from the description of the earth's past has a number of consequences. Classical earth science sought to organize earth history so as to accommodate organic evolution. Yet plate tectonics has compelled paleontologists to revise the fossil record in order to accommodate known plate motions. The result is a sequence of extinctions and radiations corresponding to the rhythms of plate motion. Graphically, this sequence makes a pattern rather like the rhythm of paleomagnetic reversals. In the same way, paleontology and stratigraphy, the core disciplines of the classical period, are more likely to reconstruct the environmental system of an organism or a rock than to be satisfied with arraying fossils and rocks into historical sequences. To put it somewhat differently, plate theory has brought the earth sciences into the philosophical world of the chance universe. The old horror of "drift" is gone; the imperative to ensure organic evolutionism, particularly after the advent of modern genetics and population biology, is no longer necessary.[20]

THE NEW THEORY HAS SHIFTED CONCERNS TO present processes rather than past ones. The sense of time that enchanted Dutton at the Grand Canyon may be likened to a pyramid, with the vast eons of the past at the base rising to a minuscule present. Indeed, the landscape of the Canyon lends itself wonderfully to this interpretation, with the dark Pre-Cambrian rocks of the gorge, its rising stack of Paleozoic strata, its neighboring Mesozoic cover, and tertiary sculpture. The work of the present is overwhelmed by the masses of past time that frame it. It was precisely such conceptions of geologic time that made classical earth science work: both evolutionary geology and geophysics could posit small forces with large cumulative effects over eons of time. The summation of infinitesimals in the integral calculus was a mathematical analogue.

Modern earth science, however, has reversed this pyramid. The earth's crust did not reach its contemporary form by the providential accumulation of strata upon strata, mountain next to mountain, fossil after fossil, but by a vast system of recycling. Old crust is continually consumed, reprocessed, and disgorged; the reworking of the crust has an input equal to its output. It is an open system existing in a steady state. The continents are a patchwork of reprocessed plates sutured by mountain ranges, so that while there is identifiably old crust, it is exceptional—the point of the inverted pyramid. Most of the earth has been remade; in particular, no oceans are more than 200 million years old. Old crust is not the basement rock of geology, but an eddy, temporarily preserved from the recycling currents of plate motion.

Instead of citing infinitesimal processes, modern earth science focuses on expressions of maximum energy to account for the shape of the earth and its landforms. It is during floods that rivers perform most of their work; similarly it is at the margins of plates that important geologic activities occur, not in the uniform processes acting on the stable interiors. Landforms are less a palimpsest of geologic relics than a shape adjusted to the processes of a dynamic present. The mathematics of modern earth science is involved with probability more than with calculus. The difference is crystallized in the two revelations of the earth—furnished, respectively, by the Grand Canyon and the Marianas Trench.

The Colorado River gorge provided a cross-section through geologic history, but the oceanic trench gives a cross-section of the mechanism of present dynamics. In the nineteenth century, as information about the earth accumulated, geologists became increasingly aware of an enormous, influential past; in the twentieth, as information has continued to increase, they are cognizant of an overwhelming present.

Plate tectonics has resolved some ancient enigmas about the fundamental asymmetries of the earth. The distribution of continental masses was a dilemma: they were bunched together far more than the contraction from a homogeneous mass would predict. Also, there was the very fact that continents existed at all. The fundamental distinction between continental and oceanic rocks was recognized at the beginning of the twentieth century, and led immediately to the question of why that continental rock was not uniformly distributed over the oceanic rocks.

The Earth Sciences After Darwin

Plate theory easily accounts for crustal and geophysical asymmetries with the same philosophical assumption that it used to critique progressive time: namely, the earth as an open system. The anomalies of heat, gravity, and crustal mass are preserved in a steady state as part of an open system into which there is a continual input of energy and matter, which approximately equal the energy lost and the matter recycled. The analogy to the operation of biological systems, from the molecular to the ecological level, is very close, and the biological clocks that govern these systems are similar to the rhythmic timepieces used by modern earth science.

The twin pillars of the American school of classical geology were the cosmology of Chamberlin and the geomorphology of Davis. Yet both fields have now become cybernetic, which harmonizes their internal developments with that of plate tectonics theory. That is, general systems theory is philosophically unifying the geology of the planet, surface, and crust. Perhaps the chief exponents of a systems approach to planetary and especially to geomorphic thought are Arthur Strahler and Richard Chorley. They have, in fact, spelled out their systems bias quite explicitly. Although they represent internal developments in fields not directly responsible for plate theory, the approach they epitomize, responding to similar stimuli, began concurrently with drift theory and reached definition at about the same time. In a recent textbook, Strahler shows how these separate fields have converged. He explains that:

The concept of open systems developed here for the earth sciences provides the foundation for an understanding of almost every conceivable event or form that may come to our attention. If our premise is correct that all natural phenomena are organized into open systems governed by a common set of general laws, we have at our disposal the necessary keys to comprehending our earth as we see it today with all its varied forms and processes.

That statement is the modern equivalent to Spencer's epigram on evolution.[21]

That the American school should collapse, and that plate theory should become practically an Anglo-American creation, marks the full triumph of the new earth sciences. That fact brings us to our final level of analysis, the cultural role of geology. During its classical period, American geology served two functions. For one, it grappled with the exponential expansion of geographic and historical knowledge accumulated between the times of Cook and Perry. Because most of that information pertained to natural history, it is not surprising that the ordering principle should be an organic metaphor.

The second function related to geology's role as an intellectual subsidy to the westward migration. Considering that most of its crucial data came from the frontier, this role was a natural one for geology to execute. But it was precisely this relationship to the frontier, both of knowledge and of settlement, that made evolutionism so attractive to geology. Consider the Currier and Ives print, "Across the Continent, 1869." It commemorates, of course, the completion of the transcontinental railroad, and one can see the West, bisected by the Industrial Revolution, proceeding as a series of stages from wilderness to civilization.

The year 1869 also commemorates John Wesley Powell's glamorous traverse of the Colorado River, and it is as easy to see in the pictures accompanying his narration a similar interpretation of the western landscape, as it proceeds stage by evolutionary stage. It is this experience that made historicism reasonable and united the thought of Chamberlin and Davis to that of Powell and of Frederick Jackson Turner.

By 1928, this experience was well over. However, just as the American school was too closely tied with the discoveries made in the American West, so was the school too intimately bound up with its historical role in the West. The modern national experience is not one of expansion and inevitable progress but of homeostasis and the potential limits to growth of all sorts. The systems approach rock-ribbing most contemporary geological theories fits this syndrome much better than evolutionism does. Similarly, the focus of geology has shifted from the scenes of the western frontier. The new imagery of nature derives from the oceanic trenches, global satellites, and even from the moon.

For geomorphology in general and environmental geology in particular, the new theories offer a model of nature that incorporates humans. In the Davis scheme, as in its biological analogue by the American Frederick Clements (biological succession), one studies the processes of pure nature. The scene is uncontaminated by the presence of man and his works. It sounds strange that an evolutionary model with a major ambition to design an earth where biological evolution could lead to man would, in turn, fail to show the action of man on the landscape. But that is exactly what occurred, and two reasons are separately noteworthy. One is that the vision of unlimited time still dazzled the minds of earth scientists, with the effect that present activity was dwarfed to a vanishing point. Here again we have the pyramid of time. Another reason is that the study of wilderness conditions conjured up the myth of a golden age before the landscape was corrupted by the arrival of European civilization—the landscape on which Daniel Boone and

Natty Bumppo manufactured an epic that became a national creation myth.

That myth has become untenable, at least as a foundation for scientific theory. The study of how a pure, wild river would behave in a natural condition over millions of years is an academic exercise when there are no wild rivers left and when, given the urgency of reclamation activities, one can hardly wait millions of years to see how the experiment turns out. It is both easier, and perhaps essential, to assume that the present is a system complete in itself, and it is extremely useful to assume, as the evidence seems to indicate, that this system exists in a steady state. One reason for the infusion of quantitative studies and engineering models into geology is simply that more and more of the landscape has been engineered. This new landscape must be understood according to the theories by which it was designed—not the laws of nature so much as the models of engineers.

Plate theory has ushered the earth sciences into the club of big sciences. The scope of investigation has swelled from International Geophysical Year through the defunct Mohole Project and that splendid surrogate, the Deep Sea Drilling Project, into a global enterprise, the Geodynamics Project, which seeks by 1979 to discover the mechanisms driving plate motion. The commitment of scientific resources and the corporate organization of the new missions are on a scale unmatched in previous earth science history. In short, as epitomized in the Geodynamics Project, modern earth science has all the trappings of big science, as manifested in the Oak Ridge National Laboratory and the National Institutes of Health. Similarly, this science promises its social payoff in the form of giving precision to mineral exploration. As an example, the specific physical chemistry of ore deposition has been known for some time, but the distribution of the general environment of precious metals and fossil fuels has baffled geology. In a paradoxical turn that would have pleased the AAPG sponsors of the 1926 symposium, plate tectonics is successfully describing those general environments.[22]

A further observation about the current revolution in the earth sciences is worth mentioning. That is the conviction that it *is* a scientific revolution, as defined by Thomas Kuhn. *The Structure of Scientific Revolutions* first appeared in 1962, a good five years prior to the acceptance of plate theory, but simultaneous with Hess's article. Several of the major advocates of the theory early referred to Kuhn's book as a vehicle for describing the upheaval taking place, and practically every commentator since then has genuflected in Kuhn's direction. In truth, the almost instantaneous conversion to plate tectonics after 1967 does conform to Kuhn's model of a psychological or gestalt change during an exchange of paradigms. But there are other aspects to note in all this. For one thing, Kuhn's anatomy of revolution preceded the acceptance of plate theory and in a sense provided a program—even a manifesto—for it. There is, then, a methodological dilemma for those who would compare Kuhn's manifesto to the result: it may seem like a self-fulfilling prophecy.

In addition, certain assumptions seem to underly both Kuhn's analysis of the history of science and plate theory's depiction of the history of the earth. Both slight historical progressivism in favor of discrete periods—whether as intellectual paradigms or geological systems. History, for one, means a succession of paradigms and, for the other, a succession of systems. General trends, such as the loss of energy or the accumulation of information, are apparent in both, but these tendencies are shorn of historicism by concentrating on the examination of the specific systems. The change in the writing of earth history is analogous to that which occurred with American historiography in general. Rather, the difference between Herbert Spencer and Thomas Kuhn is that between Clarence Dutton and Arthur Strahler.

SO IT MAY BE WELL TO RETURN TO Clarence Dutton for our conclusion. *The Tertiary History of the Grand Canyon District* was one of the great synthetic expressions for that age of discovery begun when Cook ventured into the South Pacific to measure the transit of Venus. It epitomized a theory of the earth and intersected the syndrome of the thought of an age. In a similar way, plate theory has distilled the era of oceanographic and satellite exploration.

The satellite pictures have made it possible, even mandatory, to see the earth in new ways. Aerial photography has perhaps become the dominant form of geological illustration. In trying to help the nineteenth-century reader visualize his synthetic map of the earth, Eduard Suess imagined how the earth would appear to a visitor approaching it from outer space. In opening his textbook with a similar scene, Arthur Strahler could rely on actual photographs from the moon.

Perhaps Harrison Schmitt, the geologist astronaut on Apollo 17, best summarized what this could mean when he exclaimed: "I didn't grow up with the idea of drifting continents and sea floor spreading. But I tell you, when you look at the way the pieces of the northeastern portion of the African continent seem to fit together separated by a narrow gulf, you could almost make a believer out of anybody." Yet it required a view halfway from the moon to see the obvious. The satellites and deep sea drilling ships have become the cyclotrons and electron microscopes of the earth sciences.[23]

Consider two pictures of the Grand Canyon to illustrate what this change of perspective means. First, William Holmes's drawings from Point Sublime. The sense of vast horizons of space and the infinite patience of time become a magnificent, meticulously documented ensemble. Dutton's romantic evolutionism becomes self-explanatory. A rather different perspective is demanded by the ERTS photo. The scene has been flattened somewhat like pop art; the sense of time and distance has vanished. It is a landscape visibly ready for the steady state models of systems theory. And it has made the two classical interpretations of the Canyon, rim and river, irrelevant.

Yet the earth sciences have little need to look at the Grand Canyon today. The important geologic processes occur at plate boundaries, or in the synthetic landscape of an urbanized and engineered environment. Interest in the Colorado Plateau comes from a current theory that accounts for the plateau's elevation and abnormal heat and gravity readings by hypothesizing that an entire ridge system was once subducted down a trench bordering the continent. That trench and ridge system now underly the plateau.

Even here, it is not the celebrated Colorado River gorge that captures the imagination but that deeper trench that underlies it and continues to activate the unusual geology and geophysics of the region. Nothing better summarizes the motion of two centuries of earth science as well as that juxtaposition in the Colorado Plateau, a shift of attention from the sublimest thing on earth to the great jaw crushers of the crust.

1. The best account of the western surveys is in William Goetzmann, *Exploration and Empire* (New York, 1966). I am also indebted to Dr. Goetzmann for suggesting the relationships between exploration, information, and the organic metaphor developed here for the nineteenth century. The *Trieste* episode is told in Jacques Piccard, "Man's Deepest Dive," *National Geographic Magazine* 118 (August, 1960): 224–39.

2. George Scrope, "The Origin of Valleys," *A Source Book in Geology* in ed. Kirtley Mather and Shirley Mason, (New York: 1939), p. 279.

3. Clarence King, "Catastrophism and Evolution," *American Naturalist* vol. XI (1877), p. 449.

4. James D. Dana, "On the Plan of Development in the Geological History

of North America," *American Journal of Science*, 2d series, vol. 22, no. 66 (1865): 348n.

5. John Newberry, "Report of Progress in 1870," *Geological Survey of Ohio* (Columbus: 1870), p. 8.

6. Dana, "On American Geological History," *American Journal of Science*, 2d series, vol. 22, no. 66 (1865): 306; John Newberry, "Presidential Address," *Proceedings of AAAS*, 1867, 16: 1–15.

7. Clarence Dutton: "Review of Physics of the Earth's Crust," *American Journal of Science* XXIII, no. 136 (1882): 283–90.

8. Fisher found the secular refrigeration hypothesis insufficient, agreeing with Dutton in this respect, but disagreeing with Dutton's critique on contraction per se. For an interesting monograph on the whole topic of calculating the earth's age, See Joe Burchfield, *Lord Kelvin and the Age of the Earth* (New York: 1975). A good criticism at the turn of the century is Grove Karl Gilbert, "Rhythms and Geologic Time," *Proceedings of AAAS*, 1900 49: 1–19. The major summary of a century's efforts to determine the earth's age is Joseph Barrell, "Rhythms and Measurement of Geological Time," *Geological Society of America Bulletin* 28 (1917): 745–904.

9. Bailey Willis, "Continental Drift, Ein Märchen," *American Journal of Science*, 242 (1944): 510.

10. Rollin Chamberlin in Waterschoot van der Gracht, et al., *Theory of Continental Drift* (Tulsa: 1928).

11. Clarence Dutton, "On Some of the Greater Problems of Physical Geology," *Bulletin of the Philosophical Society of Washington* 11 (1889): 51. In a footnote (p. 289) to his 1882 review of Fisher's book, Dutton says that in an unpublished paper he has used the term "isostasy." This was probably in another informal paper delivered to the Philosophical Society. From the early 1870s the problem of a contracting earth occupied Dutton.

12. For the response of geodesists to the symposium on continental drift, see William Bowie, *Isostasy* (New York, 1927). A later book that gives a fuller and more modern interpretation is Reginald Daly, *The Strength and Structure of the Earth* (New York: 1940). See also W.A.J.M. Waterschoot van der Gracht, et al., *Theory of Continental Drift* (Tulsa, 1928).

13. Bailey Willis, "Continental Genesis," *Geological Society of America Bulletin* 40 (1929): 281–336.

14. Harold Jeffreys, *The Earth: Its Origin, History, and Physical Constitution*, 2d. ed. (Cambridge: 1929).

15. Herman Fairchild, *The Geological Society of America, 1888–1930* (New York: 1932), p. 228.

16. Some of the studies worth consulting are H. Takeuchi et al., *Debate About the Earth*, rev. ed. (San Francisco: 1971); Anthony Hallam, *A*

Revolution in the Earth Sciences (Oxford: 1973); Ursula Marvin, *Continental Drift* (Washington: 1973); Walter Sullivan, *Continents in Motion* (New York: 1973); Don and Maureen Tarling, *Continental Drift*, rev. ed. (Garden City: 1974); J. Tuzo Wilson, ed., *Continents Adrift* (San Francisco: 1970). The American F. B. Taylor also invoked large continental displacements about this time (1910), but his mechanisms were no better than Wegener's and did not lead as directly into later plate theory. If Wegener is the Lamarck of drift theory, call Taylor its Erasmus Darwin.

17. Willis, "Continental Drift, Ein Märchen," p. 509.

18. Harry Hess, "History of the Ocean Basins," in *Petrological Studies: A Volume to Honor A. F. Buddington*, ed. A. E. J. Engel, et al. (New York: 1962), pp. 599, 618.

19. See Raymond Siever, "The Steady State of the Earth's Crust, Atmosphere, and Oceans," *Scientific American* 230, no. 6 (June 1974): 72–79.

20. See, for example, James Valentine and Eldridge Moores, "Plate Tectonics and the History of Life in the Oceans," *Scientific American* 230, no. 4 (April 1974): 80–89.

21. Arthur Strahler, *Planet Earth: Its Physical Systems Through Geologic Time* (New York: 1972), p. 6. A good summary of Chorley's thinking is in Richard Chorley, "Geomorphology and General Systems Theory," U.S. Geological Survey *Professional Paper 500-B* (Washington, 1962). I do not distinguish between "cybernetics" and "general systems theory" in this paper. Whether one theory encompasses the other or not is irrelevant to this subject; they represent (collectively) a whole complex of modern thought that is difficult to stabilize in one term. However the larger theory is labeled, the two chief operational concepts are open system and feedback.

22. For a prospectus on the aims of big geology, see National Academy of Science U.S. Geodynamics Committee: *U.S. Program for the Geodynamics Project* (Washington, 1973).

23. The Schmitt quote comes from the frontispiece of Marvin, *Continental Drift*.

INDUSTRIAL RESEARCH LABORATORIES

KENDALL BIRR
State University of New York (Albany)

IN 1976, INDUSTRIAL RESEARCH LABORATORIES OCCUPIED A PROMinent position on the American scientific and technological scene. In 1975, approximately 70 percent of American research and development (R & D) expenditures of more than $34 billion were performed there.[1] By that we mean that the work was performed in laboratories maintained and operated by private firms engaged in nonagricultural production.

Definitions of research and development have been much debated, but the distinctions used by the National Science Foundation (NSF) in its annual surveys are now commonly accepted. R & D includes a broad range of scientific and technological activities. It involves first what is referred to as basic research, original investigations for the advancement of scientific knowledge with no specific commercial objectives in view. Industrial laboratories characteristically perform some basic research, but more normally they are engaged in applied research and development work. The former involves investigations that seek new scientific knowledge with specific commercial objectives in view; the latter involves technical activities of a nonroutine nature concerned with translating research into products or processes. The distinctions among these activities are based on the clarity with which scientists or engineers can envision the final result of their investigations.

Industrial research laboratories, then, have become significant institutions in our modern society. We live in an age in which economic productivity is a matter of serious concern, and it is clear that such productivity in turn depends on a science-based technology. The quality of our existence is shaped by a continuous flow of scientific and technical knowledge, and industrial research laboratories play a key role in the maintenance of that flow. The history of such laboratories, then, is an aspect of the institutionalization of science and technology and is intimately involved with the growth of our modern economy.

Industrial research laboratories are a relatively recent phenomenon, barely a century old. The reasons for this are not hard to discover. Until sometime during the nineteenth century, technological innovation was the product not of scientists but of craftsmen, normally individuals skilled in and able to improve upon a particular industrial art. For the most part, scientific knowledge was irrelevant to such men, for the state of the art was far in advance of theoretical scientific understanding. Most innovation resulted from the work of individuals who not only had to originate the idea but also had to reduce it to practice and in many cases market the resulting product or process. The characteristic figure of technological innovation in the early nineteenth century was the inventor-entrepreneur, the individual who combined in his person (or sometimes in cooperation with another individual) the technical skills of the inventor and the marketing skills of the entrepreneur.

The story of Charles Goodyear is perhaps a notable case in point. He had no scientific or technical training. Instead, he spent his early years as a hardware merchant until his bankruptcy in 1830. In 1834, he chanced to see and learn about rubber products and some of the difficulties manufacturers faced in fabricating reliable goods. Subsequently he began experimenting with the substance with, as his biographer puts it, "no other tools than an unusually sanguine and determined nature and a firm belief in the future for rubber."[2]

The story of Goodyear's accidental discovery in 1838 of the process of vulcanizing rubber is well known. Experimenting with the idea of using sulfur to remove the surface stickiness of rubber, Goodyear accidentally dropped some of the mixture onto a red-hot stove where the heat, instead of melting the rubber, improved its quality and got rid of the troublesome stickiness. Goodyear spent another five years and more than $50,000 in borrowed money to perfect the process and patent it in 1844. In subsequent years, he sold patent rights in both Europe and America, but on his death in 1860 he left his wife and six children burdened with debts of nearly $200,000.[3]

The lessons were clear. Determination, some experimental skill, and some degree of luck were far more important than scientific knowledge. If an individual succeeded in generating a new idea, it was a long and expensive process to reduce it to practice. And if the inventor lacked adequate capital or entrepreneurial skills, the rewards for his innovation were at best uncertain. It was surely an unsatisfactory way of generating technical change, but it is difficult to see how it could have been improved at this stage by cooperative research and development under corporate auspices. The success stories of the nineteenth century involved either inventors of singular entrepreneurial skills (Thomas A. Edison and Elmer Sperry at the end of the century might be cited as examples)[4] or inventors who managed to find and work effectively with a skilled entrepreneur (the Scholes-Densmore collaboration on the typewriter would be a good example).[5]

Yet, as the nineteenth century progressed, conditions slowly emerged that were to generate the modern industrial research laboratory. The most important development was the steady convergence of scientific knowledge and industrial practice.[6] Through most of the nineteenth century, this convergence was most notable in chemistry, a science with strong ties to an empirical, industrial tradition. It was no accident that the great bulk of men with scientific training who entered industry had received their training in chemistry.

Scattered throughout industry appeared three types of institutional arrangements that served as predecessors of the industrial research laboratory. The first of these was the independent laboratory, usually chemical, serving a variety of individuals and firms who needed the scientific expertise of the chemist. Charles T. Jackson, best known for his experiments with ether as an anesthetic, operated a laboratory in Boston beginning in 1836. His activities included such diverse projects as geological surveys for both state and federal governments, chemical study of sorghum, and experiments seeking to use cotton seeds. Like many such laboratories, Jackson's also served as a training ground for other chemists. The noted Yale scientist Benjamin Silliman, Jr., received some of his training there.

Philadelphia in the pre-Civil War years possessed an unusual number of independent chemical laboratories. Perhaps the most important was that operated by James C. Booth, following a first-class education in Germany in the laboratories of Wöhler and Magnus. Booth opened his laboratory in 1836 and over the years worked on analyses of iron ore, tested sugar, investigated the production of gelatin, served as smelter and refiner for the U.S.

Mint in Philadelphia, and acted as a consultant for many chemical industries. Booth similarly educated a generation of chemists in his laboratory and played an important role in the rising scientific-industrial community of Philadelphia.[7]

The second of the institutional arrangements to appear in the nineteenth century was the laboratory attached to a particular industrial firm. One of the earliest and most successful was organized by Samuel Luther Dana, an M.D. who drifted into chemistry. In 1834, he began a thirty-four-year association as resident and consulting chemist with the Merrimack Manufacturing Company, a Lowell, Massachusetts, textile firm. His work there produced some important innovations in bleaching and dyeing textiles. In 1875, Charles Benjamin Dudley resigned his post as a science teacher in a Poughkeepsie, New York, military academy to join the staff of the Pennsylvania Railroad, where he established a chemical testing section. During his career there, he made important contributions to metallurgy, investigating the relations between carbon, silicon, manganese, and phosphorus in steel and their effects on steel's physical properties. Dudley was one of the founders of the American Society for Testing Materials. When he left the Pennsylvania Railroad, he left behind a well-established laboratory staffed with thirty-four trained chemists and numerous assistants.[8]

Other examples could be cited. By the end of the century, chemists had penetrated numerous branches of American industry, notably chemicals, metallurgy, meat packing, soap, photographic supplies, and other areas. Most of the activities of such individuals could hardly be characterized as research or development, but their work did slowly impress on industrialists the value of trained scientific personnel.

The third type of involvement of trained scientists with industry came as the latter found it beneficial to make use of scientific consultants, either from the academic world or from independent laboratories. Benjamin Silliman's involvement with the oil industry is well known,[9] but it should also be noted that Standard Oil turned to Herman Frasch, a German-born and trained petroleum chemist, to deal with the difficulties posed by the high sulfur content of oil pumped from the newly opened Ohio fields.[10] In the early 1870s, Western Union developed a working relationship with the rising young inventor, Thomas A. Edison, in which Edison was in a sense working for the telegraph firm.[11] Indeed, such examples were sufficiently common to lead one student of the subject to suggest that there was a kind of "putting out" system in research in which industrial firms contracted with independent scientists or laboratories for the research they felt needed to be done.¹²

By the 1870s and 1880s, the conditions which produced the modern industrial research laboratory were emerging. These conditions included, first, a felt need for scientific and technological assistance. The late nineteenth century was a period of bitter competition, deflation, and business upheaval in America as industrial leaders desperately searched for security and stability. The search most obviously led to methods of lessening competition, including inter-firm cooperation, price fixing, mergers and the other activities which so aroused the ire of the antimonopolists. But in some cases the search for corporate security led toward science and technology.

Such needs were most deeply felt in those industries characterized by rapidly changing technologies and particularly in those industries whose technologies were dependent on science. It was no accident that the modern industrial research laboratory first emerged in industries such as communications, electrical machinery, and chemicals. The first two had never had a craft tradition in advance of scientific knowledge; indeed, electricity had never been of practical use before scientific understanding of the principles on which it operated. A few people feared that technological improvement might be hampered by a lack of scientific information. Some leaders at General Electric at the end of the century feared that the industry was rapidly "using up" its scientific "capital." The organization of the General Electric Research Laboratory was one step in an attempt to rectify that situation.

In addition, the emergence of large corporations seemed to provide the financial wherewithal to support industrial research. Individual inventors earlier in the century had often suffered from a lack of capital to reduce their ideas into practice and had lacked the organization and entrepreneurial skills to convert their ideas into commercially successful products or processes. The large corporation seemed well constructed to deal with such problems and facilitate the whole process of technological innovation.

Finally, there were some successful examples at hand to encourage such industrialists. The cooperative research going on in German academic laboratories in connection with the German system of graduate education provided one example, as did the scientific research that lay at the roots of the highly successful German dye industry. The latter provided one of the most impressive nineteenth century examples of the profitability of scientific research.¹³

Surely one of the most spectacular early examples of successful organized industrial research was provided by Thomas A. Edison. The noted inventor was perhaps one of the last of the great inventor-entrepreneurs, but he spent most of his career working intimately with the emerging industrial giants of his day. An empiricist rather than a genuine scientist, he had the good fortune to work in electricity at a time when his limited scientific training was no bar to his making major contributions to the field. When in 1876 he built his new laboratory at Menlo Park, New Jersey, he proposed to devote himself to "the invention business" full time. He did so with remarkable success. From the laboratory flowed a striking series of inventions: improvements in the telephone; the early phonograph; the incandescent electric lamp; and a complete system of generating, distributing, and measuring the electric power necessary to operate the lamps. Not the least of Edison's inventions was that of the industrial research laboratory itself.[14]

There were several elements in Edison's success, factors which suggested the terms on which future research laboratories might win similar success. The first of these was Edison's leadership. The laboratory was clearly his personal instrument, and he had the genius to use it well. Strong, creative leadership played a similarly important role in other early research laboratories. Secondly, Edison showed an ability to use effectively the available scientific personnel and ideas. Note the role played in Edison's work by Francis Upton, a university-trained scientist and mathematician. Similarly, Edison quickly recognized the usefulness of the newly developed Sprengel mercury pump, a scientific instrument that permitted him to achieve the high vacuums necessary for the construction of a successful incandescent lamp.

Thirdly, the Edison experience demonstrated the power of cooperative research. Edison's organization was important, and the ability to pursue multiple lines of research played an important role in enabling Edison to beat his rivals to a successful incandescent lamp. Finally, Edison's success demonstrated the importance of firm financial backing. His support came initially from people associated with Western Union to whom he had earlier demonstrated his talents. The Edison Electric Light Company, the firm organized to pursue the development of a complete electrical system, spent nearly $650,000 between its formation in 1878 and the opening of the Pearl Street central generating station in New York, which convincingly demonstrated the practicability of Edison's system.[15] No individual could have borne such costs without meeting the fiscal fate of a Charles Goodyear.

In the first two decades of the twentieth century, the industrial research laboratory emerged in its modern form. It was a group of scientists and engineers attached to and supported by a corporation and given the specific mission of performing what we would now define as research and development work, free from concern for routine testing or day-to-day troubleshooting on the production line. The laboratories emerged most obviously in the chemical and electrical industries. The first survey of industrial research laboratories made in 1920 revealed that two-thirds of all the research workers employed in industry were to be found in the electrical, chemical, and rubber industries.[16] The pre-World War I years saw the organization of some of the most prominent and successful of research laboratories at General Electric (1900), Du Pont (1902), American Telephone and Telegraph (AT & T) (1904), and Eastman Kodak (1912).

Still, it was World War I that publicized the importance of research and assured its position on the American industrial scene. Americans were alternately impressed with the scientific expertise that seemed to make Germany strong and alarmed by evidence of the degree to which this country had become dependent on German technology, particularly in chemicals, dyes, pharmaceuticals, and optical glass. The cooperative applied scientific research of wartime also contributed to a widespread conviction that industrial research was essential to industrial progress. So powerful was the effect of the war that one historian has observed that the war produced such an "infusion of research into the economy, especially into production . . . that industrial research as a branch of the country's scientific establishment dates its rise to eminence almost entirely from the war period."[17]

THE PERIOD BETWEEN WORLD WAR I AND WORLD WAR II saw the establishment of numerous new research laboratories. Just how many is uncertain. Beginning in 1920, the National Research Council began to collect information about industrial laboratories, but coverage during the early 1920s was limited. It is clear that the 1920s were a decade of substantial expansion; by 1931 more than 1,600 companies reported laboratories employing nearly 33,000 people. The Depression brought modest cutbacks, but the expansion was quickly resumed, so that by 1940 more than 2,000 corporations reported laboratories employing about 70,000 people.[18]

Chemists and engineers were most heavily represented among the professional workers in such laboratories. In the overwhelming proportion of cases, leadership was provided by scientists or engineers with advanced degrees, frequently the doctorate.[19] There were estimates of research and development expenditures during the 1930s, but they are subject to wide margins of error. The best estimates suggest that industrial research laboratories in 1930 were spending between $100 and $160 million, and that expenditure levels had reached the $234 million mark by 1940.[20] A large proportion of the research was concentrated in industries with rapidly changing technologies. Chemicals, electrical, communications, petroleum, and rubber industries showed a strong interest in research and development work.[21]

Such gross statistics tend to hide important diversities. A look at what was happening in individual firms impresses the historian with the variety of forms, functions, and achievements that characterized industrial laboratories during these years. A brief examination of two prominent laboratories during these years—General Electric (GE) and Standard Oil of New Jersey—helps illustrate some of these diversities.

General Electric was one of the first corporations to establish an independent research laboratory with the clear mission to provide new products and ideas to the company. Under the leadership of its first director, Willis R. Whitney, the laboratory was extraordinarily successful in the pre-World War I period. Basic research paid off handsomely with impressive commercial products—most notably ductile tungsten, which produced greatly improved incandescent lamp filaments, the modern high vacuum X-ray tube, electron tubes, and the gas-filled incandescent lamp. One of GE's most distinguished scientists, Irving Langmuir, received the Nobel Prize in 1932 for his work in surface chemistry. Indeed, the General Electric Research Laboratory was often held up as a model of the productivity of scientific research in industry.

Yet a careful look at the GE laboratory's activities during these years reveals a more complex picture. Besides engaging in basic research, the laboratory performed a wide range of applied research and development work, ranging from defensive research on the incandescent lamp to applied research and development of radio vacuum tubes to pilot-plant stage operations for some of its innovations. The laboratory on the one hand gave its senior scientists considerable freedom to pursue investigations that interested them and on the other hand conscientiously met the demands for specific technical assistance coming from all parts of the company. The picture is one of a complex research organization generating new ideas, looking after the technical interests of the corporation, and interacting with other scientists, engineers, and laboratories within the firm.[22]

While General Electric undertook a research effort in order to generate new innovations within the firm, Standard of New Jersey looked to research and development to exploit ideas generated outside the company. "Up to 1919," two historians have observed, "applied chemistry had played a relatively small part in the improvement of refining processes in the industry, and theoretical chemistry almost no part at all."[23] Jersey's response was the creation of a group of technicians to improve the refining process at its major New Jersey refinery, then the organization in the fall of 1919 of a Development Department. The new department moved ahead in its early years to develop a continuous cracking process and to demonstrate the commercial practicability of continuous vacuum distillation. It also encouraged the development of laboratories in Standard's manufacturing divisions and affiliated companies, to relieve the department of some of its onerous routine duties.

In 1927, research activities were reorganized as the Development Department was transferred to Jersey's patent-holding subsidiary, Standard Oil Development Company. In subsequent years, an interesting mix of healthy rivalry and technical cooperation developed among the Development Company and the laboratories in Jersey's affiliates.

The late twenties and thirties saw some impressive innovations. Some of the ideas came from outside the organization, notably from Jersey's controverted relationship with I. G. Farben, but some came from within the corporation's laboratories. The innovations included hydrogenization processes, the improvement of lubricants, the commercial development of high-octane gasoline suitable for aircraft engines, fluid catalytic cracking, the production of toluene, and work on synthetic rubber. The R & D efforts of this major oil company presented yet another pattern of industrial research: an appropriate mix of competition and cooperation among the corporation's laboratories and a strong dependence on the exchange of technical information with a major European cartel.[24]

The growth of industrial research laboratories between the wars, then, was accompanied by surprising diversity of organizational forms, research and development goals, and research achievements. But the achievements were sufficiently striking to persuade corporate leaders to increase steadily their commitments to research and development.

World War II and the events that followed created radical changes in America's industrial research laboratories. During the conflict, America's scientific talent was mobilized on a scale previously unknown in agencies such as the Manhattan Project and the Office of Scientific Research and Development. The products of that mobilization—the atomic bomb, radar, the proximity fuse, and many others—brought to organized scientific research unprecedented prestige. Federal R & D expenditures rose from $74 million in 1940 to $1,590 million in 1945 and suggested a heretofore relatively untapped source of support for research. The imperatives of the Cold War in the 1950s and the race for national supremacy in space in the 1960s served to stimulate the demand for R & D activities in the post-World War II era.

The result was a period of explosive growth in research and development. Total national expenditures, which had been $6,182 million in 1955, rose to $34,345 million in 1975, a more than fivefold increase. Much of the increase represented inflation; R & D expenditures as a percentage of gross national product peaked at 3.0 percent in 1964 and subsequently declined to about 2.7 percent. Still, the years since 1945 have seen a new order of magnitude in research activity.

The shape of America's industrial research effort has been profoundly influenced by the size of federal funding. In the past two decades, from half to two-thirds of R & D funds have come from the federal government. Some of those funds are, of course, spent in federally operated laboratories, but much is performed on a contract basis in industry. As a result, anywhere from 40 to 60 percent of the research performed in industrial research laboratories has been federally funded. The end of the Vietnam War and cutbacks in the space program have produced a decline in federal support for industrial research, but the support remains substantial. The chief effect of federal funding was to add the aerospace industries to the traditional list of big industrial R & D spenders. In recent years aircraft, electrical equipment and communications, chemicals, machinery, motor vehicles, and transportation equipment have accounted for 80 percent of research and development expenditures in industry.[25]

The new order of magnitude of industrial research and development expenditures has generated debate over several policy questions. For example, there has been the ongoing discussion of the size of the national R & D effort. Two things are clear about this debate. First, some of the apparently unlimited confidence in the power of scientific research to achieve desirable ends has disappeared. As a result, R & D appropriations both in the federal government and in industry have responded to external events such as the end of the Vietnam War, the arrival of detente with the Soviet Union, heavy government deficits combined with high tax rates, or the vicissitudes of recession. Research and development no longer have unlimited claims on industrial or governmental funds.

Second, new questions have been raised about the way we spend our R & D funds. At one level, the debate has revolved around the distribution of effort among basic research, applied research, and development. The evidence indicates that industrial research laboratories devote only 3 to 4 percent of their effort to basic research; by contrast 77 to 79 percent of their R & D effort has gone into development. The reasons for this are fairly obvious; despite some striking examples of the payoff that can result from basic research, most corporation executives find that kind of research too risky.

To some degree, commentators may be raising a false issue. The important question may not be the allocation of R & D funds among basic research, applied research, and development but instead the effectiveness of the interaction among these three kinds of activity to maximize final results.

Other kinds of questions have been raised about allocation of R & D funds. Some critics have argued that our total R & D effort has been distorted by the Cold War and the immense resources poured into defense and defense-related activities. Other critics have noted the trivial character of much of the development work performed in industry. Both kinds of critics wonder how such emphasis can be justified when it results in the neglect of pressing social concerns, such as the long-term energy crisis or environmental issues. These are major questions of national R & D policy raised by the growing importance of research in our society and by the greatly enlarged level of R & D activity, but they are unlikely to be resolved by decisions at the level of the individual industrial research laboratory.

At that level, other kinds of questions have been raised. Many of these questions are related to the growing size of the R & D effort and of individual organizations developed to handle R & D. As industrial research laboratories have become larger, they have become increasingly bureaucratized. In 1912, when the General Electric Research Laboratory was at the peak of its productivity, its director, Willis R. Whitney, could warn one of his correspondents about the dangers of too much system. "It is perfectly conceivable to me that a laboratory could go to pieces and still have a very systematic arrangement," he wrote. "I dread organization and system so much that I want to warn others from expending much time and effort upon it."[26] Such sentiments may still elicit warm responses from individual scientists working in industrial laboratories (or university ones, for that matter). But the days in which such individualistic and antibureaucratic attitudes can be effective are long past. The first generation of R & D entrepreneurs has been replaced by a new breed of R & D managers.

The basic questions facing such managers have to do with maximizing the results from governmental and corporate investment in research and development. They include such issues as determining the level of research activity; the distribution of effort among basic research, applied research, and development; selecting the most promising areas for R & D activity; deciding how far to pursue particular projects or when to stop research; and determining how to see a particular innovation through to commercial success. The questions are no different from those faced by the first generation of laboratory directors, but they have become far more complex and the stakes are much higher.

There has long been a kind of "how to" literature designed to aid research directors in making such decisions. Until about 1950, such literature was the product of practicing industrial research people who summarized their own experiences and those of other practitioners. In the last two decades, however, the field has been taken over by individuals trained in economics and management. Although there have been relatively few comprehensive empirical studies of management practices, the growing literature of R & D management incorporates both individual empirical studies and theoretical analysis and insights.[27] How useful these analytical techniques have been is uncertain. Network techniques such as Program Evaluation and Review Technique (PERT) have proved helpful in planning and scheduling development projects. But techniques designed to improve the process of project selection seem to be less successful, probably because they oversimplify the research process and fail to deal satisfactorily with the uncertainties of research.[28] Whatever the present state of research management, an increasing dependence on such managerial methods seems inevitable as those responsible for research direction struggle with increasingly complex management problems.

Finally, recent investigations have raised questions about the ultimate effectiveness of the large industrial research laboratory. In more exuberant days of research and development, many people were convinced that in R & D, as in other varieties of production, there were economies of scale. The resources—both intellectual and financial—available in large laboratories, the power of organization and cooperation, and the superior ability of the large corporation to carry innovations through to the commercial level all seemed to suggest that research productivity would increase with the size of the research unit. But several studies have indicated that Du Pont's success with synthetic fibers or GE's with the incandescent lamp has been the exception rather than the rule among large laboratories. Instead, most important technological innovations have come from individuals or small firms operating on the edges of the industry involved.[29]

THE MODERN AMERICAN INDUSTRIAL RESEARCH laboratory is chiefly a product of the last century. It was born of the marriage between science and technology and represents the development of a particular set of institutional arrangements for the conduct of research and development within the private sector. Its successes have given it a secure niche in the institutional structure within which scientific research and technological innovation occur in America. But this laboratory is not the only or perhaps not even the most important institution. In a sense, it is suffering from an identity crisis arising from its struggle to cope with problems common to all large-scale organizations and from nagging doubts about the productivity of such laboratories as compared with other research and development units. How those problems will be resolved cannot be predicted, but it seems safe to conclude that the industrial research laboratory will continue to occupy a prominent position on the American scientific scene in the next century.

1. U.S. Bureau of the Census, *Statistical Abstract of the United States 1975* (Washington: U.S. Goverment Printing Office, 1976), pp. 546, 548.

2. Carl W. Mitman, "Charles Goodyear," in *Dictionary of American Biography*, ed. Allen Johnson and Dumas Malone, 20 vols., (New York: Charles Scribner's Sons, 1928–1936), vol. VII, p. 413.

3. Ibid., pp. 413–15; Charles Singer et al., *A History of Technology*, 5 vols., (New York: Oxford University Press, 1958), vol. V, pp. 752–69.

4. Matthew Josephson, *Edison* (New York: McGraw-Hill, 1959); Thomas Parke Hughes, *Elmer Sperry: Inventor and Engineer* (Baltimore: The Johns Hopkins Press, 1971).

5. Richard N. Current, *The Typewriter and the Men Who Made It* (Urbana: University of Illinois Press, 1954).

6. Kendall Birr, "Science in American Industry," in *Science and Society in the United States*, ed. David D. Van Tassel and Michael G. Hall, (Homewood, Ill.: Dorsey Press, 1966).

7. Howard R. Bartlett, "The Development of Industrial Research in the United States," in *Research—a National Resource*, National Resources Planning Board, 3 vols., (Washington: GPO, 1938–1941), vol. II, p. 72;

Bruce Sinclair, *Philadelphia's Philosopher Mechanics* (Baltimore: The Johns Hopkins Press, 1974).

8. Bartlett, in *Research—A National Resource*, vol. II, pp. 25–27.

9. Gerald White, *Scientists in Conflict* (San Marino, Calif.: The Huntington Library, 1968).

10. Harold F. Williamson and Arnold R. Daum, *The American Petroleum Industry: The Age of Illumination, 1859–1899* (Evanston: Northwestern University Press 1959), pp. 599–618.

11. Josephson, *Edison*, pp. 86–87.

12. I am indebted for this notion to W. David Lewis and an unpublished manuscript on the history of industrial research he prepared for the Eleutherian Mills Library.

13. John J. Beer, *The Emergence of the German Dye Industry* (Urbana: University of Illinois Press, 1959), esp. chap. 8.

14. Josephson, *Edison*, chaps. VIII–XI.

15. Harold C. Passer, *The Electrical Manufacturers, 1875–1900* (Cambridge: Harvard University Press, 1953), p. 87.

16. Bartlett, in *Research—a National Resource*, vol. II, p. 34.

17. A. Hunter Dupree, *Science in the Federal Government* (Cambridge: Harvard University Press, 1957), p. 323.

18. Compiled from various National Research Council surveys. For a detailed analysis of such statistics and other data, see George Perazich and Philip M. Field, *Industrial Research and Changing Technology*, report no. M-4 (Philadelphia: WPA, National Research Project, 1940).

19. Ibid., p. 78; statements about laboratory directors made on the basis of an analysis of backgrounds of such directors listed in National Research Council Bulletin 16.

20. Edward R. Weidlein and William A. Hamor, *Science in Action* (New York: McGraw-Hill, 1931), pp. 44–45; David Graham, "Research Expenditures and Their Effect on the General Electric Company," *Harvard Business Review*, X (October 1931): 118; John R. Steelman, *Science and Public Policy*, 5 vols. (Washington: GPO, 1947), vol. I, pp. 9–10; Frank Joseph Kottke, *Electrical Technology and the Public Interest* (Washington: American Council on Public Affairs, 1944), pp. 22, 33n.

21. Perazich and Field, *Industrial Research*, p. 21.

22. Kendall Birr, *Pioneering in Industrial Research* (Washington: Public Affairs Press, 1957).

23. George Sweet Gibb and Evelyn H. Knowlton, *The Resurgent Years, 1911–1927* (New York: Harper, 1956), p. 523.

24. Ibid., pp. 520–32; Henrietta Larson, Evelyn H. Knowlton, and Charles S. Popple, *New Horizons, 1927–1950* (New York: Harper, 1971), chap. 6.

25. Generalizations are based on analysis of annual NSF statistics on research in industry.

26. Willis R. Whitney to F. W. Lovejoy, 24 October 1912, General Electric Research Laboratory Archives.

27. C. E. Kenneth Mees, *The Organization of Industrial Scientific Research* (New York: McGraw-Hill, 1920) and C. C. Furnas, ed., *Research in Industry* (New York: Van Nostrand, 1948) are representative of the older books by practitioners. The shift symbolically occurred in 1950 with the publication of the second edition of C. E. Kenneth Mees and John A. Leermakers, *The Organization of Industrial Scientific Research* (New York: McGraw-Hill, 1950) and David Bendel Hertz, *The Theory and Practice of Industrial Research* (New York: McGraw-Hill 1950). Representative examples of the new type of analysis are Daniel Roman, *Research and Development Management* (New York: Appleton-Century-Crofts, 1968) or Edwin Mansfield, *The Economics of Technological Change* (New York: W. W. Norton, 1968). Robert N. Anthony, *Management Controls in Industrial Research Organizations* (Boston: Harvard University Graduate School of Business Administration, 1952) remains one of the few comprehensive empirical studies.

28. Mansfield, *Economics of Technological Change*, pp. 88–90.

29. John Jewkes, David Sawers, and Richard Stillerman, *The Sources of Invention* (New York: St. Martin's Press, 1958), esp. chap. vii; W. Rupert Maclaurin, *Invention and Innovation in the Radio Industry* (New York: Macmillan, 1949); Daniel Hamburg, "Invention in the Industrial Research Laboratory," *Journal of Political Economy* LXXI (April 1963): 95–115.

THE RISE AND SPREAD OF THE CLASSICAL SCHOOL OF HEREDITY, 1910-1930; DEVELOPMENT AND INFLUENCE OF THE MENDELIAN CHROMOSOME THEORY

GARLAND E. ALLEN
Washington University

THE FIRST SIXTY YEARS OF THE TWENTIETH CENTURY MIGHT WELL be called the "age of genetics."[1] Beginning with the rediscovery of Mendel's laws in 1900, through elucidation of the chromosome theory of heredity between 1910 and 1920, to the development of the Watson and Crick model of DNA in the 1950s, the field of genetics has become a dominant area of modern biology. It contributed significantly to cytology by showing that the chromosomes, whose function from 1870 until after 1910 was a much-debated issue, could be regarded as the material bearers of heredity. It gave a sound basis to Darwin's theory of natural selection, which had suffered from lack of any coherent mechanism of heredity.

Genetics also revolutionized the study of biochemistry by suggesting that genes controlled the production of proteins, many of which (as enzymes) guided the cell's myriads of metabolic pathways. It provided a significantly different viewpoint for the study of development by showing that genes influence specific changes during embryology and in some ways (still unknown) can be turned on or off at appropriate times. Finally, genetics opened up a whole host of new insights in medically related disciplines, from the study of genetic diseases to the investigation of cancer as a change in genetic control processes. Just as Darwin's theory became a focus for relating many disparate areas of biology in the nineteenth century, so genetics came to occupy a similar position in the twentieth.

Influential in establishing genetics in its central position during the early part of the century was the work of Thomas Hunt Morgan

(1866–1945) and a group of young and enthusiastic coworkers, Alfred Henry Sturtevant (1891–1970), Herman Joseph Muller (1890–1967), and Calvin Blackman Bridges (1889–1938). Between 1910 and 1915 this group, working at Columbia University with the small fruit fly *Drosophila melanogaster*, succeeded in applying Mendelian principles to the cytological facts of chromosome structure and behavior. By correlating breeding results with cytological observations, they showed that Mendel's abstract, hypothetical "factors" (or *Anlagen* as he more frequently called them), postulated in 1865, could be treated as specific points, or loci, along the length of the chromosomes. These findings, with their many extensions and elaborations, were summarized in *The Mechanism of Mendelian Heredity* (New York, Henry Holt and Co.) published in 1915 (revised 1923). It was this epoch-making book that set the Mendelian theory on a sound quantitative and experimental basis and led a number of biologists to grasp the fundamental importance of the new science of genetics.

In the present paper, I would like to discuss some of the most fundamental characteristics of the work of the Morgan group—the so-called "classical" school of genetics in the United States. The discussion will focus on three aspects of the Morgan group's work: (1) actual scientific results, (2) certain philosophical considerations behind the work, and (3) some of the sociological factors which brought the group to the forefront of the Mendelian movement in the early twentieth century.

In order to discuss these issues in some depth, the focus will be on a small, but crucial, period of time: the years between 1910, when Morgan first entered the field of heredity in a full-scale way, and 1915, the year in which *The Mechanism of Mendelian Heredity* was published. This period of time is particularly important in several ways. Although Morgan and his group continued to make outstanding contributions to the development of genetics through the late 1930s, the five years between 1910 and 1915 saw the most rapid growth of their novel and revolutionary ideas, in particular their emphasis on the material (chromosomal) basis of the Mendelian theory. It was also during this period that Morgan and his group clearly assumed leadership of the new Mendelian movement, supplanting its first champion, William Bateson.

Development of the Drosophila Work, 1910–1915

Who were the members of the Morgan group? What influence did Morgan himself exert, both through his personal life style in science and through his particular approach to the problem of heredity?

Of the four coauthors of *The Mechanism of Mendelian Heredity*, Morgan was the oldest. Born in 1866 in Lexington, Kentucky, he attended Johns Hopkins University, where he received his Ph.D. under William Keith Brooks (1848–1908) in 1890. Brooks was a morphologist who sought to discover evolutionary (primarily phylogenetic) relationships among groups of animals and plants by studying their comparative anatomy, embryology, cytology, and, to a lesser extent, physiology.

Morgan's early work was in the area of descriptive morphology of the Pycnogonids, a group of marine invertebrates called sea spiders. After he left Hopkins for a teaching post at Bryn Mawr College in 1891, his work became increasingly experimental in orientation, a fact which was of considerable importance in determining his later interest in heredity. In 1904, Morgan left Bryn Mawr for Columbia, where he remained until 1928; in that year, he accepted an invitation to organize the new Division of Biological Sciences at the California Institute of Technology (Caltech) where he remained until his death in 1945.

The remaining three members of "Morgan's group," as it was called, had all entered Columbia College as undergraduates between 1906 and 1908. In the winter of 1910 to 1911, Morgan took Sturtevant and Bridges—juniors in the college—into his laboratory and gave them desks in what came to be known as the "fly room." Muller, who had graduated the previous spring, was then a student of the cytologist Edmund Beecher Wilson (1858–1939). Muller became interested in the *Drosophila* work and took part in most of the developments carried out between 1910 and 1916, when he completed his doctoral thesis on the interrelations of linked genes in *Drosophila*.

Sturtevant and Bridges became Morgan's graduate students in 1912 and received their Ph.D.'s under his direction in 1914 and 1916, respectively. Both remained with Morgan as researchers under the auspices of the Carnegie Institution of Washington until 1928, when they moved with him to Caltech. Muller, on the other hand, left the *Drosophila* group shortly after receiving his Ph.D., assuming posts at Rice Institute (Houston), the University of Texas, the Institute of Genetics of the Academy of Sciences of the USSR, Amherst College, and Indiana University, where he remained until his death. Thus, although he was an integral part of the developments leading up to the publication of *The Mechanism of Mendelian Heredity*, Muller asserted his own independence at an early age and developed a line of work (the artificial induction of mutations) distinct from that pursued by the rest of the group.

In 1933, in recognition for his work in establishing the chromosome theory of heredity (i.e., the idea that genes are located on chromosomes), Morgan was awarded the Nobel Prize in physiology and medicine. Because Bridges and Sturtevant had worked so closely with him and had remained with the *Drosophila* studies for such an extended period of time, Morgan shared the Nobel Prize money with them equally. For reasons that are not wholly clear, but no doubt largely because of his dissociation from the group, Muller was not included in this division. He was, however, destined to receive his own Nobel Prize some fourteen years later (1947) for his work in mutagenesis.

Although trained as a descriptive morphologist, early in his career Morgan became a strong advocate of the experimental approach to biology. At Hopkins, he was increasingly exasperated with the speculative morphology of Brooks and perhaps as a result came under the influence of the physiologist H. Newell Martin, a student of Sir Michael Foster and an outspoken proponent of the experimental method. After leaving Hopkins, Morgan encountered Hans Driesch (1867–1941) at the Naples Marine Station, and also Jacques Loeb (1859–1924), who joined the Bryn Mawr faculty the same year as Morgan.

Through Driesch and the work of the Naples Station, Morgan had seen the excitement of experimentation applied to his own previously descriptive field, embryology. And through Loeb he became aware of the importance of trying to put all areas of biology on an equal footing with the exact sciences (i.e., physics and chemistry) in terms of the validity of their generalizations. Morgan conveyed his enthusiasm for experimental work to his younger colleagues Sturtevant, Bridges, and Muller; they were strongly attracted to Morgan and Wilson, both of whom were outspoken critics of the old-style descriptive biology and currently involved in the newer experimental work in heredity (Morgan on *Drosophila*, Wilson on the cytology of sex inheritance).

There is irony in the fact that although his own work did so much to establish the validity of, and provide a material basis for, the Mendelian theory, Morgan was for a number of years prior to 1910 openly hostile to the factor hypothesis. The various aspects of his opposition have been detailed elsewhere,[2] but one aspect deserves mention here. Along with the descriptive and often highly speculative phylogenizing of the morphologists had come a whole series of attempts to deal with the problem of heredity in terms of hypothetical units or particles. Darwin's pangenes, Weismann's biophors, and Nägeli's micellae were only a few of the almost endless number of speculative theories of heredity and embryology current between 1870 and 1900. To Morgan, the Mendelian theory, as it was understood between 1900 and 1910, seemed to share many characteristics with these hypothetical particulate theories, remnants of what he considered an outmoded approach to biological problems. In 1901, Morgan had written:

A favorite method of biological speculation in the last forty years has been to refer the properties of the organism to invisible units and to explain the action of the organism as the resultant of their behavior.... Elements are endowed by their inventors with certain properties and these give the appearance of an explanation to organic phenonema.[3]

Morgan was opposed to the Mendelian theory because of what appeared to be its hypothetical nature. It was useful, perhaps, as a mathematical theory, he once said, but it must not be mistaken for reality. Because of his emphasis on the experimental method, Morgan did not see the Mendelian theory as having any greater foundation in fact than Weismann's or Nägeli's.

In a recent study, Edward Manier has stressed the relative importance of experimentalism, as compared to simple empiricism, in Morgan's methodology, particularly as it related to his acceptance or rejection of the Mendelian and chromosome theories.[4] Manier points out that the empirical approach is characterized by concern with a large amount of basically similar kinds of evidence (for example, testing a new mutation against a number of different known stocks by a series of cross-breeding experiments). On the other hand, the experimental approach requires at least two independent types of evidence—the method of obtaining one kind being different from that for obtaining the other (for example, determining the existence of a chromosomal deletion both from breeding data and from cytological examination of chromosome preparations).

On these grounds, Morgan was clearly motivated more by the experimentalist than purely empiricist considerations. By 1909, considerable empirical evidence was available that Mendel's laws had wide application—i.e., they held for a large number of species. Yet Morgan remained skeptical. What began to change his mind was not the fact that he could apply the Mendelian theory to yet another organism such as *Drosophila*, but that he could test the Mendelian theory (studied by breeding experiments) with evidence from a wholly different area, namely cytology, specifically in the observed behavior of chromosomes during gametogenesis. As soon as Morgan saw that the white-eye mutation acted *as if* it were somehow part of the X-chromosome, he began to view the Mendelian theory in a completely different light. From that point on, his reservations vanished, and he began the full-scale investigations of *Drosophila* which comprise the major results described in *The Mechanism of Mendelian Heredity*.

The fact that Morgan saw Mendel's "factors" as having a possible material basis on the chromosomes does not mean that he automatically accepted the idea that genes were physical entities, the reality of which was necessary for the validity of the Mendelian theory. In the preface to *The Mechanism of Mendelian Heredity*, he suggested that Mendel's laws and the superstructure of hereditary theory based upon them could be viewed independently of chromosomes, although that was not necessarily the point of view favored by his group:

> We have ... put our own interpretation of the facts, and while this may not be agreed to on all sides, yet we believe that in what is essential we have not departed from the point of view that is held by many of our co-workers at the present time. Exception may perhaps be taken to the emphasis we have laid on the chromosomes as the material basis of inheritance. Whether we are right here, the future—probably a very near future—will decide. But it should not pass unnoticed that even if the chromosome theory be denied, there is no result dealt with in the following pages that may not be treated independently of the chromosomes; for, we have made no assumption concerning heredity that cannot also be made abstractly without the chromosomes as bearers of the postulated hereditary factors.[5]

Morgan goes on to pose the logical question:

> Why then, we are often asked, do you drag in the chromosomes? Our answer is that since the chromosomes furnish exactly the kind of mechanism that the Mendelian laws call for; and since there is an ever-increasing body of information that points clearly to the chromosomes as the bearers of the Mendelian factors, it would be folly to close one's eyes to so patent a relation.[6]

This point illustrates one important feature, often overlooked, in the development of classical genetic theory during the twentieth century. Although it is true that the establishment of the chromosome theory provided both a material basis for Mendel's postulated factors and a specific mechanism for understanding how segregation, assortment, and recombination took place, the theory of the gene as developed by the Morgan school was often interpreted as a highly formalistic structure. The breeding results, genetic maps, concepts of multiple alleles, position effect, or lethal genes were able to stand on their own, even without the sophisticated cytological work that related these ideas to the physical structure of chromosomes. Historian Donald Fleming has pointed out that some workers even felt a certain pride in the formalism of genetics as it developed through the 1920s and 1930s. For the theory was rigorous and logically consistent on its own. Yet there was little question in the minds of Morgan and his colleagues that genes were real entities, and that they corresponded to specific points on chromosomes.

Although Morgan and his group were not unaware of questions about the reality of genes or how they functioned, they focused their attention almost exclusively on the physical relationship between genes and chromosomes. Questions regarding the nature of genes or how they functioned physiologically were considered too difficult to get a handle on at the moment. For this reason, a gap existed until the late 1930s and early 1940s between the classical gene theory of the Morgan school and areas such as biochemistry and cell physiology. But one cannot ask for everything, and there were good reasons why Morgan and his group stayed away from questions of gene function in the period between 1910 and 1940. They restricted their work to areas where actual experiments and observations could be made. This was perhaps one of the most important and brilliant aspects of their methodology.

Morgan appears to have begun his studies with *Drosophila* somewhere around 1908 or 1909.[7] His original interest in breeding *Drosophila* was equally as much for the light it would throw on the problem of evolution as on that of heredity. He tried to induce large-scale variations, or mutations as he called them, in a variety of organisms. *Drosophila* was the only one in which he actually found a particularly striking mutant—the white-eyed male which turned up in December 1909.

When Morgan began to breed his white-eyed mutant, he noted that in successive generations the mutation appeared almost exclusively in males; only occasionally would he see a white-eyed female. It was then, in early 1910, that he saw the possible parallel between the segregation of Mendel's factors and the observed movements of the accessory chromosomes that determined sex. Even in his first paper[8] of 1910, however, Morgan refrained from speaking of the white-eye factor as a physical part of the X-chromosome. The white-eye factor, he pointed out, could be treated *as if* it segregated along with the X-chromosome, and he indicated this by labeling the eye-color and sex-determining factors separately (i.e., RX-RX stood for red-eyed female, WX-X for white-eyed male).

Once he had established the factor-chromosome relationship in his own mind, somewhere between 1910 and 1911, another idea occurred to Morgan. He was a close associate and friend of Edmund Beecher Wilson, who was then (1910) head of the zoology department at Columbia. Since the early 1900s, Wilson had been increasingly involved in demonstrating and championing the chromosomal theory of sex determination. In 1910, Wilson was regarded as the foremost cytologist in the United States, if not the world. In 1909, F. A. Janssens had published a careful series of cytological observations on what was called chiasma—the intertwining of homologous chromosomes during metaphase I of meiosis. In observing this phenomenon, Janssens believed he could show that occasionally homologous chromosome strands exchanged parts.[9]

Either through Wilson, or through his own work on cytology in aphids in 1909 and 1910, Morgan had become familiar with Janssens's concept, and applied it to the conception of genes as parts of chromosomes. He reasoned that the strength of linkage between any two factors must be related in some way to the distance between them on the chromosome. The farther apart any two genes, the more likely that a break could occur somewhere between them, and hence the more likely that the linkage relationship would be disturbed.

In a conversation with Morgan one day in 1911, Sturtevant, then still an undergraduate at Columbia, suddenly realized that the variations in strength of linkage could be used as a means of determining the relative spatial distances apart of genes on a chromosome.[10] According to his own testimony, Sturtevant went home that same night and drew up the first genetic map for the sex-linked genes Y, W, V, M, and R.[11] The order and relative spacing which Sturtevant determined at that time are essentially the same as those appearing on the recent standard maps of *Drosophila*'s X-chromosome.

All of the early work on *Drosophila* between 1910 and 1915 was carried out in the winter in Morgan's small laboratory at Columbia and during the summers at the Marine Biological Laboratory in Woods Hole, Massachusetts. Although Morgan was considerably older than his coworkers, there was a give-and-take atmosphere in the fly room that precluded the existence of formal barriers and rigid distinctions between teacher and student. Age and past experience (or lack of it) were no obstacles to the effective functioning of the group as a research team. According to A. H. Sturtevant, there was little consideration of priority in new ideas or discoveries, and all were free to criticize one another openly and sometimes vehemently:

> There was a give-and-take atmosphere in the fly room. As each new result or new idea came along, it was discussed freely by the group. The published accounts do not always indicate the sources of ideas. It was often not only impossible to say, but was felt to be unimportant, who first had an idea. A few examples come to mind. The original chromosome map made use of a value represented by the number of recombinations divided by the number of parental types as a measure of distance; it was Muller who suggested the simpler and more convenient percentage the recombinants formed of the whole population. The idea that "crossover reducers" might be due to inversions of sections was first suggested by Morgan, and this does not appear in my published accounts of the hypothesis. I first suggested to Muller that Lethals might be used to give an objective measure of the frequency of mutation. These are isolated examples, but they represent what was going on all the time. I think we came out somewhere near even in this give-and-take, and it certainly accelerated the work.[12]

The group worked as a unit, each carrying out his own experiments but keeping abreast at all times of what the others were doing. According to Sturtevant, the atmosphere of excitement and enthusiasm was a long-standing and persistent one. There was no race against time or against another group for the "right answer" or the "prize." The fly room group was a model for the value of cooperative rather than competitive attitudes in scientific research.

All was not as idyllic in the fly room as Sturtevant's picture implies, however. Muller, for example, felt farther from Morgan than either Sturtevant or Bridges, claiming that Morgan played favorites. As E. A. Carlson has indicated,[13] Muller and his close friend Edgar Altenburg both felt that Morgan would do anything to keep Sturtevant and Bridges involved with the work, but was much less interested in involving either Muller or Altenburg. In later years, Muller was even to feel that Morgan played a role in preventing him from being appointed to a permanent position at Columbia in 1920.[14]

Muller continued to feel that Morgan used his students' ideas "too freely," and often did not give others the credit they deserved. In addition, Muller and Morgan were quite different sorts of people. Morgan came from a low-key, genteel background which reflected itself in his unassuming manner, as well as a strong dislike of personal controversy and of outside intrusions into his scientific work. Muller was aggressive and persistent, involved in radical politics, and likely to be grating to the more aloof Morgan. Whatever the exact reasons, and they are far from clear, the members of the fly room interacted in ways that were not always so harmonious as common myth has it. Like most social situations, the *Drosophila* group had its share of personal likes, dislikes, and conflicts.

Despite some of these internal problems, the group worked more or less harmoniously, especially in the years between 1910 and 1915. This was partly because the younger men got along reasonably well and enjoyed one another's company. In terms of their working relationships, each member of the *Drosophila* group had an area of speciality upon which he concentrated. Muller was particularly adept at building stocks of flies for various gene or chromosomal combinations, Bridges was the cytological expert, and Sturtevant worked a good deal on the mapping problem.

Morgan kept up with all these efforts, suggested further experiments, developed ideas for new breeding schemes, and learned from the others what their newest findings or ideas were. Morgan's mind was far reaching, with an incisive ability to see the theoretical implications of a new set of observations or experimental results. Sometimes called a romantic in his intellectual orientation, Morgan was capable of enormous enthusiasm. He often jumped from one idea to another, and the breadth of his research interest through a long career attests to the many subjects that attracted his attention. He moved easily from embryology to heredity, from evolution to cytology, often seeing the theoretical relationships among them all, yet being able to discern where, at any particular time, the most fruitful questions could be asked.

Of the others, Muller was probably intellectually the most similarly endowed to Morgan in imagination and enthusiasm. Sturtevant appears to have been highly reflective, thoughtful, less intense than Morgan, but with a quick sense of the specific implications of any new theory. Bridges was apparently the most meticulous, able to spend hours at a time at his microscope pouring over anesthetized flies, counting phenotypes, and searching for new mutants. He was also the most highly skilled in manipulative techniques, a feature that led him into the highly important cytological work which formed a basic aspect of the group's approach.

These four were by no means the only ones to frequent the fly room, especially after 1915. As Sturtevant points out, there was a "steady stream of other students using the fly room also—rarely were there fewer than five people working here at any one time."[15]

Included in this group were foreign visitors, postdoctoral fellows, and Morgan's own graduate students, names now legendary in the history of modern genetics: Curt Stern, Theodosius Dobzhansky, Edgar Altenburg, Alexander Weinstein, and Edgar G. Anderson, among many others. Behind-the-scenes support for the rapidly expanding *Drosophila* work was provided by Morgan's close friend and chairman of the department, Edmund Beecher Wilson. Sturtevant and Muller have both paid considerable tribute to Wilson for his help to individual students, as well as to the group as a whole during these early years.[16]

The chromosome theory of heredity did not convince all biologists in the period following 1910. The Danish botanist Wilhelm Johannsen, who had introduced the term "gene" in 1909 in his *Elemente der exakten Erblichkeitslehre* to replace the words *Anlagen* or "factors," wrote in the 1913 edition of that book that the chromosome theory was "a piece of morphological dialectic."[17] In a review of *The Mechanism of Mendelian Heredity* in 1916, the English Mendelian William Bateson wrote:

> ... it is inconceivable that particles of chromatin or of any other substance, however complex, can possess those powers which must be assigned to our factors (i.e., genes).... The supposition that particles of chromatin indistinguishable from each other and indeed almost homogeneous under any known test, can by their material nature confer all the properties of life surpasses the range of even the most convinced materialist.[18]

Somewhat later, in the 1920s and 1930s, Richard Goldschmidt, director of the Kaiser Wilhelm Institut für Biologie in Berlin-Dahlem, became a strong opponent of the chromosome theory.[19] Johannsen and Bateson eventually came to accept the theory, although the latter had continued misgivings. Goldschmidt, however, never accepted the idea and proceeded to develop his own alternative conception in which individual genes did not exist, the chromosome functioning rather as a whole, or what he termed a "continuum."

To a certain group of biologists, of which Bateson and Goldschmidt are good examples, the insistence on drawing a relationship between Mendelian factors and chromosome structure smacked of unnecessary materialism. Although Morgan and his group cannot be called materialists in the strict philosophical sense of the word, they were distinctly not idealists. To the extent that they considered explicitly such philosophical implications of the gene-chromosome theory, however, they showed a marked preference for viewing heredity as having a material basis in chromosome structure and function. They were not willing, because of their view of proper method in science, to let any theory stand as a pure formalism, as an idealized conception having no basis in concrete reality.

More important than the question of philosophical materialism or idealism (Morgan and his group were not prone to much philosophizing on these matters), however, was the issue of experimentalism. To Morgan and his group, the establishment of the chromosome theory of heredity marked a triumph for the application of quantitative, rigorous, and experimental methods to an area of biology that had largely been qualitative and descriptive. Loeb, the dean of mechanistic philosophers, hailed the *Drosophila* work as the most revolutionary development in biology in the twentieth century;[20] even Bateson was moved to add to his basically hostile review of 1916 that "not even the most skeptical of readers can go through the *Drosophila* work unmoved by a sense of admiration for the zeal and penetration with which it has been conducted, and for the great extension of genetic knowledge to which it has led—greater far than has been made in any one line of work since Mendel's own experiments."[21]

The ultimate importance of the chromosome theory of heredity was, to some extent, not its substance, but its method. To Morgan and others, modern biology was beginning to extricate itself from the domination of morphology and was becoming an exact science. Morgan was not a simple reductionist. He did not want to reduce genetics *per se* to physics and chemistry, though he thought that the methods of physics and chemistry were important experimental tools. What he did want was to put biology "on the same footing" as physics and chemistry—i.e., to make it exact, rigorous, and quantitative. This could be accomplished only by insisting on an experimental approach, of which the *Drosophila* work was a prime example.

Influence of the Drosophila Work, 1915–1930

The widespread acceptance of the Mendelian-chromosome theory by the middle or late 1920s was in no small part due to the direct influence of Morgan and his group. That influence came from two sources. Morgan himself played a key role in the dissemination of the new genetics. Through his voluminous writing, his influential position in various societies and on the editorial boards of important journals, and through lecturing in the United States and Europe, the new discoveries of the *Drosophila* group became known. A second source of influence came from the work of many investigators trained in the Morgan lab over an eighteen-year period. Through the graduate students and postdoctoral fellows who worked in the fly room the wonders of the new genetics were transmitted to students all over the world.

One of the most direct means Morgan used to get before the scientific (and to some extent general) public was through his writing. A quick survey of Morgan's bibliography shows that between 1910 and 1915, he published two books dealing solely with the new work in genetics,[22] and fifty journal articles. In addition, he also published a number of articles on embryology and other subjects, including the cytology of chromosomes in parthenogenetic forms. Between 1915 and 1930, he published an additional two books devoted solely to the new genetics[23] and one on the relationship between the new genetics and evolution.[24] He also wrote, as author or coauthor, another fifty articles dealing either with genetics alone or with the relationship between genetics and evolution or embryology. The *Drosophila* group as a whole (Morgan, Muller, Sturtevant, and Bridges) published a laboratory manual on *Drosophila* experimentation for college genetics courses.[25]

Through his influential position in various professional societies and on the editorial boards of several journals, Morgan made the work of his group well known from the very first. Morgan's influence was especially important in the early years before his younger associates had established their own reputations and developed their own channels for quick and easy publication. For example, when the National Academy of Sciences decided to begin producing a *Proceedings* in 1915, Morgan made certain that a report of the group's findings was published in the first issue. As a founder and member of the editorial board of two very important journals, *The Journal of Experimental Zoology* (founded 1904) and *Genetics* (founded 1916), Morgan saw to it that the *Drosophila* work received its fair share of space. Morgan's personal connections also helped in this regard. He was a close friend of J. McKeen Cattell, a long-time professor of psychology at Columbia and throughout the 1920s and 1930s the editor of *Scientific Monthly, The American Naturalist*, and *Science*, in all of which Morgan and his students published frequently.[26]

The influence of Morgan's students, subsequent to leaving the Columbia laboratory, was also an important factor in making known the new work in genetics.[27] It should be pointed out that the influence which graduate students and postdoctoral investigators carried away from Columbia was not only Morgan's; it was also E. B. Wilson's. A basic aspect of the *Drosophila* work, the unification of the breeding tradition and the cytological tradition, was personified in the warm and close relationship between Morgan, the experimentalist and breeder, and Wilson, the cytologist. The interaction of these two men, personally and scientifically, impressed upon students the unique combination of ideas being forged in the Columbia laboratory.

Those who worked in the fly room subsequently took positions in diverse geographic regions of the United States and Europe, thus ensuring that a wide range of new students were brought into contact with the *Drosophila* work.[28] Muller, for example, was probably the most widely traveled. In each place he visited or taught, he helped to form a group of students devoted to the *Drosophila* work. Fernandus Payne (Ph.D. 1909) spent his entire teaching career at Indiana University, where he was dean of the faculty and, in 1945, instrumental in bringing Muller to Bloomington. Charles W. Metz (Ph.D. 1917) spent fifteen years as a researcher at Cold Spring Harbor, and subsequently taught for a time at Johns Hopkins. H. H. Plough (Ph.D. 1917) spent his entire career at Amherst; during the summers, however, he taught in the courses offered at the Marine Biological Laboratory and thus transmitted the *Drosophila* work to many graduate students in this way. Donald E. Lancefield (Ph.D. 1921) taught at Columbia from 1922 until 1938, and at Queens College 1938 to 1963.

Others were strongly influenced by the *Drosophila* work, though they were not technically graduate students of Morgan's. Franz Schrader (Ph.D. with Wilson, 1919), was a cytologist who taught subsequently at Bryn Mawr and in 1930 returned to Columbia to replace Wilson. Charles Zeleny (Ph.D. Chicago, 1904) had worked with Wilson at the Stazione Zoologica in Naples, Italy, and through Wilson met Morgan. Zeleny became a champion of the *Drosophila* work at the University of Illinois from 1911 onward. Leslie C. Dunn wanted to study with Morgan in 1914, but Morgan's facilities were already crowded; so Dunn went to Harvard instead and took his degree with W. E. Castle studying not *Drosophila*, but mammalian genetics. In 1928, however, Dunn returned to Columbia to replace Morgan and carry on the work in genetics there.

Among others who worked in the fly room were three of international importance: Otto L. Mohr from Norway, Theodosius Dobzhansky from the Soviet Union, and Curt Stern from Germany. All came to Morgan as postdoctoral investigators. Mohr and Stern returned to their native countries and carried the new genetics along with them. Dobzhansky remained in the United States and carried the *Drosophila* work in new directions, especially the relationship between genetics and evolution. In Norway, Mohr extended Mendelian-chromosome principles not only to animal and plant breeding, but also to human genetics. After returning to Germany in the early 1930s, Curt Stern carried out work on the Mendelian and chromosomal nature of genetic mosaics.

Another vehicle by which the new genetics was propagated was through the influence of Morgan and his group at the Marine Biological Laboratory at Woods Hole, Massachusetts, where they spent most of their summers. The Marine Biological Laboratory was, from the turn of the century onward (and still is today), the most prominent summertime gathering place in North America for biologists from the United States and Europe. Morgan was one of the original corporation members of the laboratory, and he spent virtually every summer of his active career (from the late 1880s until the early 1940s) in Woods Hole. Prior to 1910, Morgan worked primarily on embryological problems at Woods Hole. Between 1910 and 1925, he regularly packed his entire group, bottles of *Drosophila*, assistants, graduate students, and family, off to Woods Hole for the months of June, July, and August. The group and its research became the center of much attention at the laboratory, and elicited considerable interest and excitement.

It is important for historians not to minimize the importance of such institutional settings in trying to assess the factors influencing the spread of ideas. Woods Hole was not necessarily a better place to raise and breed *Drosophila* than New York (except that it was more pleasant), but it was through the gathering together of many biologists, with differing interests and approaches to problems, that new ideas arose and were transmitted. Morgan frequently paid much tribute to the role of the Marine Biological Laboratory in fostering a spirit of research and collaboration among scientists who were often semi-isolated from one another during the rest of the years.

National Influences and Reception of the Drosophila Work

If we confine our investigation to the Mendelian-chromosome theory as enunciated by the Morgan group in 1915, some interesting patterns emerge regarding the acceptance of the *Drosophila* work in various countries. In science, as in any other aspect of culture, national traditions may contribute strongly to the acceptance or rejection of particular ideas.

Between 1915 and 1930, the Mendelian-chromosome theory gained acceptance not only in the United States, but also Norway,

Sweden, Denmark, the Soviet Union, and to a lesser extent, in Germany. It did not take hold strongly in England or France. What were the factors which contributed to either acceptance or rejection?

First, was the philosophical mood prevalent among scientists in general, and biologists in particular, in the period between 1910 and 1925. Where idealistic philosophy held greater influence, the Mendelian-chromosome theory fared less well; where it held less influence, the theory fared better. The philosophical mood in much of American biology around 1910 or 1915 was decidedly materialistic; that in England was decidedly idealistic.[29] In Germany, there was a strange mixture, but the emphasis was on the materialistic.

Second, and in many ways related to the first factor, is the role which an existing or potential cytological tradition played in acceptance or rejection of the Mendelian-chromosome theory. Where there was a strong interest in cytology, as in the United States, Belgium, Scandanavia, and Germany, the chromosome theory was viewed more favorably. Where there was no cytological tradition, as in France, and open hostility to it, as in England, there was strong skepticism about finding a material basis for Mendel's *Anlagen*. It is obvious how a relationship could well exist between philosophical position and cytology: materialists would tend to emphasize cytology as the study of material components of the cell; idealists would put less emphasis on rigorous scrutiny of the cell's minute structures and take more pride in the formalistic structure of Mendelian (minus the chromosome) theory.

A third factor influencing the acceptance or rejection of the Mendelian-chromosome theory is the personal element. In Germany and England, two powerful and influential men, Goldschmidt and Bateson, dominated most of the work in heredity. Both were strong opponents of the Mendelian-chromosome theory, largely on philosophical grounds (i.e., both were philosophical idealists). Yet their positions of power and prestige within the scientific community of their respective countries allowed them to set directions and priorities for research. Because both were directly skeptical of the Morgan school's whole approach, they did not encourage such research among their own students.

A fourth factor has to do with the calamitous, but unevenly distributed, effects of World War I. Germany and France were most devastated, economically and psychologically, by the war, and they were in no position to take up and devolop a new science—especially one in which they did not already have a good start. England was also caught in the general depression which afflicted all of Europe (and the United States to a lesser degree) in the aftermath of the war. Combined with Bateson's recalcitrance about chromosomes, it was unlikely that English workers would take up a wholly new field at this time.

Aside from the fact that the *Drosophila* work was well underway in the United States by the outbreak of the war, and that the United States did not enter the war until late, it may not be a coincidence that the United States was the country in which development of the Mendelian-chromosome theory was most possible in the 1920s and 1930s. Although the *Drosophila* work was not expensive by modern standards, it did require constant funding from the early 1920s onward—to the tune of about $5,000 per year. While European scholars were having trouble raising funds even to get their research results published, Morgan and his group were enjoying a level of financial support which was indeed unusual for non-medical sciences in those days.

This brief analysis of the early years of the Morgan group illustrates several points about the development of biology in the twentieth century. The first is the importance of group, cooperative efforts in the development of a research project. Although Morgan's name has been the one most frequently associated with the *Drosophila* work, in the early years of the century, the role of his constant collaborators cannot be overemphasized. Historians of science have too often echoed the prevalent view that science is a lonely, creative effort. In the development of scientific research, from theory building to bottle washing, there is an interdependence among co-workers without which research would be impossible. Recognition of this fact is commonplace at one level of study in the history of science. Too often, however, there is a tendency to focus on the contribution of only one member of what is actually a team of contributors, though many "teams" are not as tightly knit as was the *Drosophila* group between 1910 and 1915.

A second generalization which the above analysis emphasizes is the role of philosophical biases in determining the approach of a group of workers to a particular problem. The inclination of all members of Morgan's group toward one or another form of mechanistic materialism was an important factor in their determination to forge a connection between Mendelian and cytological evidence. Without this bias they, like Bateson, Goldschmidt, or Johanssen, might have remained content with Mendelian theory as a pure formalism, and cytology as just a microscopic branch of comparative anatomy. Further, their philosophy predisposed them to emphasize the experimental, over the purely descriptive, side of biology. At a time when biology was under much attack for its speculative, nonrigorous theories, the Morgan group's insistence upon rigorous canons of experimentation brought much acclaim and attention to their work in genetics.

A third generalization relates to the roles of institutions in the development of scientific work. Without Morgan's specific connections to professional societies, his role as a professor at Columbia, and his connection with the Marine Biological Laboratory, dissemination of the *Drosophila* results would have been far less extensive. This, in turn, would have slowed down the number of workers brought into the field, and thus the general rate of progress of the work. More is not always better in terms of advancing research programs, but a critical level of exposure appears to attract the attention of others who can carry the work in a variety of divergent but nonetheless related directions.

A fourth generalization derives from the role which certain intellectual and national traditions play in furthering research work. In the United States, the existence of a strong cytological *and* breeding tradition made it easier for Morgan and his coworkers to forge a link between the two. Lack of one or the other tradition, or both, could have made such a unity of disciplines much more difficult. Furthermore, since intellectual traditions often have certain national associations, it is not surprising that certain lines of research develop more rapidly in some countries than others. The fact that the *Drosophila* work developed rapidly in the United States and certain Scandanavian countries, less rapidly in Germany, and least of all in England and France, can be related to philosophical predispositions and the existence of cytological and breeding traditions (or the lack of them) in those countries.

There is much fruitful research awaiting historians investigating the interrelationships between these various factors as they affect the development of certain fields of science. Historical research along such lines may help to eliminate the artificial dichotomy frequently encountered in the history of science between "internal" and "external" historiography.

1. A substantial part of this paper was originally published as an introduction to the reprint edition of T. H. Morgan, H. J. Muller, A. H. Sturtevant, and C. B. Bridges *The Mechanism of Mendelian Heredity* (New York: Henry Holt, 1915). Another part is modified from the author's *Thomas Hunt Morgan: The Man and His Science* (Princeton: Princeton University Press, 1978).

Allen

2. G. E. Allen "T. H. Morgan and the problem of sex determination, 1903–1910." *Proceedings of the American Philosophical Society*, 1966, 110: 48–57.

3. T. H. Morgan, *Regeneration* (New York: Macmillan, 1901), pp. 277–78, quoted from Edward Manier, "The experimental method in biology," *Synthese* 20 (1969): 185–205; esp. p. 189.

4. Manier, "Experimental method in biology," p. 202.

5. Morgan et al., *Mechanism of Mendelian Heredity*, p. viii.

6. Ibid., pp. vii–ix.

7. For more details, see Garland E. Allen, "The introduction of *Drosophila* into the study of heredity and evolution, 1900–1910," *Isis* 66 (September 1975): 322–33.

8. T. H. Morgan, "Sex limited inheritance in *Drosophila*," *Science* 32 (1910): 120–22.

9. F. A. Janssens, "La théorie de la chiasmatypie," *La Cellule* 25 (1909): 389–411.

10. A. H. Sturtevant, *A History of Genetics* (New York: Harper and Row, 1965), 47.

11. These are symbols for the following characteristics in *Drosophila*: y = yellow body, w = white eye, v = vermilion eye, m = miniature wing, and r = rudimentary wing. All are located on the X chromosome.

12. Sturtevant, *A History of Genetics*, pp. 49–50.

13. Carlson, E. A., "The *Drosophila* group," *Genetics* 79, supplement, part II (June 1975): 15–27; Carlson, E. A., "The *Drosophila* group," *Journal of Histology and Biology* 7 no. 1, (1974): 31–48. See also the exchange of letters in *Genetics* 81, no. 1, (September 1975): 222 a–d (between Carlson and Moti Nissani).

14. Sturtevant to T. M. Sonneborn, 5 May 1967, Sturtevant Papers, California Institute of Technology, box 4.

15. A. H. Sturtevant, "Thomas Hunt Morgan," *Biographical Memoirs, National Academy of Sciences* 33 (1959): 294–95.

16. Personal communications.

17. Wilhelm Johannsen, *Elemente der exakten Erblichkeitslehre*, 2d. ed., (Jena, Gustav Fischer, 1913).

18. William Bateson, "The Mechanism of Mendelian Heredity (a review)," *Science* 44 (1916): 536–43.

19. G. E. Allen, "Richard Goldschmidt's opposition to the Mendelian-chromosome theory," *Folia Mendeliana* 6 (1971): 299–303; esp. pp. 300–301.

20. Jacques Loeb, *The Mechanistic Conception of Life* (Cambridge: Harvard University Press, 1964), reissue of the 1912 book with an introduction by Donald Fleming. Quoted from the introduction, p. xxviii.

21. Bateson, "Mechanism of Mendelian Heredity," quoted from Sturtevant, *A History of Genetics*, p. 49.

22. T. H. Morgan, *Heredity and Sex* (New York: Columbia University Press, 1914); Morgan et al., *Mechanism of Mendelian Heredity*.

23. Morgan, *The Physical Basis of Heredity* (Philadelphia: Lippincott, 1919); idem, *The Theory of the Gene* (New Haven: Yale University Press, 1926).

24. Morgan, *A Critique of the Theory of Evolution* (Princeton: Princeton University Press, 1916), revised later as *Evolution and Genetics* (Princeton: Princeton University Press, 1925).

25. Morgan et al., *Laboratory Directions for an Elementary Course in Genetics* (New York: Henry Holt and Co., 1923).

26. For a discussion of Cattell's life, see Michael Sokal, "The Unpublished Autobiography of James McKeen Cattell," *American Psychologist* 26 (1971): 626–35.

27. Sturtevant has drawn up a "scientific genealogy" for the Morgan group, [from *A History of Genetics* (New York: Harper and Row, 1965); pp. 140, 142]. This genealogy indicates to some extent the pathway of influences affecting Morgan and through which he affected others.

28. Morgan did not, of course, seek to place students in jobs based upon achieving a wide geographic distribution. Many of the "fly room" students took jobs where they were available, which also happened to take them far and wide.

29. William Colman, "Bateson and chromosomes: conservative thought in science," *Centaurus* 15, nos. 3 and 4, (1970): 228–314.

AMERICAN FOUNDATIONS AS PATRONS OF SCIENCE: THE COMMITMENT TO INDIVIDUAL RESEARCH

STANLEY COBEN
University of California (Los Angeles)

THE NEARLY SIMULTANEOUS DECISION, SHORTLY AFTER THE END OF the First World War, by a number of philanthropic foundations to support on a large scale the research of individual scholars—which in practice meant primarily natural science—was one of the most significant events in the history of learning in the United States. This decision coincided with several other events that magnified its importance and hastened developments already well under way.[1]

The proportion of the population of the United States between eighteen and twenty-one years of age enrolled in institutions of higher education increased from 4.0 percent in 1900 to 12.42 percent in 1930. Enrollment of graduate students rose from a total of 6,000 in 1900 to 47,000 in 1930.[2] American universities in the 1920s responded to this sudden acceleration of the long-term increase in graduate and undergraduate enrollments by appointing new teachers at roughly the same rate as the increase in the number of students.[3] This upsurge in enrollments, especially in the sciences, occurred in part as a response to increased professional opportunities for scientists and highly trained scientific technologists. The largest American industrial corporations expanded enormously the size of their research staffs in this period, and the number of industrial research laboratories grew rapidly to about three hundred in 1920 and to more than a thousand in 1927.[4]

The enlargement of teaching staffs in universities and colleges supplemented those opportunities. A trend towards professionalization in all sectors of knowledge made academic degrees nearly essential for employment in such positions. The ambitions of parents in the middle class, who believed that higher education for their children would improve the younger generation's lot in life, also contributed to the growth of the student body, and a larger proportion of these students sought advanced training.[5] The development of interesting new ideas, novel techniques, and new or drastically improved equipment also attracted students and patrons to scientific research. The emergence of certain scientists who proved adept—especially during the war—at dealing with government and foundation officials and with one another, was important in stimulating and guiding the flow of funds from the foundations.[6]

Fellowships

The freeing of hundreds of scientists from the routines of undergraduate teaching and academic administration during the 1920s alone was much more important to the development of science in the United States than the equipment provided by foundations, even though the latter included large telescopes and—beginning in the early 1930s—cyclotrons that were unavailable elsewhere. Each year, starting in the 1920s, scores of young scientists, who might otherwise have been overwhelmed by the heavy tasks of teaching the introductory courses then characteristic of early academic careers, received foundation grants. These grants enabled the scientists to master their own fields and to learn the relevant concepts and theories from other closely related fields of science and mathematics; they were then able to use this knowledge to develop their own ideas.

Fellowships that freed young scientists from other obligations for one, two, and sometimes three or four years immediately after completion of their doctoral dissertations had an especially salutary effect, and not only because these fellowships encouraged original thought at especially fertile periods in the scientists' careers. J. Robert Oppenheimer commented on the stimulating effects on their teaching of the work of young American theoretical physicists like himself in quantum mechanics. This comment was true also of young American astronomers, astrophysicists, geneticists, chemists, geologists, and mathematicians: "Some of the excitement and wonder of the discoverer was in their teaching."[7] This exhilarating process of exploration and discovery within a rapidly developing science was an experience seldom available to the ordinarily overburdened junior teachers of undergraduate students in the sciences.

As the universities enlarged the size of their staffs, clusters of scientists thoroughly educated in the most advanced theories and techniques—which many of them had helped to develop—collected at the leading institutions. Consequently, thousands of engineering students, as well as students in scientific disciplines, received a quality of education available only to a carefully chosen few in other countries. A large proportion of these university students received degrees and then entered industrial research, joining in that work a significant proportion of the first generation of postdoctoral students who had benefited from the largesse of the foundations. For example, about 20 percent of National Research Council Fellows during the 1920s were employed by industrial firms within a few years after receiving their awards.

American business firms increasingly became able to manufacture superior products of a type made possible by basic scientific knowledge. Such technological innovations included condenser microphones for telephones, circuit breakers, push-button elevators, FM radios, teletype machines, neoprene synthetic rubber, nylon, and lead-ethyl gasoline—all developed in the 1920s. Other American laboratory discoveries during the period, such as radar, the rocket engine, and electronic television, were not yet developed commercially or on a large scale. In fact, although benefits were obvious immediately, the greatest effects of this improvement in the training of industrial research workers occurred during and after the Second World War.[8]

Moreover, even before the emigration from National Socialist Germany, scientists from all over the world, particularly from Europe, flowed to the American universities and to other scientific institutions and business firms. They were attracted by the superior facilities (which foundations had helped to construct), by concentrations of specialists in their particular fields, fellowships from American foundations, and higher salaries in both business and the universities. Some of the best of these European scientists, particularly ambitious young men whose advancement in their academic careers was obstructed by the small number of professorships in European universities and the fact that below that level it was difficult to have a successful academic career, accepted offers of permanent employment in the United States.[9]

The wealth of the foundations, no matter how carefully or widely disbursed in fields prepared to make good use of it, could not have stimulated the extraordinary development of science that began after the First World War without the prior presence of many talented and experienced scientists; numerous well-equipped laboratories; professional associations and scientific journals; a large network of universities, including about a dozen with demanding intellectual standards; and a large group of very wealthy businessmen who believed that fundamental scientific research would further technological progress.[10]

The businessmen and politicians who had created the large foundations did not share all the values of academic intellectuals, and they did not invariably appreciate academic ways of doing things. Nevertheless, the promotion of research and training by these foundations served a crucial function for the development of science in the United States.

The Great Foundations

Before the 1920s, the private philanthropic foundations supported little research by individuals in universities. With the exception of a handful of grants to already eminent scientists, their funds supported large-scale projects with practical applications consistent with their various general policies, or the funds were contributed to university endowments to be used any way the chosen institution preferred. The shift began in 1919 when the National Research Council introduced a program for the award of postdoctoral fellowships on the basis of a gift of $500,000 from the Rockefeller Foundation. Foundation officials had been considering creation of an institute for research in the physical sciences, similar to the Rockefeller Institute for Medical Research (now the Rockefeller University). Leaders of the major scientific professional associations persuaded those officials to support a program that could be controlled by academic scientists. George Ellery Hale, the astrophysicist, Robert A. Millikan, the physicist, and other influential scientists used the argument that the success of research in industrial products depended on the acquisition of new knowledge attained through pure scientific research. American industry, they claimed, might be endangered by postwar competition unless aid were given to the basic science on which new products depended.

It was also funds provided by the Rockefeller Foundation that enabled the International Education Board in 1923 to begin awarding fellowships similar in most respects to those offered by the National Research Council. Money from the same source allowed the General Education Board, beginning in 1923, to subsidize research with multimillion dollar grants to particular scientific departments of universities. The Rockefeller Foundation itself increased the research funds it provided for projects too large or unusual for its subsidiary foundations. Large amounts of money from the Rockefeller family fortune were donated by the Laura Spelman Rockefeller Memorial Fund to individuals and to intermediary organizations to aid various types of research; this branch of the Rockefeller's system of philanthrophy had been established in 1918 by John D. Rockefeller, Sr., in memory of his wife. During the next ten years, before its amalgamation with the Rockefeller Foundation in 1928, the Laura Spelman Fund received about $74 million for distribution. Its gifts helped both the Social Science Research Council and the American Council of Learned Societies to establish, during the 1920s, programs of grants to individual scholars for study and research in the social sciences and the humanities.

The John Simon Guggenheim Foundation, commencing in 1925, supported the work of talented individuals in virtually every field of scholarly, scientific, and artistic activity. In 1926, thirty-eight individuals received grants from the Guggenheim Foundation and the number of fellowships granted by the foundation rose annually. Most of the recipients were thus given a year of freedom to work at whatever intellectual task they chose. Despite publicity emanating from the foundation suggesting an emphasis on aid to the arts, humanities, and social sciences, from the start of the Guggenheim Foundation program, physical and biological scientists annually received between one-third and one-half of the foundation's fellowships.[11]

A number of other foundations, inspired by these programs, or perhaps moved by the same considerations, began aid to research on a smaller scale during the 1920s. These included the Charles A. Coffin Foundation and the American Institute of Architects. Large philanthropic operations already in existence, such as the Carnegie Corporation, the Carnegie Institution of Washington, and the Russell Sage Foundation, redirected their policies somewhat towards the support of research by individuals.[12]

Three Maecenases: Consensus on the Value of Science and Scholarship

The foundation officials most responsible—among many who shared responsibility—for the new types of policy originating at about the same time within a variety of philanthropic organizations were Wickliffe Rose, Beardsley Ruml, and Henry Allen Moe. Each possessed great administrative talents; in their outlook on learning and society, they resembled the leading scholars with whom they frequently dealt. They were not distinguished scholars themselves but had begun their careers as academics of considerable promise. As a result of that experience, they shared the outlook of their contemporaries among the leading academic intellectuals.[13]

Rose was a professor of philosophy and occasionally of history at Peabody College, Nashville, Tennessee, for fifteen years, until his uncle and godfather, Wallace Buttrick, director of the General Education Board, persuaded him to leave the academic life he enjoyed for broader responsibilities as head of a large medical project for the Rockefeller Foundation.[14] Rose gradually developed a system for combating specific diseases in various parts of the world, the success of which affected the approach to intellectual matters of many Rockefeller Foundation officials and his own subsequent projects as well. The policy Rose found most successful was to identify research centers where the understanding of a disease and its treatment was most advanced, then to enlarge those centers and support their activities with additional financial contributions. Physicians and paramedical workers from areas especially afflicted with that disease were then paid to come to these hospitals and medical schools to participate in research and learn about the most advanced means of treatment. When these medical specialists returned to their homes, they were encouraged further with funds for medicine and other assistance and were expected to train others. In this fashion, diseases that had decimated populations for centuries were brought under control and sometimes practically eliminated.

As Rose advanced through the hierarchy of administrators within the Rockefeller foundations, he decided that the most rapid way to develop any field of knowledge was to strengthen and enlarge the foremost centers in that field, and then to arrange for trained persons and the most advanced ideas to radiate from these. While directing medical programs, he concluded also that the future welfare of mankind depended on advances in the physical and biological sciences. "This is an age of science," he recorded in his private notebook early in the 1920s. "All important activities from the breeding of bees to the administration of an empire call for an understanding of the spirit and techniques of modern science."[15]

Once John D. Rockefeller and his closest advisers became aware of Rose's gifts as an administrator, he was placed in a position to carry out his schemes in what he considered the critical fields of science. These included certain special fields within physics, astrophysics, chemistry, and biology; they also included mathematics, which he believed supplied concepts of critical importance to each of the other fields. He received a great deal of freedom to initiate policy and was placed in charge of the General Education Board as well as a supplementary foundation, the International Education Board, which he himself conceived to aid foreign scientific centers and students being trained in Europe.

In the United States, Rose halted the General Education Board's long-standing policy of contributing to the general endowments of universities, which had reached a total of $60 million when he assumed presidency of the board. He inaugurated instead a program to subsidize research within the few strongest university departments of science, in the expectation that "the high standards of a strong institution spread across oceans." Money from the foundations under Rose's control also flowed to Europe's major scientific institutions, in accordance with recommendations made after careful study by experienced American scientists (especially Augustus Trowbridge, who resigned as professor of physics at Princeton University to serve as Rose's chief investigator and adviser) and a subsequent five-month tour of European scientific institutions by Rose himself. A system of fellowships to promising European and American scientists brought them to the centers expanded with the aid of Rose's donations, where they were welcomed. National Research Council fellows also found facilities and professors available. Funds had been disbursed only after assurances were received that aid to a center would enable it to accommodate these additional advanced students.

Rose acted in the conviction that the fields he assisted not only stood at crucial points in their own development, and that the assistance would be of decisive importance, but that the example of their achievements would inspire work in other areas of scholarship. He especially wished to encourage use of mathematical and quantitative techniques. Scientific leaders at the California Institute of Technology, the University of Chicago, and Princeton University obtained millions of dollars by showing the close relationship between strong mathematics departments and the science departments of their respective institutions. The University of Göttingen, traditionally renowned both for the ability of its mathematicians and their cooperation with other scientists at the institution, probably received less money than it might have because scientists there failed to persuade Trowbridge and Rose that this tradition would be maintained. Rose believed that all sciences would benefit from the study of accomplishments resulting from use of mathematical and quantitative techniques, and he specifically stated that he expected the social sciences to learn a salutary lesson from achievements of physical and biological scientists that depended on mathematical concepts.[16]

Rose devised his programs largely in anticipation of their long-term effects, and therefore their full results cannot be calculated easily. Almost everyone concerned, however, agreed that their impact was considerable. The experience of Princeton University is illustrative. The leading physicists and mathematicians there decided in the autumn of 1929 to use part of the income from $1.5 million in endowment funds given to the two departments by the General Education Board, and by donors who provided matching funds required under terms of the grant from Rose, to bring to the University two of the most brilliant young mathematician-physicists in Europe: Eugene P. Wigner and John von Neumann. At that time, both were research and teaching assistants at the University of Berlin. When they arrived in the United States, Wigner, then twenty-seven years of age, and Neumann, who was twenty-six, were already recognized as leading contributors to the development of quantum mechanics. Within a few years, the graduate students whose training Wigner directed included John Bardeen, Frederick Seitz, and Conyers Herring, who were among the founders of modern solid state physics. Bardeen shared one Nobel Prize for his part in developing the transistor, and subsequently won another. Von Neumann, best known to theoretical physicists of his generation as the author of several papers and a monumental book (which remain among the most comprehensive mathematical explanations of quantum mechanics) gained a wider reputation for his part in developing the electronic computer.[17]

Beardsley Ruml received his doctorate from the University of Chicago in 1917, after conducting research in the new field of psychological testing. His talents attracted attention when he served the federal government during the First World War as codirector of the Division of Trade Tests. Philanthropic organizations took further note of his performance as assistant to the president of the Carnegie Corporation in 1921–22. In 1921, Raymond B. Fosdick, trustee and high official of the Rockefeller Foundation and later its president, asked his associate Abraham Flexner, head of the foundation's vast scheme of aid to medical education, to recommend an able young man to help manage the new programs on which the General Education Board planned to embark. Flexner sought advice from President James Angell of Yale, and the two agreed that Ruml stood out as the ideal candidate. Ruml, then twenty-seven years old and apparently on the verge of entering a business career with the prospect of high financial rewards, was persuaded that the opportunities that philanthropic work presented for service to mankind outweighed the smaller income he would earn, and he accepted Fosdick's offer.

Within two years, Ruml was made director of the Laura Spelman Rockefeller Memorial Fund and was given responsibility for aid to the social sciences and humanities. Ruml, helped by advice from his friend and former teacher Charles E. Merriam, chairman of the political science department of the University of Chicago, attempted to further the development of the social sciences.

When Ruml arrived on the scene, the leading figures of the disciplines he sought to aid were suffering "withdrawal" pains as wartime activities such as various forms of economic control, testing of recruits, and propaganda were concluded. Assured of his support, the recently organized and financially destitute Social Science Research Council (SSRC) called a meeting of leading scholars in each field represented within it. This meeting was held at Dartmouth College in the summer of 1925; Ruml arranged for the Laura Spelman Memorial Fund to pay all expenses. A series of annual conferences followed, attended by the two delegates to the council already selected by each professional association, by other influential scholars, and by representatives from several foundations and from federal, state, and municipal research bureaus. Conversations at these meetings produced not only financial results but agreement among academic representatives on research programs and methods of distributing the steadily increasing funds available. Throughout the 1920s, Ruml continued to act as the chief and most certain financial patron of the SSRC and as an instigator of more effective organization. After the Rockefeller Foundation absorbed the Memorial Fund, Ruml accepted a post as dean of the division of social sciences at the University of Chicago.[18]

Like Ruml, Henry Allen Moe appeared headed for almost certain business success, when the intellectual values he had assimilated at a great university were appealed to on behalf of service to a philanthropic organization. A student of mathematics as an undergraduate, he studied law at Oxford University. While a Rhodes Scholar, he began a life-long interest in old English and Elizabethan literature and, throughout his career as a foundation official, he continued to publish scholarly articles on these subjects. Moe was appointed lecturer in law at Brasenose and Oriel Colleges. In

1927–29, he was a lecturer at the Columbia University School of Law.

When Moe returned to the United States in 1923, at the age of twenty-eight, he received an offer to join a New York City law firm. As he prepared to accept, former U.S. Sen. Simon Guggenheim of Colorado—who had gained a fortune as head of the Guggenheim family mining interests in the western United States—requested his assistance in drawing up plans for a projected major foundation. Moe postponed his entry upon a legal career to participate in this evidently irresistible project. Moe's partner in the formation of the new foundation, Frank Aydelotte, a professor of English at several universities before his appointment as president of Swarthmore College in 1921, joined him in months of intensive study on the practices of other foundations and the amounts of money available from all sources for fellowships. They interviewed scores of scholars, artists, college and university presidents, physicians, and businessmen, seeking suggestions about fields within the arts and intellectual disciplines that the Guggenheims' money might affect most beneficially.

Finally, Moe drew up plans for the foundation, which was named after the senator's son, John Simon Guggenheim, who had recently died of pneumonia just before entering Harvard. During a luncheon meeting with Simon Guggenheim to discuss these plans, Moe was offered the opportunity to direct the foundation and to serve as its secretary. He claims that he, like Ruml, weighed the opportunity to affect the course of American intellectual and artistic life against the certainty of a larger income in business. The decision required, he said, half a second of serious consideration for him to accept the offer.

Moe described the donor of the funds he disbursed in terms similar to those used about John Rockefeller by Rose and Ruml: "always interested . . . , never obtrusive." These successful industrialists having learned, as their enterprises grew, to delegate responsibilities to carefully chosen experts, applied the same principles as did Andrew Carnegie—another important instance of the class—to the foundations they established. Moe created a system of unpaid referees, juries, selection committees, and an advisory board. They read applications and letters of recommendation, assessed works submitted by artist-applicants, and then at each level selected from the host of applicants those most likely to use a grant for creative accomplishment.

Moe almost always accepted the recommendations made by the committees of specialists he had appointed; among the information considered by these committees were analyses of the more likely candidates by Moe himself. Moe's criteria for these evaluations may be inferred from some of his informal statements. He quoted with approval the declaration by one of his fellows that the ideas underlying what appeared to the public as great scientific achievements came from the minds "of a handful of men, scattered over a continent and a century—men who were willful, uncompromising, quarrelsome, arrogant, and creative."

Moe attempted to support scholars who could be described as "uncompromising," "creative," and "really good." His attitude can be discerned in a case described by a historian of the foundation. A scientist whose name was entirely unfamiliar to Moe and his staff requested a fellowship for what seemed unusual research in immunology. Preliminary inquiries revealed that the applicant had only one arm. "This gave me pause," Moe confessed. "Here was a man who did his research in a laboratory where a fellow certainly needs two arms, and could use three. Moreover one of the referees submitted by the applicant had written us that his ideas were old hat." Just to be sure, Moe requested a few of the scientist's papers and brought them to the Rockefeller Institute's experts in immunology. They reported that the man's research was greatly in advance of anything else then in progress. Apparently the selection committees accepted this opinion, for the applicant received a fellowship. About twenty years later, the research he had described in his application earned the scientist a Nobel Prize. "That's how it goes," Moe commented revealingly. "We hooked that one and he did us proud, but I get nervous when I think how close we came to losing him. We made mistakes."[19]

The generation of foundation officials to which Rose, Ruml, and Moe belonged, and others with similar intellectual training, values, and objectives—such as Flexner, Trowbridge, and Aydelotte—carried out programs never imagined by those whose fortunes provided the necessary funds, nor in most cases by the original advisers of these businessmen about how their philanthropy should be exercized. Rockefeller, for example, had started his foundations after persistent urging by Frederick T. Gates, a former Baptist clergyman, who was serving as head of the American Baptist Education Society when Rockefeller asked him to join his personal staff in 1892. Gates envisaged the Rockefeller foundations as contributing to the spread of Christian ethics and "Christian civilization" throughout the world and to the development of "civic virtue," particularly in the United States. Building centers for study and research in theoretical atomic physics and sponsorship of academic research in the social sciences critical of American institutions and traditions did not seem to Gates consistent with God's wishes for use of Rockefeller's wealth.

In a memorandum about the General Education Board distributed in January 1927, Gates complained: "The business of the Board has passed into the hands of salaried officers, and in that business the members of the Board meeting statedly only three times a year, have of late years taken little part. Indeed they know little about it." Referring explicitly to Rose's board program in which the General Education Board and International Education Board supplemented each other, Gates protested that "International boards are obviously undesirable. Each board should be fully made up of men specially fitted for the work of that board." He acknowledged that "I have wholly disapproved very much of the work of the present chief officers, as of course they know full well." Gates asserted, however, that his chief concern was with the future: "The administration of the present officers has disclosed clearly," he proclaimed, "the future possibilities of evil [the last word was crossed out and 'bureaucracy' substituted] to which those defects might lead."[20]

Fortunately for Gates, and possibly Rockefeller's peace of mind, they lacked the time and information needed to evaluate the tendencies Beardsley Ruml was indirectly encouraging through the support of research formally assisted by the Social Science Research Council and its subsidiaries. At that moment, for example, the council was supporting research by Robert Redfield, Melville Herskovits, Otto Klineberg, and Ruth Bunzel and was soon to act favorably on applications from Harold Lasswell and Margaret Mead—whose research disclosed aspects of society that the traditional moral outlook did not provide for.[21]

What Gates failed to articulate and perhaps understand, was that the basic source of his discomfort did not lie in the foundations' governing bodies or in a few unfortunate appointments. The best administrators who could be drawn into philanthropic activity by the 1920s had assimilated certain intellectual values during their advanced training at universities such as Oxford and the University of Chicago. The policies of the foundations were deeply affected by the development of an intellectual and moral culture in the United States, whose members and values had thoroughly infiltrated the faculties of the major universities and the officers of the foundations by the early 1920s. Consequently, Ruml drew up plans most comfortably in consultation with his friend Charles Merriam, head of the SSRC, and kept in contact with Charles Merriam's brother John, a prominent geologist and paleontologist appointed in 1920 to direct the Carnegie Institution. Rose depended heavily on detailed advice about European scientific institutions and American students there, given him by the eminent physicist Trowbridge. Rose's judgment also was valued by most of the widely respected scientists at major universities. His attempts to aid thoretical physics, and to do so in part by subsidizing mathematics, were suggested first as if to a fellow scientist by Oswald Veblen, one of the foremost mathematicians of his time. Moe too clearly enjoyed cooperation with creative artists and scholars in choosing applicants with exceptionally fertile minds for the financial aid he could offer.[22]

Fellowships and the Advancement of Science

Close to a majority of John Simon Guggenheim Foundation fellowships were awarded to scientists, particularly those working in new fields, or with new ideas, who were not supported ordinarily by other foundations. In the area of physical chemistry for example, which developed in its modern form during the 1920s, nearly every pioneer in the field was aided by Moe's decision to award a fellowship that gave time for study, research, thought, and writing. First if not foremost among them came Professor Linus C. Pauling. The year Pauling received a doctorate from the California Institute of Technology in 1925, at the age of twenty-four, he requested a Guggenheim Fellowship for the following year, stating in his application: "The subject of mathematical physics has been in existence for more than 100 years. A system of mathematical chemistry . . . is only on the point of being created." The new quantum mechanics, he declared, would increase chemists' ability to identify the position of electrons in an atom and to explain "how this qualifies the atom to form molecules and to enter into chemical compounds." The nature of chemical bonding thus could be understood with greater precision.[23]

Although two German physical chemists, Walter Heitler and Fritz London, wrote the first important paper leading towards a new theory of molecular structure in 1927, a group of young American scientists advanced the field rapidly late in the 1920s. The most renowned among them included Professor Robert S. Mullikan of the University of Chicago—who like Pauling won a Nobel Prize for this work—John H. Van Vleck, John C. Slater, and the somewhat older chemist, Richard C. Tolman. Each of these scientists received a fellowship of the Guggenheim Foundation enabling him to continue his research shortly after receiving his doctorate—Van Vleck held his in 1930. All four had studied quantum mechanics while it was being developed at leading centers of theoretical physics, Pauling and Mullikan with the aid of grants by the National Research Council. Moe and his committees were deliberate in promoting the new physical chemistry; similar instances could be cited from many other fields.

The most important item purchased with the foundations' money was time for undisturbed research, study, and intellectual speculation. Scholars who discussed the subject agreed almost unanimously with the view expressed by Oswald Veblen in a conversation with Rose on 1 November 1923: "Almost all of the men now engaged in productive scientific work in the United States have to devote most of their time to routine teaching. Most of them are indeed so burdened with teaching responsibility that but little time is left for research." An example Veblen chose illustrated another burden inhibiting scientific research and thought, namely, increasing administrative duties in universities growing rapidly in size and increasingly committed to departmental autonomy with all the benefits and burdens this entailed. Veblen pointed to what he considered the tragic story of one of Harvard's most brilliant mathematicians, who was appointed to the chairmanship of his department. "Veblen regards him as a genius as a mathematician," Rose noted. "He must do so much routine teaching, however, that he has relatively little energy left for the work he ought to be doing."[24]

Although established scientists and scholars might be burdened with teaching and administrative work, those who suffered most from the division of labor in American universities were the imaginative, ambitious younger teachers who performed the most time-consuming and uninspiring duties. After years devoted to teaching the rudiments of their subjects to undergraduates, and assessing and annotating essays and examinations dealing with elementary problems, few of these teachers retained much energy or enthusiasm for original thought in the advanced areas of their field, even if they continued to follow the literature that explored those areas.[25]

The demands of routine teaching and administration, so deleterious to creative activity, continued throughout the 1920s for the majority of college and university teachers. But the number of those who escaped, and thus were able to achieve significant work of science or scholarship, multiplied during the decade. In this achievement and its consequences, the fellowships grants for one year (sometimes even two or three), with their concomitant release from teaching and administration and the opportunity for travel, undistracted research, and reflection, played a very great part. This remarkable institution for the support of learning was largely the work of a small number of persons who shared the values and aspirations of the best academic minds of their time, and who could use the funds at their disposal as officials of philanthropic foundations.

1. A case history of the concurrence during the 1920s of these conditions in one important scientific field can be found in my essay on the rapid development of theoretical atomic physics in the United States: Stanley Coben, "The scientific establishment and the transmission of quantum mechanics to the United States, 1919–32," *American Historical Review* LXXVI, 2 (April 1971): pp. 442–66.

2. U.S. Department of Commerce, Bureau of the Census, *Historical Statistics of the United States, Colonial Times to 1957* (Washington, D.C: U.S. Government Printing Office, 1960), pp. 210–11.

3. U.S. Department of the Interior, Office of Education, *Biennial Survey of Education, 1926–1928* (Washington, D.C.: U.S. Government Printing Office, 1928), p. 698; U.S. Department of Commerce, Bureau of the Census, *Biennial Survey*, pp. 207–11; Paul Forman, John L. Heilbron, and Spencer R. Weart, "Physics circa 1900: Personnel, Funding, and Productivity of the Academic Establishment," *Historical Studies in the Physical Sciences*, vol. 5 (Princeton: Princeton University Press, 1975), esp. tables II, A.2 and A.7, on pp. 8, 13, 34–35; George J. Stigler, *Employment and Compensation in Higher Education*, National Bureau of Economic Research, Occasional Paper no. 33 (New York: National Bureau of Economic Research, 1950), p. 210.

4. Joseph Ben-David, *Fundamental Research and the Universities* (Paris: Organization for Economic Cooperation and Development, 1968), esp. pp. 29–34; Ronald C. Tobey, *The American Ideology of National Science, 1919–1930* (Pittsburgh: University of Pittsburgh Press, 1971), pp. 6–7; Spencer R. Weart, "The Physics Business in America, 1919–1940: A Statistical Reconnaissance," in this volume; Kendall Birr, "Industrial Research Laboratories," ibid.; W. H. G. Armytage, *The Rise of the Technocrats: A Social History* (London: Routledge and Kegan Paul, 1965), p. 246.

5. For representative evidence of this change in attitudes, see Robert and Helen Lynd, *Middletown: A Study in Modern American Culture* (New York: Harcourt, Brace, 1929), pp. 182–87. The Lynds summarized the replies of informants among ambitious elements of the working class. "Over and over again one sees both parents working to keep their children in college. 'I don't know how we're going to get the children through college, but we're *going* to. A boy without an education today just ain't *anywhere*!' was the emphatic assertion of one father" (p. 187).

6. For discussions of this factor, see Barry D. Karl, "The Power of Intellect and the Politics of Ideas," *Daedalus* XCVII, 3 (Summer 1968), pp. 1002–35; Daniel J. Kevles, "George Ellery Hale, the First World War, and the Advancement of Science in America," *Isis* LXIX, 4 (Winter 1968), pp. 427–37; S. Coben, "Quantum Mechanics," pp. 446–47.

7. Robert J. Oppenheimer, *Science and the Common Understanding* (New York: Simon and Schuster, 1953), p. 36.

8. Arthur D. Little, *The Relation of Research to Industrial Development: An Address before the Canadian Manufacturers' Association, Toronto* (Boston: A. D. Little, 1918). Again, see the case study in S. Coben, "Quantum Mechanics," and the broader analyses in J. Ben-David, *Fundamental Research and Universities*, and S. R. Weart, "The Physics Business in America." One of the most important discussions of the relationship that developed between fundamental research and industrial technology is United States: Stanley Goldberg, "The Concept of Basic Research and the Rise of the Industrial Laboratory in America," address delivered at Harvard University, May 1964.

9. Charles Weiner, "A New Site for the Seminar: The Refugees and American Physics in the Thirties," in *The Intellectual Migration: Europe and America, 1930–1960*, ed. Donald Fleming and Bernard Bailyn (Cambridge, Mass.: Harvard University Press, 1969), esp. pp. 32, 196–200; S. Coben, "Quantum Mechanics," pp. 452–53, 460–65.

10. See Nathan Reingold, "Definitions and Speculations: The Professionalization of Science in the Nineteenth Century," in *The Pursuit of Knowledge in the Early American Republic*, ed. Alexandra Oleson and Sanborn Brown (Baltimore: Johns Hopkins University Press, 1976), pp. 33–69.

11. John Simon Guggenheim Memorial Foundation, *Directory of Fellows 1925–1974* (New York: Lind Brothers, 1975). During the period from 1925 to 1932, physical and biological scientists and mathematicians received slightly over 30 percent of the foundation's awards. Furthermore, some individuals whose work overlapped these fields to some degree—such as geographers and psychiatrists—were placed in other categories. According to Milton Lomask's semi-official history of the Guggenheim Foundation, a survey of past grants made in 1953 showed that "almost one half of the John Simon's expenditures for fellowships went to the support of scientific research"; Milton Lomask, *Seed Money: The Guggenheim Story* (New York: Farrar, Straus, 1964), p. 256. Foundation authorities, Lomask reported, had considered complaints that large sums of money were available from other sources for scientific research and had decided to continue giving almost one-half of their grants to scientists because the John Simon Guggenheim Foundation was one of the few organizations that supported individual scientists "under conditions that leave them free to pursue objectives they themselves have set up": ibid., pp. 256–57.

12. Money transferred from the Rockefeller Foundation, which also subsidized research, or from the Rockefeller family fortune, supported the patronage of individual research by the National Research Council, Laura Spelman Rockefeller Memorial Fund, Social Science Research Council, American Council of Learned Societies, General Education Board, and International Education Board. Therefore, the best single source for following these changes in the history of American philanthrophy is the correspondence and reports in the Rockefeller Foundation Archives, Rockefeller Archive Center, Hillcrest, Pocantico Hills, North Tarrytown, New York (hereafter referred to as Rockefeller Archive). Every general history of American philanthropy suffers from the previous inaccessibility of these records. The John Simon Guggenheim Foundation's records remain closed, although individual files sometimes are made available upon request. More detailed accounts of the shifts mentioned in this account can be found in S. Coben, "Quantum Mechanics"; Barry D. Karl, *Charles E. Merriam and the Study of Politics* (Chicago: University of Chicago Press, 1974), pp. 118–39, 153, 182–84, 201–14; Raymond B. Fosdick, *The Story of the Rockefeller Foundation* (New York: Harper, 1952); M. Lomask, *Seed Money*; and John Higham, "The Schism in American Scholarship," *American Historical Review* LXXII, 1 (October 1966): 1–21.

13. On the development of the intellectuals' subculture in the United States, see Henry F. May, *The End of American Innocence: A Study of the First Years of Our Time, 1912–1917* (New York: Knopf, 1959); Richard Hofstadter, *Anti-Intellectualism in American Life* (New York: Knopf, 1962); Laurence R. Veysey, *The Emergence of the American University* (Chicago: University of Chicago Press, 1965); and Alfred Kazin, *On Native Grounds: An Interpretation of Modern American Prose Literature* (New York: Harcourt, Brace, 1942).

14. Buttrick to Rose, 8 May 1909; 11 June 1910; 2 March 1912; Rose to Buttrick, 31 October 1912; E. C. Sage to Rose, February 1921; all in General Education Board Collection, sub-series 2, Wickliffe Rose papers, Rockefeller Archive (hereafter referred to as Rose papers).

15. Rose's comments about the importance of science were presented to a meeting of the International Education Board; see "Excerpt from Docket for Meeting of April 30, 1923," p. 13, International Education Board Reports, Series 1, Box 5; also Raymond B. Fosdick, *Adventure in Giving: The Story of the General Education Board* (New York: Harper and Row, 1962), p. 229.

16. The details of Rose's policies can be traced in the general records of the General Education Board, and of the International Education Board for the period during which he led these organizations, as well as the Rose papers and Rockefeller Archive. For summaries, see S. Coben, "Quantum Mechanics," pp. 448–50; R. B. Fosdick, *Adventure in Giving*; George W. Gray, *Education on an International Scale: A History of the International Education Board, 1923–1938* (New York: Harcourt, Brace, 1941).

17. Veblen to von Neumann, 26 November 1929; Hibben to Wigner, 10 December 1929; von Neumann to Veblen, 13 November, 16 December 1929, 2 January 1930, Box 9, papers of Oswald Veblen, Library of Congress, hereafter referred to as Veblen papers. Interviews with Eugene P. Wigner by the author, 10 October, 30 November 1966; Wigner to the author, 13 January 1972; interview with Wigner, 21 November 1963, Archive for the History of Quantum Physics (hereafter AHQP), American Philosophical Society Library, Philadelphia, copies at the University of California, Berkeley, and the Institute for Theoretical Physics, Copenhagen. Max Jammer, *The Conceptual Development of Quantum Mechanics* (New York: McGraw-Hill, 1966), pp. 343, 367, 376. Laura Fermi referred to John von Neumann as "the only Hungarian of his generation to be recognized as a genius by his countrymen (at least by all with whom I spoke)," and to Wigner as "generally considered a colossus in theoretical physics, the living physicist whose over-all performance has been the greatest . . .": Laura Fermi, *Illustrious Immigrants: The Intellectual Migration from Europe 1930–1941* (Chicago: University of Chicago Press, 1968), p. 9. Wigner also is a Nobel laureate in physics. Herring eventually was employed by the Bell Telephone Laboratories; Seitz is now president of Rockefeller University (formerly the Rockefeller Institute).

18. Ruml's policies can be followed in the manuscripts available in the Laura Spelman Rockefeller Memorial Archives, series 3, esp. boxes 58 and 67, Rockefeller Archive; also in the papers of Beardsley Ruml, University of Chicago, for the period during which Ruml directed the Laura Spelman Rockefeller Memorial.

19. Because the John Simon Guggenheim Memorial Foundation declines to open its records to scholars, anyone seeking to write about the foundation's history and Moe's career must depend heavily upon Lomask's virtually uncritical *Seed Money*. Lomask evidently received permission to view at least a selection of documents in the foundation's files, and he reprinted valuable manuscripts and anecdotes, such as Moe's tale about the one-armed scientist (p. 261). To some extent, this situation may be alleviated when the manuscript division of the American Philosophical Society completes the task of processing Moe's personal papers for use by scholars.

20. Frederick T. Gates, "The General Education Board," 8-page memorandum dated January 1927, Frederick T. Gates papers, Rockefeller Archive.

21. G. Stocking, *Race, Culture, and Evolution*, pp. 299–300; Social Science Research Council, *Fellows of the Social Science Research Council* (New York: Social Science Research Council, 1951), pp. 54–55, 225–26, 262–63, 329.

22. For example: Merriam to Ruml, 24 October 1924; Ruml to Merriam, 27 October 1924; Merriam to Ruml, 29 October, 21 November, and 16 December 1924; Laura Spelman Rockefeller Memorial Archives, series 3, box 67, Rockefeller Archive; B. D. Karl, *Charles E. Merriam* passim; "Record of Interviews, with Doctor Rose, November 27, 1922, Rose

papers, May 22, 1928," Rose diaries, Rose papers, interviews with John C. Merriam summarized on pp. 134, 160, 181, 214, 252, 258, 278, 327, 329, 330, 337, 340, 357, "Record of Interviews with Doctor Rose." Trowbridge, who resigned from Princeton to investigate conditions at European and American universities for Rose, sent such detailed and voluminous reports that they could be of significant use to any student of twentieth-century science: see Records of the International Education Board, file 71, "Interviews." Rose's conversations with both Veblen and Trowbridge about his plans to promote science in the United States are summarized in "Record of Interviews with Dr. Rose," p. 117, all Rose papers. Also see "letters and memorandum for Dr. Wickliffe Rose...," boxes 5 and 29, Veblen papers.

23. Interview with Pauling, 27 March 1964, pp. 12, 22–23, 28–29; session two, pp. 1, 3, 19–20; interview with Van Vleck, 2 October 1963, AHQP, pp. 25–28; J. H. Van Vleck, "The New Quantum Mechanics," *Chemical Review*, 4 (December 1928), pp. 467–506; M. Jammer, *The Conceptual Development of Quantum Mechanics*, p. 343.

24. "Record of Interviews with Dr. Rose," p. 117, Rockefeller Archive.

25. For examples in the field of physics, see S. Coben "Quantum Mechanics," especially pp. 459–65. For more general discussion, see L. Veysey, pp. 77, 358, 388. Typical complaints not cited in these works include that of Leonard B. Loeb, son of Jacques Loeb, who began his distinguished career as a biologist in Europe before migrating to the United States. The younger Loeb was familiar with conditions for university professors both in Europe and the United States in the period before postdoctoral foundation grants became widely available: "Those few who had gone abroad to study or achieve a Ph.D. returned to instructorships and the drudgery of laboratory assisting and lecturing to lower division students. Being kept at low rank for many years before promotion, with many class-room hours, much work on up-keep of apparatus, reading problems, etc., and doing routine committee work, they had little time and less energy for research." Leonard B. Loeb, "Autobiography of Leonard B. Loeb," p. 19, prepared for the Center for the History of Recent Physics in the United States, American Institute of Physics, New York City, March 1962.

Frank Hoyt, a young professor at the University of Wisconsin in 1921–22, recalled his own difficulty in finding time for research: "Very heavy teaching schedules, as you know, were quite common in those days." Hoyt received a National Research Council Fellowship for research in Copenhagen the following year. Interview with Frank C. Hoyt, 28 April 1964, AHQP, 1. Ruth Benedict's inability to find the time necessary to write about her research among the Zuni was mentioned in a letter by Franz Boas; "Of course Ruth has her hands full (teaching) this winter and I do not suppose she will make much headway." Boas to Reichart, 4 November 1926, Boas papers. Benedict published her first article about research accomplished during 1923 in 1928.

WARREN WEAVER AND THE ROCKEFELLER FOUNDATION PROGRAM IN MOLECULAR BIOLOGY: A CASE STUDY IN THE MANAGEMENT OF SCIENCE

ROBERT E. KOHLER
University of Pennsylvania

IT HAS BECOME INCREASINGLY CLEAR TO HISTORIANS OF SCIENCE that patronage is a strategic area for understanding the structure and dynamics of science. So long as the development of science was seen as depending only on an intellectual dynamic, the question of where the money for research came from was as peripheral as discovering who paid for a great scientist's shoes. But when intellectual priorities in science are seen to depend on a complex set of relations with clients and markets in other sectors of society —education, industry, government, private patrons, etc.—then patterns of patronage became crucial indicators of influence and control.

In a market system, customers and entrepreneurs may be as significant in determining the quantity and kind of scientific work as those who produce it. I say *may* be, because the degree to which patronage actually influences science depends on the particular ways in which relations between patron and client institutions are handled. In the United States prior to World War II, the large private foundations, such as the Carnegie or Rockefeller Foundations, were the most significant "external" patrons of science. (Government and industry outspent the foundations but did so "internally," within the various agencies and corporations; they affected university science indirectly, as markets for university-trained scientists.) Foundations supported university science directly, and they were thus a significant source of ideas and policies regarding the role of science in society.[1]

Even within the Rockefeller Foundation (RF), however, styles of patronage were not uniform. From 1913 to about 1921, policy was to aid college teaching and practical applications of science, such as public health programs. From 1922 to 1929, emphasis was put on scientific and professional education, especially medical; in the 1930s and 1940s, a new policy focused on aid to individual research; and from 1950, emphasis was put again on a more practical aim in agricultural science—breeding miracle grains for the "green revolution."[2]

This study is concerned with the period from 1929 to 1939, when new programs supporting research came into being, and with the policies of the RF's Natural Sciences division, run from 1932 on by Warren Weaver. The significance of Weaver's program in supporting basic research in molecular biology is not that it was representative of the period—it was not. Its significance lies rather in two points. First, its effectiveness: Weaver's program developed the idea of molecular biology (Weaver coined the phrase in 1938),[3] and provided much of the support for the applications of new physical and chemical techniques to biology in the 1930s (isotopes, ultracentrifuge, X-ray crystallography, etc.). Weaver's policies deeply influenced several disciplines, notably biochemistry.

Second and more important here is Weaver's program itself. In 1932–1939, Weaver developed a special role, which I will call the *science manager*. Whereas foundations had generally avoided making decisions on allocations to basic research, Weaver selected projects according to a well-developed plan; he took an active role in locating and developing suitable projects. He made it his business to envision a large area within the chemical and biological sciences as a whole, to set priorities, strengthen weak points in the system, identify strategic opportunities, and plan for future growth. He exercised a managerial role, of the sort pioneered about the same time in Britain by Walter Fletcher, secretary of the Medical Research Council.[4] In some ways, Weaver's managerial style looked forward to the patrons of big science, such as the National Science Foundation (NSF) or National Institutes of Health (NIH), whose directors had in theory similar opportunities (though in practice the vaster scale of these agencies and the prevalence of peer review cramped an effective managerial style).

The question addressed in this study is how and why Weaver was able to function as a manager of science. Evidence in the RF archives[5] suggests that several subtly interlocking factors were crucial: the ideals of the Foundation regarding science and its own institutional responsibilities for managing rational social change;

Weaver's personal ideals of a new biology reformed by methods of physical science; the administrative structure of the Foundation, which permitted a degree of specialization that could accommodate Weaver's plan; the administrative strategies of Foundation President Raymond Fosdick, which maintained a balance of power between officers and trustees; economic ups and downs, and changing opportunities within various sciences—all these shaped and gave substance to Weaver's program.

Foundation Ideals and Reorganization, 1913–1928

Since the role of science manager was guided by the Foundation leaders' conception of science and of its legal responsibilities as an institution, we must first understand some basic points about the history of the Rockefeller Foundation in the 1920s.

First, Foundation leaders combined a deeply idealistic enthusiasm for science and research with an abiding skepticism of the academic institutions in which most research was carried out. The large foundations were creations of the Progressive era, and their goals reflected the Progressive ideal that human welfare was best promoted by the systematic and rational application of "objective" knowledge. Their institutional raison d'être was to provide a model of disinterested behavior in ". . . promoting procedures in the rationalization of life."[6]

As exemplars of objective knowledge, the sciences were highly regarded: the physical sciences for their rational perfection, the social sciences for their potential utility in fostering a rational social order. But in Foundation circles, "science" had a broad meaning: it was "organized knowledge," not science in the narrow sense of compartmentalized academic disciplines. "Research" for Foundation leaders meant striving to be well-informed on socially important issues, not the overly specialized research of academic scientists, who tended to be guided by purely internal criteria of relevance and by academic rewards (e.g., prestige, tenure) rather than social improvement. Thus, early Foundation leaders made a sharp distinction between research and education and were skeptical that the university was the proper institution for large-scale scientific research.

Rockefeller money did support research, but in the Rockefeller Institute—a special institution having ultimate reference to medicine rather than academic disciplines and run by Simon Flexner in the German style—department directors had complete freedom to pursue basic problems free of external policy, the constraints of artificial discipline boundaries, and the demands of teaching.[7] The mission of the other Rockefeller Boards—the General Education Board, the International Education Board, and the Foundation—was to support college teaching and the demonstration and application of knowledge to improving the human condition. In 1916, the Rockefeller Foundation declined to support academic research through James McKeen Cattell's Committee of 100.[8] The leaders of the 1920s, such as Wycliffe Rose and Abraham Flexner, opposed grants to individual research projects as academic charity: they felt that foundations had a higher responsibility to all of society and a broader conception of science as a social activity.

The second characteristic of Foundation policy was an extreme sensitivity to the issue of intervention. On the one hand, the wealth and power of the large private foundations could only be justified if they had a large influence on national institutions. On the other hand, the foundations were legally only quasi-public institutions; they were not accountable to the political process, and their right to influence national institutions was unclear and, prior to World War I, highly controversial. Like other quasi-public institutions of the Progressive period, such as the National Research Council or the National Civic League, the Rockefeller Foundation was caught in a perpetual bind: it was set up to be the disinterested manager of social institutions, yet it was vulnerable to criticism as being simply a powerful vested interest in disguise. As a result, Foundation executives were very sensitive to the criticism that they were using great wealth to promote special groups or to tell individuals what to do. Thus the trustees, conservative men of affairs, opposed planning in general, aid to individuals, and any action that might be construed as "dictating the course of scientific research." Fear of public criticism reinforced their unquestioning faith in the virtues of laissez-faire and the invisible hand of scientific progress.[9]

This conflict between the need to exercise influence and the fear of dictating behavior was resolved in the 1920s by adopting the following policies regarding aid to colleges and universities: (1) giving only to institutions, not individuals, and in the form of capital grants allocated by the donees, not the Foundation, and (2) giving to the "best" institutions, according to prevailing standards and on a regional or national scale. Foundation policy was to "make the peaks higher," helping strong institutions provide regional models and allowing normal competitive pressures to reform the national system by eliminating the weak and stimulating the more progressive.[10] Money was provided only for institutions as a whole—bricks and mortar or endowment, not for specific research.

Although the Foundation thus avoided dictating priorities within universities, science departments were being enormously stimulated by general capital grants.[11]

Grants to individuals took the form of fellowships, awarded on the basis of merit by committees of the National Research Council.[12] In this way, the Foundation managed to externalize all decisions regarding allocation of money to individuals or to individual institutions. It thus operated like an investment banker or a manager, helping the system do more efficiently and even-handedly what it was doing anyway, without actually seeming to dictate policy.

Prior to about 1923, the Rockefeller Boards concentrated on endowment of general education and public health. From 1923 to 1929, however, Rose and others began to focus their interest more on scientific and medical education. In this way, huge sums were expended on all areas of science: $45 million on the natural and medical sciences from 1913 to 1933, all but a tiny fraction on plant and endowment and most of it to a few top institutions.[13] Under Rose's direction, the International Education Board (IEB) invested $16 million in the natural sciences during 1923–1929; 1.6 percent went directly for research, and 96 percent of funds spent in the United States went to two institutions. In the same six years, the General Education Board (GEB) invested $12.2 million on the natural sciences in the U.S.; 1.1 percent was earmarked for research, and 98 percent went to nine institutions. The Medical Education Divisions of the RF and GEB between 1914 and 1932 put $28.1 million into medical schools; prior to 1929, $60,000 went to research.[14] Comparable sums were spent on the social sciences by the Laura Spelman Rockefeller Memorial.

The policies of the 1920s had intrinsic limitations, however, and their very success created problems. American universities in the 1920s, stimulated in part by the Rockefeller Boards, simply outgrew the resources of the Foundation. The market for higher education boomed; university science became a favored object of alumni giving; the boom in industrial research provided a large market for university scientists; rising costs of scientific education and research, stimulated by the high standards of the Foundations, outgrew endowments. All this meant that by the mid-1920s the Rockefeller Foundation, with its fixed endowment, was simply no longer able to have an impact on the system of university science as a whole, at least not by capital investment. For the Foundation to remain a significant influence it would have to concentrate its efforts on a narrower range of activities.

Second, the entry of the Foundation into support of science had occurred by the individual initiative of active entrepreneurs like Rose, Beardsly Ruml, or Richard Pearce, rather than by a planned and coordinated effort. The five Rockefeller Boards staked out and defended independent territories that often overlapped or left vital areas unrepresented; there was confusion and dismay among potential clients, and the administrative mare's nest in the New York offices became more and more obvious. In 1928, a complete reorganization of the Boards was effected by Raymond Fosdick, John D. Rockefeller's chief counsel, a League of Nations advocate and Progressive reformer. A single Foundation was created, with five divisions: the Natural Sciences (NS), Medical Sciences (MS), Social Sciences (SS), Humanities, and Medical Education. These divisions took over appropriate activities from the Boards.[15]

Fosdick's strategy for reorganization included two main policies: First, a policy of *concentration* of effort on science, and moreover on one aspect of science—the "advancement of knowledge." Although Fosdick shared with his fellow trustees the broad conception of research as including demonstration and application of knowledge (especially in the social sciences), his strategy meant that the Foundation would henceforth concentrate on grants to university research. Second, Fosdick did not elect to concentrate on one area of science, such as physical science, but the whole range of learning. The divisional structure reflected the divisions of knowledge in a university.

Although the divisional structure implied acceptance of the academic disciplines, Fosdick did not have in mind support of disciplinary science for its own sake, but rather the application of disciplines to a large central problem defined by the Foundation—namely, "the science of man." Fosdick envisioned each division contributing, through support of appropriate sciences, to human problems in all aspects: natural, biological, medical, social, and cultural. Relevance to this theme was the touchstone of relevance for the activities of all divisions. Thus, Fosdick avoided locking the Foundation into a highly defined, specific mission, or into a passive role of doling out discretionary funds to university departments. Yet he internalized decision making in the Foundation by providing a coordinating rationale.

However, Fosdick's strategy brought the submerged conflict between Foundation ideals to the surface. Concentration on "the advancement of knowledge" raised troubling questions about the stated mission of the Foundation to promote "the welfare of mankind." Fosdick's broad definition of "research" did not mollify critics such as Rose and Abraham Flexner, who felt that the traditional ideals of aiding education and rational behavior had given way to charity for a highbrow academic elite.[16] Both Rose and Flexner retired in 1928, bitterly opposed to the new policy. The trustees remained apprehensive about that policy, and social relevance remained a controversial issue into the 1930s. Moreover, Fosdick's broad plan for the "science of man" left a good deal of room for the officers in charge of each division to exercise individual initiative. Concentration entailed making choices, setting priorities among disciplines and even within disciplines on the basis of Foundation priorities. Officers became, in effect, policy makers for scientific research. This entailed assembling a staff with sufficient technical competence to judge the merit of research proposals and administer a large number of individual projects.

The departure from the traditional Foundation role of a "neutral" investor caused a good deal of uneasiness, especially among the trustees. In developing and managing the program, Weaver had to be acutely sensitive to these inherent contradictions and ambiguities regarding the proper limits of Foundation activity as patron and visible hand in scientific research.

Interim: 1929–1932

The period from 1929 to 1932 was marked by uncertainty as to how to put the new policies into effect. The habits of the 1920s were continued, at least in the natural sciences. Of the $11.9 million spent on the natural sciences, $7.98 million (67 percent) was for capital investment. Most of the $1.63 million spent on research was in the form of "fluid" (i.e., discretionary) grants to universities. Like the NRC fellowships ($1.86 million), these grants were allocated by recipient institutions.[17] Individual research grants, $0.368 million, went mainly to Europeans, thanks to Rose's network of IEB contacts and deference to European science. The fields favored also reflected prior interests: marine biology and oceanography ($3.14 million or 26.5 percent) and biology ($1.25 million or 10.5 percent), with the rest scattered over eight other fields.[18] There were few clear precedents as to how programs should be developed and administrated. Should the Foundation support mainly training or bench research, institutions or departments? Should it continue the policy of making the peaks higher, or support worthy individuals wherever they were found? These issues were discussed in a two-day staff conference in October 1930, but the only clear resolution was to continue the NRC fellowships.[19]

The ability of the divisions to put together a program depended a great deal on the officer in charge. Edmund Day, a veteran of Ruml's program, rapidly developed a broad program in the social sciences, based on Social Science Research Council surveys. Alan Gregg, a lieutenant of Pearce in the old Medical Education Board, had a thorough familiarity with European medical science and began at once to construct a program in brain and neurological research and psychiatry, which was to be the main theme of the MS division.[20]

The natural sciences were more problematic, and there was little in Foundation experience to guide the way. Rose's enthusiasm had been limited to the safer physical sciences. There had been in the mid-1920s a nascent program in "human biology," organized by Edwin Embree in a new Division of Studies.[21] Following the model of Foundation programs in mental hygiene and public health, Embree focused on areas of biological science that impinged on social concerns: human genetics, race biology, physical anthropology and race, brain research, and (by way of general development of biology) experimental and marine biology. A large grant had gone to Raymond Pearl for work on mammalian genetics and race hybrids; capital grants were made to Woods Hole and Pacific Grove.[22] For various reasons, however, Embree's program did not survive the reorganization. Besides the fact that it concentrated on the more controversial fringes of human biology, it represented an organizational model that Fosdick had explicitly rejected—namely, a program organized around medical problems. Fosdick's aim was the broad development of biology along with the other natural sciences.

But while it was agreed that the remnants of Embree's program be terminated, there was no agreement on what was to take its place. There had been uncertainty in the last phases of the reorganization over the NS division. A suggestion for a separate division of biology was rejected. A division of agriculture and forestry survived up to the very last moment, and Gregg continued to think of a future division of biology with medicine and agriculture attached to it.[23]

The difficulty of deciding between these various options is reflected in the lack of a full-time director for the NS division. However, Max Mason, the new president, gave some direction to the NS. Mason was a mathematical physicist, and his projects reflect his background: they included the application of X-ray crystallography to chemistry and perhaps in time, he thought, to biology.[24] For a few years Richard Pearce and William Carter, from the Medical Education division, served as acting directors of the NS.[25] In September 1930, Mason persuaded Herman Spoehr to take the job. Spoehr was a professor of plant physiology, and his tentative proposals for an NS program in October 1930 pointed to the needs for developing research in the basic sciences underlying agriculture and forestry, especially forestry, a vital national resource in need of basic science.[26] He envisioned a range of activities from physics to biology, dealing with energy and photosynthesis and centering on the study of enzymes, vitamins, and other accessory factors in cells: "This may be taken as an example of a . . . large problem from which we can concentrate to bring correlated studies."[27]

Spoehr resigned in August 1931, however, and in April 1932 Mason visited Harvard and MIT in search of a new director from the physical sciences. He consulted Carl Compton, A. A. Noyes, and Arthur Lamb, an organic chemist who expressed some interest in the job but was unwilling to leave research altogether.[28] Other possible candidates included Floyd K. Richtmeyer, professor of organic chemistry at MIT. Finally, in the fall of 1931, Mason tapped his former colleague at the University of Wisconsin, Warren Weaver.

Like his teacher, Weaver was a conservative classical physicist, firm in the belief that the new quantum mechanics was a flash in the pan.[29] By 1932, however, quantum physics was firmly established in American universities,[30] and it is perhaps not surprising that Weaver came to New York with the conviction that the long-range future of physical science lay in its application to biology. This was what Mason and the trustees wanted to hear. Overcoming his diffidence that a physicist would be able to develop and run a program in biology, Weaver accepted the job.

When Weaver moved to New York, the policy options that had been open in the years 1928–1931 closed. In this fluid situation, the shape of divisional programs depended a great deal on individual preference. Weaver's presence ensured that the preference of the NS division would not be forestry, agriculture, or medicine, but the physical sciences.

"Psychobiology" and "Vital Processes": 1932–1934

The hallmark of Weaver's program from 1932 to the late 1940s was the idea of a "new biology," reformed and inspired by the application of techniques of molecular physics and chemistry and the standards of experimental rigor of the physical sciences. Weaver's program took various names—"vital processes," "psychobiology," "experimental biology," and "molecular biology"—but the idea behind it was constant.

There was nothing novel about Weaver's conception. The notion that "progress" in biology meant a closer approximation to abstract physics and mathematics was a textbook commonplace. The 1920s and 1930s were a period of particular enthusiasm for the "reductionist" program. Within many biological disciplines, reformers were promoting the promise of physical science. In genetics, Thomas Hunt Morgan and his school anticipated that the next major advance would be the chemical understanding of the gene and gene expression. In embryology, the early 1930s was the period of the greatest enthusiasm for the "organizer" theory and chemical embryology. In endocrinology, sensational chemical and biochemical discoveries regarding hormones were bringing the field from the clinic to the chemical laboratory.[31]

These programs were widely circulated in pronouncements of professional societies and in the popular press. Throughout the 1920s, chemists were increasingly eager to promote biology as a source of research opportunities. The late 1920s and early 1930s saw the famous prophecies by Max Delbrück, Niels Bohr, Pascual Jordan, and other physicists of a biological quantum revolution.[32] Weaver's expectations were somewhat more mundane than "new laws" of living matter, but he shared with the physicist-biologists an outsider's naive, rather condescending view of biology as an underdeveloped country, rich in potential but wrapped in unscientific habits and tradition. Professional biologists knew well that the application of physical and chemical techniques had long been a recurrent ideal in biology; they also knew the difficulties of actual research. Weaver saw biology from the warped perspective of the upper rungs of the Comtean ladder of the sciences.

Although Weaver's program for a reformed biology was not a novelty for biologists, his outsider's perspective was critically important for his new role as science manager. Because he was not trained in a biological discipline, he tended not to conceive biology in terms of disciplines but rather in terms of large problems to be attacked from many points of view. In formulating his program, Weaver did not think in terms of aiding separate disciplines for their own sake, as a biologist might have done. He saw the disciplines as providing opportunities for selective application of mathematical, physical, and chemical techniques and technologies to biological problems. His criteria for selecting projects was not relevance to particular disciplines, but relevance to his plan. The outsider's perspective lent itself to a cross-disciplinary program.

This approach was also congruent with Fosdick's and the trustees' distrust of academic disciplines and their conception of science as the application of knowledge to human problems. For, example, one trustee in 1930 asked Alan Gregg: "What impression do you get of the artificial division we have made of physics, chemistry, bacteriology and pathology? Are they in the way now or not? Isn't it about time we forgot those names and scrambled the whole thing to see if we cannot get some new terms? Haven't we interfered with our development?" Gregg assured him that he saw no advantage in holding strictly to conventional academic categories.[33] Weaver too tended to think in terms of problems rather than disciplines as such. In retrospect, the difference between Weaver and the trustees over basic and "relevant" research was less important than the similarity of their views on the role of the Foundation as manager. For Weaver, biology was an area to be developed by application of physical science; for the trustees, society was an arena to be improved by the application of science. They thought on different levels but in analogous modes.

The important point here is not the novelty of Weaver's conception of "vital processes," but rather how this mission became an ongoing program; how selection criteria were articulated and relations with other divisions settled; and how the whole package was sold to the trustees.

Weaver's first detailed proposal for the NS program, prepared in the fall of 1932, was a broad program of support for the whole range of physical sciences:

Major: 1. Mathematics, physics, and chemistry of vital processes
2. Mathematics, physics, and chemistry of the earth and atmosphere
3. Genetic biology
4. Quantitative psychology
Minor: 5. Fundamental construction problems
6. Physical and colloidal chemistry
7. Theory of probability and statistics[34]

(A slightly earlier verson did not have psychology and in place of statistics had studies in filterable viruses, a legacy of Fosdick and Mason's earlier plan.)[35]

The principle behind Weaver's scheme is clear. The "vital processes" area, the application of physical science to biology, was to be the contribution of the NS division to the overall Foundation program of a new science of man.[36] In support of this core program were more basic areas of physical science: earth and atmosphere—man's physical environment; fundamental construction problems, i.e., the structure of inorganic matter, from atoms to galaxies; colloidal chemistry, then in vogue in biology and medical fields as an explanation of the special properties of living matter; and the mathematical techniques that underlay all sciences. In keeping with Weaver's and Mason's prior interests, physics was prominent in most areas. It was an ambitious plan—too ambitious for a year as lean as 1933.

Weaver's proposal emerged from Mason's office shorn of psychology and all the minor areas in physical science. In addition to vital processes (including genetics), only earth science remained, and Mason made it clear that he saw earth science as marginal to Foundation interests.[37] No project would qualify for support simply because it was intrinsically interesting; it would have to be directly related to the core program.[38] Only a few small grants were made in earth science before the trustees pronounced it stillborn in 1934.[39] In effect, then, the NS program was concentrated from the outset solely on "mathematics, physics and chemistry of vital processes."

In Weaver's original plan, "vital processes" was the interface with Gregg's MS program in psychiatry and neuroscience. An unintended result of Mason's shearing away all but "vital processes" was that Weaver's program overlapped in virtually every area with Gregg's. The point of reference of Weaver's program shifted away from physical science and toward "psychobiology." Moreover, since Weaver had lost those fields in which he was most at home, he tended at first to rely rather heavily on Gregg's greater experience and expertise and on justifying his program in terms of Gregg's mission. As a result, the most striking feature of Weaver's program from 1933 to about 1935 was its close relation to psychobiology.

In the report to the trustees in December 1933, for instance, the MS and NS programs were presented as a single unit comprising these subfields:

1. Psychobiology (psychiatry, neurophysiology)
2. Internal secretions (hormones and enzymes)
3. Nutrition (vitamins)
4. Radiation effects
5. Sex biology
6. Experimental and chemical embryology
7. Genetics

8. General and cell physiology (nerve conduction, osmosis)
9. Biophysics and biochemistry (spectroscopy, microchemical analysis, and basic studies)[40]

The supporting role of the NS was made quite explicit: "Psychobiology is the single principal topic, all the other subjects being viewed as contributory. A large percentage of the activity in the MS will, from the outset, fall into this first category; while the NS will be more concerned with the immediately contributory studies."[41]

The division of labor between Gregg and Weaver was not, however, a division between pure and applied fields. Gregg conceived of "psychiatry" not as clinical specialty, but as including the sciences of mental development and behavior.[42] His mission was to bring the methods and rigor of biological sciences to psychiatry and the behavioral sciences; Weaver's was to develop the biological sciences by introducing methods of physical science into them. The division of labor was not along discipline lines; it was a functional division within each area. Gregg was to take research projects that had already progressed to the point where they could be applied clinically (mainly in areas two to five). Weaver was to take those that required further development as basic science before being ready for medical or psychiatric use (mainly in areas six to nine).[43] In areas such as endocrinology or nutrition, which covered both biochemical and clinical studies, Gregg took the former, Weaver the latter.

Foundation thinking did not encourage a sharp distinction between pure and applied science but between applicable and not yet applicable. All science was ultimately useful, but to be used it had first to be perfected as a science, for its own sake. This point is vital, because this conception of science made it possible for Weaver to develop the basic biological sciences without stepping beyond the bounds of "relevance" set by Foundation ideals of social utility.

In the short term, an intimate relation to psychobiology was a useful shelter for Weaver's nascent program, providing it with ideological justification at a time when Weaver was still feeling his way into the role of science manager. So long as Weaver and Gregg worked together, both basic and clinical aspects were developed, and the functional division of labor allowed Weaver to build a program of support for the basic biological sciences under the umbrella of psychobiology. The division of labor between Weaver and Gregg established the NS as the one division that could legitimately foster academic science for its own sake. Had Weaver originally been more independent of the MS program, ideological pressure could have led him to design a program with "relevant" aspects built in, such as Embree's human biology program; or he might have emphasized the practical sides of the biological disciplines. (It seems unlikely that an attempt to foster the physical sciences as such could have succeeded in 1933–1934 any better than the earth sciences scheme.) In short, the umbrella of psychobiology was critical for Weaver's ability later to nurture selectively those sides of biochemistry, genetics, and cell physiology that became identified as molecular biology.

Pressures for "relevance" were particularly intense during the Great Depression. The euphoric celebration of physical science as the source of industrial productivity and social progress—a celebration that had accompanied the 1920s boom—suffered a sudden reversal. Witnesses to economic collapse, rising crime, and social disruptions at home and the rise of dictatorship and political violence in Europe began to doubt that science solved all social evils. Physical science was blamed for causing the industrial collapse, and the idea of a "science holiday" was widely discussed—a moratorium on scientific research to give society a chance to catch up and cope with a disaster brought about by too rapid technical progress.[44]

No one seriously proposed a science holiday, but one concrete result of the crisis in faith was a marked shift of goodwill toward the biological and behavioral sciences. D. C. Broad's contrast between "physics and death" and "psychology and life" was an extreme form of a common theme.[45] Some physical scientists, alarmed by talk of a science holiday, turned to biology and psychology for problems that might recoup declining prestige and ensure the survival of threatened programs. Foundation directors, pressed by the rising tide of criticism of capitalist institutions and demands for constructive action by the nation's leaders, likewise looked with greater favor on socially "relevant" programs.

Weaver was sensitive to these currents of feeling and their implications for his own plans:

There is ... a lesson to be learned from our present situation: ... our understanding and control of inanimate forces has outrun our understanding and control of animate forces. This, in turn, points to the desirability of an increased emphasis, within science, on biology and psychology, and on the special developments in mathematics, physics, and chemistry which are ... fundamental to biology and psychology.[46]

For Fosdick, Mason, and Weaver, attacks on reason reinforced their faith in the need for science to bring release from social superstition and irrationality, to bring the individual into ". . . a more intelligent, a more accurately adjusted and a happier relationship with our modern scientific civilization."[47] For Weaver in particular, the application of physical science to biology was a counter to the irrationality of the science holiday, and he presented psychobiology to the Foundation trustees with an almost messianic fervor:

> There is a strong and growing belief, held by many thoughtful scientists—even by many of the ablest specialists in the physical sciences—that the past fifty or one hundred years have seen the supremacy of physics and chemistry, but that hope for the future of mankind depends in a basic way upon the development during the next fifty years of a new biology and a new psychology. As one views the present state of the world, with its terrific tension, its paradoxical confusion of abundance, and its almost uncontrollable mechanical expertness, one is tempted to charge the physical sciences with having helped to produce a situation that man has neither the wits to manage nor the nerves to endure. One should be critical in distinguishing between basic pure science and the inventive and technological activity that is often incorrectly referred to as science: and yet the fact must be faced that no one hopes or expects that technological advances will not continue.
>
> The challenge of this situation is obvious. Can man gain an intelligent control of his own power? Can we develop so sound and extensive a genetics that we can hope to breed, in the future, superior men? Can we obtain enough knowledge of the physiology and psychobiology of sex so that man can bring this pervasive, highly important, and dangerous aspect of life under rational control? Can we unravel the tangled problem of the endocrine glands, and develop, before it is too late, a therapy for the whole hideous range of mental and physical disorders which result from glandular disturbances? Can we solve the mysteries of the various vitamins . . . ? Can we release psychology from its present confusion and ineffectiveness and shape it into a tool which every man can use every day? Can man acquire enough knowledge of his own vital processes so that we can hope to rationalize human behavior? Can we, in short, create a new science of Man?
>
> This point of view has recently been realized by various scientists, philosophers and statesmen; many of the techniques are at hand; but direction, stimulation, support and leadership are for the most part lacking. The Foundation has a unique chance to correlate and direct existing forces and to stimulate the creation of new forces for a coherent and strategic attack. The proposed program recognizes here one of the most inspiring opportunities with which science has ever been faced.[48]

The Rockefeller Foundation Program in Molecular Biology

The science holiday mood certainly did not give Weaver's program its substance. But it certainly *did* heighten Weaver's sense of its urgency, adding ideological to tactical reasons for a close relation between his program and Gregg's, and thus aiding its survival in a period of volatile ideological sensibilities.

The close association between Weaver and Gregg was temporary. By 1935, Weaver was breaking away from psychobiology and developing a program for support of the biological and allied physical sciences. This trend was given implicit approval by an official appraisal of Foundation policy in late 1934.

Appraisal, 1934

Although the Foundation trustees approved the divisional programs in December 1933 without apparent controversy, they were anxious in general about the future of the Foundation, and at the same meeting appointed a Committee of Appraisal, chaired by Raymond Fosdick, to consider whether new "sailing directions" were called for that were more responsive to the social crisis.[49] The economic forecast was bleak. Yield on Foundation securities had dropped from 6.59 percent of book value in 1929 to 4.21 percent in 1933, and their greatly diminished market value following the crash precluded selling capital.[50] Even more worrisome, termination costs of old programs from the 1920s were reaching a peak in 1934–1935, just as demands for new programs were rising.[51] (See table 1.)

TABLE 1. Expenditures in Natural Sciences, 1933–1938

	Concentrated Program	General Program	Exceptions to Program	Total New Program	Old Program	Total for Year
1933	223,925	175,000	11,100	410,025	340,020	750,045
1934	501,457	150,000	24,035	675,492	328,201	1,003,963
1935	1,236,804	75,000	17,000	1,328,804	1,089,870	2,418,674
1936	880,351	277,500	174,000	1,331,851	10,000	1,341,851
1937	1,461,875	none	12,750	1,474,575	none	1,474,575
1938	2,690,650	none	7,000	2,697,650	none	2,697,650

The trustees were in an anxious and stringent mood, unlikely to respond enthusiastically to large new schemes. Moreover, the pressure of public opinion for social reconstruction was mounting. Although the archives give little detailed evidence on trustee opinions (detailed minutes of their meetings were not kept), it seems unlikely that the trustees would have seen the advancement of knowledge as the Foundation mission had a reorganization occurred in 1933. One trustee, for example, wrote that he was "definitely agnostic" about the wisdom of Foundation policy of investing in new knowledge for the future when the very present survival of civilization was in the balance. The wave of juvenile crime, unemployment, bread lines, and "guerilla warfare" against all established institutions told him that the Foundation should abandon all its planning and pour everything into immediate countermeasures to stave off collapse.[52]

From my point of view, the Rockefeller Foundation's work has become too largely an investment in remote futures with an attendant policy of ignoring the present to such an extent that civilization may never reach the future. I am quite convinced personally that a large proportion of all the money that we have available for appropriation might be turned into a study of the crime situation.[53]

It would be a mistake, however, to suppose that the Committee of Appraisal was expected to alter Foundation policy radically. Although the committee prepared its report in an atmosphere of tension and uncertainty on the part of the officers, there is no evidence in the record that Fosdick ever considered major changes in policy, such as dropping research or removing divisions. As I read it, the appraisal was a move by Fosdick, the architect of the reorganized Foundation, to reassert the 1928 consensus: to defuse trustee mistrust by providing an occasion to lay all doubts and differences openly on the table and to establish communication between officers and trustees. The result of Fosdick's report was, in general, to reinforce existing policy: concentrate on core programs and give officers more power to design and manage support for scientific research. The constancy of Foundation programs owed a great deal to Fosdick's energetic and continuing support.

The burden of Fosdick's report in December 1934 was: *economize* and *concentrate*. As to the need to economize there was no dispute, but Fosdick's strategy of doing so by concentrating on special programs was controversial, so much so that discussion was tabled and taken up by the trustees after the other items had been approved. The strategy of concentration reinforced the officers' powers of planning and influencing the course of scientific research. This was precisely what made the trustees most uneasy: the prospect of an administrative enclave intervening in the internal affairs of academic science but not responding to large social needs.

Fosdick met this issue by embracing Foundation influence as inevitable and by enjoining the officers to be circumspect and sensitive to social needs:

> We do not have to be cynical to admit that if a foundation announces an interest in anthropology or astronomy or physiochemical reactions, there will be plenty of institutions that will develop a zeal for the prosecution of these studies. The responsibility which this inescapable fact throws upon a foundation is enormous. The possession of funds carries with it power to establish trends and styles of intellectual endeavour. With the best will in the world, the trustees of a foundation may select unwisely. . . . To guard against these evils requires critical judgement, common sense, wide understanding and eternal vigilance; and frankly, in this matter of promoting research, your committee is inclined to believe that the Foundation has followed its enthusiasms too far.
>
> We are by no means suggesting that research be omitted from the Foundation's activities. . . . But in our opinion we should avoid research for the sake of research without regard to its relevance. Moreover, these should be no exclusive interest in research as an end and aim. Indeed we would strongly advocate a shift of emphasis in favor not only of the dissemination of knowledge, but on the practical application of knowledge in fields where human need is great and opportunity is real. As a means of advancing knowledge, application can be as effective an instrument as research.[54]

Fosdick's message sounds like a sharp limitation of the officers' prerogative to plan programs and to support basic science. (Fosdick's remark about overenthusiastic selling of programs was aimed at Weaver's promotion of psychobiology.) Yet by urging the officer's to exercise judgment and wisdom, Fosdick was acknowledging that they had the power to judge and plan. Fosdick's report was intended, in short, to make clear the limits both to the trustees' demands for social "relevance" and the officers' liberty to pursue their special programs.

The burden of the committee's criticisms fell unequally on the different divisions. The social sciences were hardest hit, and Fosdick's remark about the need for more socially relevant work and more application was mainly directed toward Day's program. The social sciences were regarded by Fosdick and the officers themselves as the sciences closest to social problems, in which testing of theory meant application in the field. On the point of relevance, the natural sciences came under little criticism, as Weaver noted:

> Resolutions 1 and 2 indicate a conviction that the Foundation has, in some instances, chosen a cloistered, academic type of research on problems which have sometimes been pretty far removed from any conceivable contact with present and pressing human needs. The discussion of these resolutions made no special reference, as far as I recall, to our division; although the Trustees would doubtless have judged (rightly or wrongly) that research in topology, in quantum mechanics, in atmospheric elctricity, etc., would not fall in the territory where "human need is great" even though "opportunity" may be "real".[55]

It was implicitly accepted that the natural sciences were not expected to have the same degree of social applicability as the social sciences. Again, it is important to keep in mind the division of labor within the divisions. For the social sciences, Fosdick's report meant more practical work on social problems; for the medical sciences, it meant more psychiatry. For Weaver, it meant more concentration on the vital processes projects and basic biology.

Fosdick's role in continuing the strategy of concentration on planned programs was critical. Some trustees we know were skeptical; so too were some members of his subcommittee of technical experts appointed to appraise Weaver's vital processes program.[56] Henry Dakin, who operated a private laboratory in physiological chemistry, argued strongly against any interference with individual scientific genius: ". . . to sum up. Less plan, less emphasis on the future coordination of scientific knowledge and its human implications, and more scientific opportunism."[57] Dakin shared with George Ellery Hale and others a conception of science as high culture, as a calling which could not be planned or managed, and he saw with dismay the rise of science in government agencies and industrial research laboratories.[58] William Howell, emeritus professor of physiology at Johns Hopkins, shared Dakin's fear that the promotion of favored programs by the Foundation would damage the "idealism and independence" of science, but he did not think the Foundation should rely on scientific fashion in selecting projects as it had in the past. He favored the strategy of setting long-range goals and endowing a few strong university institutes to provide leadership in carrying them out.[59]

The most vigorous opponent of planning and of Weaver's program in particular was the chairman of the subcommittee, Simon Flexner, who argued that any Foundation plan must inevitably prohibit some areas of research and tempt scientists into others out of a desire for grants, rather than intrinsic interest.[60] (In this regard he was preaching what he practiced as director of the Rockefeller Institute, a bastion of scientific individualism.) Flexner opposed the vital processes program on the grounds that physical and chemical methods were general tools, not a "new" program, and he questioned Weaver's competence to administer a biology program:

"I am . . . not sure that the officers, captivated by their own notions, may not have imposed their ideas on individual laboratories. You and I discussed this point briefly. The power of the Foundation is so great that I doubt whether an entirely neutral attitude on either side can be maintained. There is also something anomalous in mathematicians and physicists dominating in a wide way research in biology and medicine. . . . A disturbing element is that the chief men in charge are so completely "sold" to the program, and as I gather from a long talk last spring with President Mason, the "program" is looked upon as a significant innovation, which it can scarcely be said to be."[61]

(These remarks, in a personal letter to Fosdick, were not expressed in Flexner's subcommittee report.) Asked if, as a trustee, he would have supported Weaver's program, Flexner replied: "I should have been able, I think, to point to examples in which the 'project' given seemed to me to be framed not so much on its feasibility as because it fitted into the Foundation scheme. I should not have disapproved of parts, and not have approved of the program as a whole."[62]

The attitudes of Dakin, Flexner, and Howell toward science are those of an older generation of scientists, imbued with the German ideal of "pure science" as high culture, in conflict with a new generation, imbued with the managerial ideal of science as a resource. One sees their disdain for Weaver's promotional style and the hint of professional competition.

Other university scientists on the subcommittee saw the strategy of concentration as adaptable to their own interests. W. B. Cannon, professor of physiology at Harvard Medical School and a leading expert in the physiological basis of emotion, applauded Weaver's conception of psychobiology, only suggesting that it include more neurophysiology and animal behavior.[63] Frank R. Lillie, leading experimental embryologist and dean of the biology division at the University of Chicago, also embraced the policy of concentrating on strategic areas: "Seeing that influence is unavoidable, it should be intentional."[64] However, he felt that "vital processes" was a less appropriate term than "experimental biology," and he suggested that the Foundation consider endowing institutes of research in experimental biology in universities or, in any case, that the selection of projects be placed in the hands of "qualified" scientists.

None of this was what Fosdick wanted to hear: "Frankly, I got very little from their reports except, perhaps, their general feeling that if properly limited the [Natural Sciences] program was good. Their irrelevancies seem to center about two points: (1) The old row between university and institute research . . . (2) Planned research versus general research. . . ."[65] For Fosdick, these issues were "old rows" settled in 1928, when it was resolved that the Foundation support research in universities and develop planned, concentrated programs—in short, when the Foundation adopted a managerial role in science patronage. Fundamental policies were not in question for Fosdick. The question was not whether Weaver's program would continue, but how fast and how far.

Flexner's criticism of the psychobiology program struck a more responsive chord. Fosdick had had similar doubts. He worried that Weaver was too much the advocate, too oversold himself and therefore too much the promoter and salesman of a particular scheme to prospective clients. In March 1934, he had invited Weaver to his home for a personal tête-à-tête and warned him against being too rhetorical and zealous in presenting his program to the trustees. The Foundation must not seem to be intervening in scientific priorities. Fosdick also worried that Weaver was supporting "bizarre" or "esoteric" projects (a project for the spectroscopic analysis of body fluids was mentioned.)[66] These were the two points on which Fosdick's policies were most vulnerable to criticism (as the Committee of Appraisal showed), and he was anxious to prevent excesses by the officers.

Fosdick's and Flexner's misgivings about the psychobiology program were confirmed by David Edsall, dean of the Harvard Medical School and an active trustee, whom Fosdick asked for an informal opinion following the inconclusive subcommittee report. Edsall was sympathetic to Fosdick's policies, having chaired a committee in 1928 that approved a shift from medical education to research.[67] He had been a constant supporter of Weaver's program and had himself advised the new president of Harvard, James B. Conant, that the application of physical science to biology and medicine would in the next generation produce as many important advances as had occurred in the past generation in such sciences as biochemistry.[68] Yet the burden of Edsall's advice was caution and skepticism about psychobiology. He too had been struck by what seemed to him overly optimistic claims by Mason and Weaver, and on consulting informed colleagues he was still skeptical:

The Rockefeller Foundation Program in Molecular Biology

> There is apparently not any dependable evidence as yet . . . to arouse confidence that large efforts would be rapidly so productive as to justify great expenditures. I believe that . . . there will be a slow, painstaking accumulation of knowledge that in the course of a few decades is likely to be of profound importance, but I question very much whether there would be any prompt solution to any very important problems.[69]

Edsall favored support for promising projects, even if they were expensive or calculated risks, but he discouraged rushing into a large and systematic program in the hope that money alone would produce large results.

What Edsall apparently had in mind was the psychobiology programs, specifically endocrinology and sex biology. The isolation of sex hormones in the early 1930s had touched off a craze of hormone cures. Drug industries rushed to exploit them commercially, and there was a wave of publicity campaigns and claims of miracle cures.[70] By the mid-1930s, a reaction had already begun to set in against excessive claims for psychobiology, and Weaver's enthusiastic promotion smacked to many solid citizens of pseudo-scientific faddism. Fosdick's appraisal of Weaver's program relied heavily on Edsall's expert advice:

> Our recommendation would be that this program in experimental biology be conducted on a modest, tentative, and opportunistic basis. . . . The strategy would be to feel out the area, to proceed cautiously, to be misled by no preconceived hopes, and to maintain a detached and healthy kind of skepticism in relation both to the program as a whole and to its constituent parts.[71]

But again, Fosdick's admonition was a virtual acknowledgment of Weaver's role as research manager.

Every economizing measure the trustees approved increased the officers' control over program policy. They were directed by the trustees to draw up schedules for the fastest retreat from general programs "consistent with Foundation obligations and dignity." Areas cut back were pre–1928 programs unrelated to new areas of concentration.[72] The general fellowship program was cut back in areas not related to special programs—that is, the NRC fellowships in mathematics, physical science, and medicine. Fluid research grants to universities were abolished, as were all grants given on the basis of geography or to underdeveloped countries to build up general scientific institutions. The officers were directed to make maximum use of grants in aid for individual research projects related to special programs.[73] The trustees urged Weaver to consider changing the name of his program from "vital processes" to "experimental biology."⁷⁴ They shelved the earth sciences.⁷⁵

As Gregg later quipped, "The Scylla and Charybdis of foundations is on the one hand doing small things in a big way and on the other doing big things in a small way. The policy of 1933 was to do everything in a small way . . ."⁷⁶ But the ban on budget increases and large projects should not obscure the more important point: the strategy of concentration on special programs, adopted for the sake of economy, gave Weaver and other divisional officers a larger, not a constricted, role in planning and managing their program. Fosdick's strategy was not Flexner's policy of less management but Edsall's policy of more stringent and judicious selection of projects. Having ratified the officers' role as science managers, what else could the trustees do but enjoin them to be wise and discreet?⁷⁷

From Psychobiology to Molecular Biology, 1934–1938

The most important changes in Weaver's program from 1934 to World War II were a shift away from psychobiology and toward molecular biology; a decline in the rhetoric of utility; a decline in the importance of areas in endocrinology, sex research, and nutrition associated with clinical application; and an increasing reference to the physical sciences, notably organic chemistry. The question is: what structural and political circumstances permitted Weaver to develop a highly selective program of cross-disciplinary research in genetics, cell biology, and biochemistry, which reflected his vision of a "new biology"?

The trend away from psychobiology was implicit in the directives of Fosdick's report in 1934, and the officers realized it. Weaver appraised the appraisal more or less correctly for his associate in the Paris office, W. E. Tisdale:

If I were to sum up in a very few words what appears to be the spirit of the whole decision, these words would be a higher degree of concentration, especially as regards the use of all our mechanisms such as fellowships and grants-in-aid to serve concentrated program interests; and an increased emphasis upon the application as well as upon the discovery of knowledge to pressing modern problems. I think it can be honestly said that, measured with this kind of a yardstick, the work of the N.S. division came in for a relatively small amount of negative criticism, and also that our program requires a relatively mild amount of readjustment.⁷⁸

In fact, the trustees' directive to avoid the more dubious areas of psychobiology and to focus on experimental biology was an invitation to Weaver to concentrate on more basic fields, such as biochemistry or genetics, which were further from clinical application but more certain to produce solid scientific results.

W. E. Tisdale, more insulated from the immediate pressures in Paris, spotted the fundamental issue most clearly:

> The "pointedness" of the MS division in psychiatry is clear, and some parts of the NS program clearly relate to that: but others, such as biophysics and biochemistry do not. Therefore it seems logical to conclude that the psychiatrical direction is unique to the medical field, and that the NS have another kind of pointedness. This pointedness seems to me to be the limitation of fields in which we act, but that within these fields it is to take the form of a general development, with perhaps a leaning towards the mammalian side, rather than a pointedness directed at any given objective.[79]

In other words, the touchstone of relevance was the division plan, and the NS division plan was to develop the biological sciences as such. Weaver concurred, only emphasizing that the aim was not the development of eight separate academic disciplines but a single plan drawing upon all:

> I hope . . . that it is not true that we have eight or more separate objectives. I would say, rather . . . that the direction of our activity results from the fact that we are attempting to sponsor "the application of experimental procedures to the study of the organization and reactions of living matter."[80]

More and more, Weaver found the most appropriate projects for the application of new physical methods in the fields of biochemistry, cell physiology, and genetics—i.e., molecular biology.

Weaver's emerging managerial role was reflected not only in a sharpened sense of policy, but also in the development of new administrative patterns. To evaluate and select research that fit his program in experimental biology and to identify the most able researchers required building up an administrative staff and a network of contacts in eight different disciplines. He had to know who was who, whose opinions might be relied upon, how to make contacts and evaluate clients in a way that did not seem overly promotional.

Weaver tended at first to rely on external professional agencies, such as the NRC committees and fellowship boards,[81] but as he became more confident and experienced, he took a more active role, seeking out clients and selecting projects himself. A fellowship program in "vital processes" was developed as the NRC fellowships for general development of science and medicine were phased out.[82] The NRC committees on sex research and radiation effects continued to administer these two areas, but they became ever smaller parts of the program. When Weaver hoped to create six more NRC committees in endocrinology, nutrition, genetics, and so on, he envisioned not administrative but advisory bodies to carry out surveys and locate promising individuals and projects for Foundation support.[83]

Perhaps the most significant change came in the administration of grants. A great deal of energy was expended in the first few years in debating the advantages and disadvantages of various forms of patronage. The individual fellowship to aid recruitment and training was best administered by expert committees. The individual grants-in-aid, small grants ($100 to $1,000) for specific and limited pieces of research, were seen as follow-up grants for Foundation fellows in their first jobs. These had the disadvantage of being difficult to coordinate and cumbersome to administer. The institutional grants, on which Weaver had first hoped to rely, were designed to be administered by university or department "research committees," to strengthen university science as a whole.[84] Like the fellowship program, these had the advantage of not requiring administrative staffs and technical know-how, but they did not permit exercise of policy. Weaver's general plan was to use all these forms of grants in a balanced way. But in the crisis of 1933–1934, institutional grants and general fellowships were ruled out, and Weaver was forced to develop a middle-level "project grant."

The project grants were a most significant development. They were grants to individuals or groups of people (often in different disciplines), averaging about $6,700 per year and usually given for three years. They were given for planned or programmatic projects, usually involving application of some physical technique to biological problems and often involving collaboration of physical scientists and biologists. This form of grant is mentioned in Weaver's earliest reflections on administrative policy, where it seems to have been conceived as a makeshift compromise between capital grants and grants-in-aid.[85] But from a minor form, the project grant quickly became the favored instrument for Weaver's program in experimental biology.

It was an ideal form: large enough not to attract complaints from the trustees of "scatteration" and small enough not to appear extravagant. Aid to interdisciplinary groups rather than individuals parried criticisms of charity to academics and fit the ideological

The Rockefeller Foundation Program in Molecular Biology

style of Foundation leaders. Most important, the project grant gave Weaver control over policy. Since each grant was given for a specific project, with an eye to long-term potential, it could be carefully selected, and projects were often discreetly shaped by Weaver himself.

Project grants were most effectively used to develop Weaver's idea for a new biology. Typical projects were the grant in 1934 to Harold Urey and a group of Columbia University biologists and biochemists for biological investigations with heavy water;[86] or the grant to Niels Bohr, George V. Hevesy, and August Krogh, for use of radioactive isotopes in physiology; or to The Svedberg, to develop the ultracentrifuge for biochemical work; and so on. By the end of 1933, five such projects were funded in NS and three in MS.[87] By the end of 1934, Weaver alone had thirty-nine projects: sixteen in biochemistry and biophysics, ten in physiology and embryology, six in genetics, and seven in endocrinology and nutrition.[88]

Weaver was explicit about the advantages of project grants: they were selective, helping the best (eight Nobel laureates to date); they were efficient—universities carried the overhead for research facilities; and they enabled the Foundation to foster the "natural and genuine common interests" between biologists and chemists.[89] In short, the project grants, with their associated network of fellowships for recruiting and small grants-in-aid to test investment potential, were the institutional body to the intellectual soul of Weaver's vision of a new biology.

The shift from psychobiology to molecular biology was reinforced by structural changes within the Foundation and by external circumstances. The division of labor between the NS and MS divisions was officially recognized by Fosdick in 1937, and the transfer of endocrinology and sex to Gregg permitted Weaver to move into the physical sciences. Improving economic conditions also encouraged expansion in that direction. Finally, changing investment opportunities in the various fields drew Weaver increasingly to "successful" disciplines such as biochemistry. It is no mystery that Weaver's plans took the turn they did; his personal interests were always with the physical sciences. Institutional and economic factors permitted him to act upon his preferences.

Relations between Weaver and Gregg were cordial and cooperative. There was at first no formal policy covering jurisdictional problems, owing to the considerable overlapping of interests. Borderline projects, such as Einar Lundsgaard's work on the biochemistry of muscle contraction or Henry H. Dales's work on acetylcholine and nerve transmission, were allocated ad hoc by Weaver and Gregg.[90] In practice, Gregg took those projects that were directly related to psychobiology and Weaver took the more basic studies. Both tended to see their interests in terms of missions rather than disciplines or problems[91]—a mode that calmed the trustees' fears of academic disciplines. This intimate and cooperative arrangement was also useful for Mason in reassuring the trustees' misgivings about artificial boundaries within the Foundation. As he assured them: "If there is such complete understanding, it makes little difference where a project ends up."[92] Divisional cooperation was welcomed by Fosdick and the trustees as a sign that the divisional structure was not impeding interdisciplinary, problem-oriented projects.

This informal handing of jurisdictional problems led to confusion when Gregg and Weaver were not on hand, especially in the Paris office, and in 1934 more explicit guidelines were set. Weaver took all of embryology, biochemistry, and biophysics, the two NRC grants in sex and radiation, and all biological or biochemical projects in the other fields. Gregg took the more clinical projects in endocrinology and nutrition, human genetics, neuro- and electrophysiology, and most of pharmacology.[93] This division by fields rather than functions was the first step toward separation.

Fosdick's decision in 1937 to separate the MS and NS by subject areas seems to have been motivated not by actual administrative problems, but by a long-standing difficulty in the relationship between the officers and the trustees. Prior to the 1928 reorganization, the trustees had to make decisions about institutions and social or public health policies, i.e., issues familiar to them in their experience in law and business. But when the goal of the Foundation became scientific research, the trustees found themselves responsible for appropriating large sums of money for technical projects in which they had no expertise. Moreover, they were provided little advanced explanation, no alternatives to what the officers put before them as being the best proposals, and no evaluation of the success of past projects. Their enthusiasm was aroused by the officers for projects of which they never heard again.[94] As Weaver quipped, the trustees used to come out of meetings rubbing their hands; now they came out scratching their heads.[95] The trustees' natural distrust of academic science was heightened by lack of regular communication with the officers, and criticism of administrative inefficiency and duplicated programs were used to oppose divisional programs and policies generally. This was the weakest link in the Foundation structure, and the overlapping interests of the NS and MS divisions were especially vulnerable to such criticism.

In 1936 Mason retired as president and Fosdick succeeded him. The switch from a scientist to a trustee as president was a strategic move, intended to bridge the gap between the officers and trustees. In April 1936, Weaver wrote Tisdale to expect more frequent reviews and appraisals of projects.[96] In an extensive report in May 1936 to the incoming president, Weaver attempted for the first time a systematic evaluation of project success.[97] In November 1936, Fosdick invited Weaver to prepare a defense of "small projects," i.e., grants to individual research projects, in anticipation of trustee criticism.[98] Fosdick encouraged more open discussion of sensitive policy issues between officers and trustees, and Weaver seized every opportunity to explain his plans and procedures.

One of Fosdick's first moves as president was to rationalize administrative structure. In a letter to Jerome Greene in March 1937, Fosdick mentioned he had decided to relocate endocrinology, embryology, nutrition, and sex research in Gregg's division, leaving Weaver with biochemistry, biophysics, general physiology, genetics, and radiation effects.[99] The main advantage of a division of labor along recognized discipline lines was, of course, that Weaver and Gregg would not appear to be competing for the same territory. Weaver responded with a less radical proposal, assigning only endocrinology and sex to the MS division, and this resolution was approved by the trustees in late 1937.[100] Fosdick thus brought the NS division out from under the psychobiology umbrella.

This change was in keeping with Weaver's own interests. He was anxious to jettison sex biology anyway, and endocrinology and nutrition were beginning to lose their fashionable appeal.[101] Weaver wrote in 1938 regarding endocrinology: ". . . it is a field which lends itself to hurried work, exaggerated claims, unfortunate publicity, and premature and dangerous applications."[102] Even by 1937, he had little need for psychobiology to legitimize his program. Moreover, loss of endocrinology and sex research provided an occasion for Weaver to press for expanding his program into basic sciences such as organic chemistry, which had previously been defined as too distant from experimental biology. Knowing that it was unofficial policy to maintain more or less parity between the divisions, he was able to argue that the new arrangement with MS ought not to be construed as a contraction of NS: "I have hoped that when our program was narrowed on the one flank, by excluding endocrinology and sex research, it would be widened on the remainder."[103]

In December 1937, one week after the trustees' approval of the separation, Weaver pressed Fosdick to acknowledge organic and physical chemistry as being an official "minor" area of NS interest.[104] He asked Fosdick for a new staff member, preferably an academic biochemist: "Activities in this field constitute a considerable fraction of our program, and it is the one field in which we are least well prepared."[105] A year later, Weaver reported that he had not located a suitable candidate, and he proposed instead a series of surveys by eminent experts. His order of preferred fields reveals the new shape of his program:

> If we were to try out this procedure in one field, my choice definitely would be biochemistry: and this field is so large and so important to us that I would be inclined to suggest two rather heavily overlapping surveys,—one by a biochemist and one by an organic chemist.... If the survey in biochemistry proved useful, I would suggest genetics as a second choice and general and cellular physiology as a third choice. A somewhat briefer survey would suffice for the field of embryology and developmental mechanics; and we might eventually wish to have a general study made of biophysics.[106]

In short, Weaver used the loss of the psychobiological fields to justify moving into areas in the physical sciences that he had been unable to cultivate in the stringent years.

The assignment of psychobiology to Gregg also made it official that the special function of the NS division was to support basic science. In a conference of the Foundation and GEB in October 1938, for example, the officers of all divisions agreed that it was proper and necessary that each develop a different conception of "pure" and "applied" research. The social sciences were by their nature less abstract and more a matter of concrete social and economic problems. The natural sciences rightly focused on more abstract disciplinary research. The medical sciences lay somewhere in between, with research and application needing to be judiciously mixed to keep each from getting stale.[107] It was noted that the International Health Division, when it was limited to applied work, had suffered from the lack of input from basic research. Its director, W. A. Sawyer, pointed to the special position of the NS in providing this input: "... the NS should continue to concentrate on pure research because of the ability of other divisions to make use of the material which resulted, as, for example, in the field of nutrition, where discoveries were applied as rapidly as they were found."[108]

This division of labor was crucial for Weaver's program in

molecular biology. The Foundation program as a whole had to be relevant to social or medical problems; its public responsibility was the "welfare of mankind." But specialization of functions within the Foundation made it possible for one division to pursue goals that were in themselves not directly relevant to the larger goal of all. So long as Weaver's program was seen as feeding into the more utilitarian programs of other divisions, Weaver was able to concentrate on molecular biology without stirring up divisional rivalries or transgressing larger institutional ideals. The specialization of functions was implicit from the start in the Foundation leaders' conceptions of the different sciences; it was made explicit when Fosdick untied the knot between the MS and NS.

Weaver's move into organic and biochemistry was also facilitated by the improving economic conditions in the late 1930s. As old programs were phased out and the economy turned up, pressure on the budget was relieved, and the trustees again began to murmur about "scatteration" and the desirability of undertaking larger and more significant projects.[109] Weaver was prepared to take full advantage of this return of confidence. In November 1936, he wrote Tisdale that the stringent policies of 1934–1936 were a passing phase and urged him to look for opportunities in institutional grants, such as the grant to the departments of organic and physical chemistry at Oxford then pending before the trustees.[110] Tisdale replied that David Keilin at Cambridge had just been telling him of his need to enlarge his laboratory for studies of the cytochromes.[111] These projects could not have been thought of in 1934, but in 1937 they were just the thing to tempt the trustees. Tisdale worried that the trustees' dislike of small projects endangered the grant-in-aid program, but Weaver saw more clearly that the NS program could only benefit from an increase in larger "project" grants.

"Group grants" also began to play an important role in Weaver's program. These were clusters of related projects to groups within a single university such as organic chemists, biochemists, and biologists. The biological division at Caltech, with Linus Pauling, Thomas Hunt Morgan, and Henry Borsook in chemistry, genetics, and biochemistry, was a favored recipient of this form of grant.[112] Group grants were obviously related to Weaver's increasing ability to develop the physical sciences as such, and he saw them as pilots for future capital grants to university departments for the purpose of institutionalizing the application of physical science to biology on a permanent basis.[113]

The timing of economic ups and downs worked for Weaver in an interesting way. The selective pressure of hard times forced him to eliminate all but the essentials of his program in vital processes, pruning away what probably would have been diversionary interests in earth science, colloid chemistry, etc. Thus a coherent program evolved, clustered around a single theme, and when growth resumed it was these areas that grew most rapidly. Economic exigencies thus promoted Weaver's role of science manager.

The third factor in Weaver's shift from psychobiology to molecular biology was the changing opportunities within the fields of science themselves. I have already mentioned the decline in enthusiasm for the more dramatic aspects of clinical endocrinology by 1937. Since grants in this area were divided rather neatly between clinical studies (mainly American) and biochemical studies of hormones and enzymes (mainly European), the partition of 1937 left Weaver with the sounder investments. There are hints that psychobiology lost its appeal even within the MS division. Max Mason, as a member of the executive council of Caltech, tried in 1938 to hatch a program linking biochemistry, biology, neurophysiology, and psychiatry. A letter to Fosdick proposing Foundation support suggests he was not getting an enthusiastic response.[114] Sex research, on the other hand, was moving in the opposite direction, from physiology and toward behavioral studies of human sexuality, a controversial area that Weaver was anxious to avoid.[115]

Scientific opportunity was not a property of science alone, however, but of the fit between Foundation ideals and scientific promise. In his role as science manager, Weaver had to assess appropriateness to Foundation interests and his need for visible results, as well as the importance of the project by his client's standards. The ideal investment, from the Foundation point of view, was one with a certain degree of past success, but with great future promise—not too needy as to require a great deal of development capital before any results would show, nor so well provided for that more support would have little impact: in short, a good scientific growth stock. Investment was the business of many of the trustees, and a sound investment policy was something they appreciated. Since the application of new physical techniques to old problems in biology or biochemistry was almost bound to yield novel and often unexpected results, it was an ideal theme for an expanding program.

The success of Weaver's strategy is exemplified by the outcome of a second review of Foundation policies in 1938. Weaver managed the committee of experts smoothly; despite continuing uneasiness with the term "experimental biology," they put an enthusiastic stamp of approval on his program, policies, and procedures. The trustees were delighted, and from then until the outbreak of World War II, nothing disturbed the functioning of Weaver's program.[116]

Weaver's reports clearly reveal his policies regarding scientific futures. The NRC radiation effects area was a catchall, unrelated to Weaver's program, and Weaver felt that opportunities for investment there were few.[117] Experimental embryology he saw as a small but healthy field, needing no special stimulation or long-term development funds. He saw real opportunities in biochemical and X-ray analysis of developing embryos, but not in the foreseeable future.[118] This area remained small. Nutrition, on the other hand, was a vast and rapidly expanding field of great public concern, but it was also heavily supported by government agencies and the pharmaceutical industry. With such competition there was no room for private investment, and Weaver confined his interest to the biochemistry of vitamins.[119]

On the other end of the scale, biophysics illustrated the problem of an underdeveloped discipline without a core of coherent problems and methods, recognized academic standing, or university departments.[120] "Biophysics . . . is still for the most part an orphan subject. Able young physicists, however genuine their interest, hesitate to devote themselves to a profession which is insufficiently recognized to offer a reasonable chance for a permanent job."[121] To build biophysics as a field would require endowment of chairs, construction of laboratories, and systematic recruitment and training, a task the Foundation was unwilling to take on in 1938.[122]

Increasingly it was biochemistry, cell physiology, and genetics that provided the most attractive investment opportunities. They were well-established disciplines with well-defined problems and methodologies, notable records of success, available facilities and recruits, and most important, undeveloped areas within them of great promise, especially through the application of new physical chemist techniques. In cell physiology, there were Robert Chambers's studies of cellular microstructure, with the micromanipulator[123] and spectroscopical studies of cellular oxidation and reduction systems by Otto Warburg and others. In genetics, the chemistry and physiology of mutation, gene structure, and gene expression provided immediate opportunities.[124] All the early work by George Beadle and Boris Ephrussi and by Alfred Kühn on eye color "hormones" was supported by the NS, along with Beadle and Tatum's development of biochemical genetics in the early 1940s. (Mammalian genetics Weaver saw as a long-range opportunity, requiring investment in facilities, programs, and recruiting.)[125]

Biochemistry was perhaps the most promising field for application of physical techniques, and the majority of Weaver's projects in molecular biology were in this field, as were most of the projects involving cooperation of biologists and physical scientists.[126] There was The Svedberg's work on proteins using the ultracentrifuge; a number of the first ultracentrifuges in the United States were funded by the Rockefeller Foundation. The application of isotopes to biochemistry by Hans Clarke and Rudolf Schoenheimer and other groups was particularly successful, and William Astbury's work on the X-ray crystallography of macromolecules found special favor in Weaver's eye. Weaver also emphasized the critical importance of chemistry:

... developments within the United States of the whole divisional program are being and will continue to be critically limited by the weakness in this country of those fields of chemistry which should contribute most directly to biological studies. This remark applies mildly to physical chemical studies of high molecular weight compounds and to surface chemistry; but this remark applies with full force to the organic structural chemistry of natural substances. This field has been notably developed in Europe ... the leadership in American organic chemistry can be counted on less than the fingers of one hand.[127]

Organic chemists such as Robert Robinson were critical links in Weaver's program for the development of molecular biology.

Weaver's investment priorities can be seen in his aggregate expenditures for 1933–1938.[128] (See table 2.)

Weaver's increasing focus on a few academic disciplines did not impede his active role as a manager of science. His aim remained to further projects that were not confined to any one discipline. In 1936 he wrote:

A considerable part of the support given to date admittedly has added to the quantity of research in the chosen fields but without changing, in any significant way, the nature or quality of such research; but there are underway certain researches, certain general developments, certain reorientations of interest which would not have occurred if this program had not been followed. The major success of the program rests, although as yet potentially rather than actually, in these deeper influences.[129]

Weaver identified the work of Pauling, Hogness, Wrinch, Robert Robinson, Astbury, Rünnstrom, Bohr, Krogh, and Hevesy as the best exemplars of this deeper influence.[130]

Because Weaver's conception of the scope and direction of aca-

TABLE 2. Expenditures in Natural Science Fields, 1933–1938
(Percentages figured on total of items 1–9)

			Percentage
1. Application of Physical Techniques to Biology			
a. Math., physical investigations of tissues, cell, molecules	$175,904		2.5
b. Spectroscopy and biology	168,786		2.4
c. Isotopes and cyclotron	309,604		4.3
d. Physical chem. and biology	356,137		5.1
e. Organic chem. and biology	635,794		9.0
f. General biochemistry	369,375		5.2
g. General biophysics	141,290		2.0
Subtotal "vital processes"		$2,156,890	(30.5)
2. General physiology	651,052		9.3
3. Nutrition	290,153		4.1
4. Genetics	422,365		6.0
5. Embryology	131,026		1.9
6. Endocrinology	326,248		4.6
7. Radiation effects (NRC)	234,340		3.3
8. Sex research (NRC)	530,620		7.5
Subtotal (2–8)		2,585,804	(36.7)
9. Aid to groups	2,305,485		(32.8)
Subtotal concentrated program (1–9)		7,048,179	(100.)
10. Exceptions to program	245,835		
11. General program	677,500		
12. Old program	1,956,159		
Total (9–12)		$9,927,673	

demic disciplines was broader than those of his clients, his management resulted in real changes of direction within certain disciplines, biochemistry providing the clearest example. Looking down a list of recipients of project grants in biochemistry/biophysics, one sees only a few who were professional biochemists (Vincent du Vigneaud, Hans Clarke, Hans Krebs). The older leaders of American biochemistry are conspicuously absent, and physical and organic chemists, physiologists, physicists, mathematicians, and biologists make up the majority of recipients. Weaver's priorities reflect his implicit sense that the most promising work in biochemistry was being done by outsiders to the discipline. As a result, a new generation of biochemists emerged in the late 1930s, many out of Rockefeller Foundation programs. These new biochemists had assimilated a broader sense of the discipline; they knew how to use isotopes and the ultracentrifuge, and they combined a knowledge of organic chemistry with a sensitivity to physiological process.

The term molecular biology, first used by Weaver and Fosdick in 1938,[131] suggests this subtle combination of mission and discipline. Weaver noted that most of his projects in biochemistry/biophysics and all the cyclotron and isotope projects fell under molecular biology.[132] He hailed that biology as a new discipline in the academic order: ". . . a new branch of science . . . an advance which may prove as revolutionary . . . as the discovery of the living cell. . . . A new biology—molecular biology—has begun as a small salient in biological research."[133] This intersection of traditional disciplines with a cross-disciplinary mission, guided by institutional ideals of the Foundation and an active manager, had profound effects on the structure and research programs of several disciplines.[134]

Conclusions

The most striking result of this case study is the way in which Weaver's program and policies are expressions of the Foundation as an institution. The point is not that Foundation ideology determined the choice of fields patronized: that choice seems to have depended mainly on Weaver's preference and his reading of current scientific fashions. Options were open in 1929–1932, and the opportunity for individual influence correspondingly great. The point is rather that the Foundation determined Weaver's *role*. Here individual preference plays a relatively minor part. The resilience of Weaver's program and of his activities as science manager reflects the fact that it was a social role, i.e., a mode of behavior in keeping with the internal structure and external policies of the institution as a whole, supported by other individuals and seen as serving everyone's interests.

No single determinant was critical for Weaver's development of this role; it was the result of a subtle orchestration of structural factors shaping strategies and strategic choices exploiting structural opportunities. First is the institutional ideals of the Foundation, rooted in the managerial ideals of the Progressive period. More than either industry or government in the 1920s, the large foundations reflected the idea that science was a cultural resource to be managed in the interest of the nation as a whole. Trustee misgivings regarding the responsiveness of academic science to national needs should not obscure the agreement on essential points: science was central to national culture, and the RF had a special responsibility for developing this resource.

Second is Fosdick's role in maintaining the policies he initiated in 1928. Foundation ideals of science were consistent with many strategies for promoting science; the particular strategies that were used in the 1930s were the result of individual action. Moreover, Weaver's development as science manager depended on institutional stability. The new policies were controversial and vulnerable; directives from the trustees that fluctuated with the economic and political tides would have made it difficult to frame and execute a coherent long-term plan and to hold the trust of his clients. For instance, Fosdick's constant support for Weaver in the 1934 appraisal was crucial for the success of the officer's managerial role.

Third, the organization of the RF, encompassing all branches of science in its divisional structure and following academic rather than problem lines, permitted a flexible division of labor. This specialization of function enabled Weaver to develop a program in basic science even though it was further from the Foundation ideal of service than the other divisions. The potential for this divergence of style was implicit in the reorganization plan of 1928, but the division of labor was developed by Weaver as his plans matured.

Finally, there are the external circumstances of the Depression and the science holiday "Zeitgeist." It is difficult to demonstrate direct influence here. The trustees, representing the public face of the Foundation, were most sensitive to external pressures, but they were not actively involved in initiating policy and served to set economic and political limits to the officers' initiatives. On occasion economics played a determining role, as in 1934 when the trustees approved for economic reasons a managerial role for the officers that they opposed on policy grounds. But in general, economic and political factors were limiting conditions for the exercise of strategies within the divisions.

In conclusion, I would like to relate this case study to larger trends in the history of American science. In his study of American agricultural research stations, Charles Rosenberg suggests that the domestication of academic science in that new institutional niche was made possible by the emergence of a new social role, the scientist-entrepreneur.[135] With sympathetic understanding of both the academic scientist and the practical service needs of the host institution, the scientist-entrepreneur created a new market for science by mediating between groups with quite different conceptions of science and its social uses.

Since science grew in the twentieth century by the creation of many such markets for scientists, the conception of the scientist-entrepreneur has broad explanatory value.[136] In the rise of industrial research laboratories, for instance, one finds scientist-entrepreneurs such as Willis Whitney (General Electric) and Kenneth Mees (Eastman Kodak). In government bureaus, scientist-entrepreneurs such as Harvey Wiley and Gifford Pinchot pioneered a managerial role for scientists in government regulation. Weaver too may be seen as a variety of scientist-entrepreneur, not mediating a new institutional niche for academic scientists but rather a relationship between academic scientists and their new patrons, the private foundations. Weaver was special in that he was not the manager of an in-house research establishment but of a far-flung system of academic clients.

Weaver's role as patron and manager invites comparison with other attempts at managing science. George Ellery Hale's conception of the National Research Council as a mediating organization, locating opportunities for research and allocating government funds to the scientific community, reflected the managerial ideal. (As Elihu Root once put it, the idea was to apply the scientific method to science itself.) But Hale's plan foundered on his own unwillingness to accept a degree of accountability along with patronage, and on government preference for the more tightly controlled scientific bureaus.[137] The Science Advisory Board (1935–1938) was likewise seen as representing the estates of academic science to a potential federal patron. But it failed to be truly representative (social sciences were excluded) and thus was left politically vulnerable; it was not able to formulate a workable set of guidelines for the rights and responsibilities of patron and client.[138]

Perhaps because Weaver was neither an official representative of professional science nor of the public, he was more successful in striking a balance between the partly overlapping, partly conflicting needs of patron and client.

The structure and institutional ideals of the large foundation were perhaps more suited to the creation of a gentle but effective managerial role than were either professional bodies or government bureaus. With its traditional links to higher education and its quasi-public status, the foundation was in a unique position to mediate between scientific and public interests. The same factors within the RF that made it possible for the role of science manager to be institutionalized may also be invoked to explain why that role remained peculiar to the large foundations. Although it is premature to come to any general conclusions, I hope that this case study may begin to provide a picture of the increasingly diverse and intimate linkages between science and specific institutions—links that characterize the history of science in the twentieth century.

1. On foundations in general, see Ernest V. Hollis, *Philanthropic Foundations and Higher Education* (New York: Columbia University Press, 1938); Eduard Lindemann, *Wealth and Culture* (New York: Harcourt Brace, 1936).

2. Raymond B. Fosdick, *The Story of the Rockefeller Foundation* (New York: Harper, 1952); idem, *Adventure in Giving The Story of the General Education Board* (New York: Harper and Row, 1962).

3. Warren Weaver, "Molecular Biology: the Origin of the Term," *Science* 170 (1970): 581–82.

4. A. Landsborough Thomson, *Half a Century of Medical Research*, vol. I (London: H. M. Stationery Office, 1973).

5. Address, Rockefeller Archives Center, Hillcrest, Pocantico Hills, North Tarrytown, N.Y., 10591. I would like to thank Dr. J. William Hess for his gracious and expert help in using these archives. Documents cited here are identified by the following shorthand notation: RF, series number, box number, folder number.

6. RF.900.22.168. Agenda, meeting 11 April 1933, pp. 61–65.

7. George Corner, *The History of the Rockefeller Institute* (New York: Rockefeller University Press, 1964).

8. RF.915.1.1. Correspondence, 1913–16. Nathan Reingold, "The Case of the Disappearing Laboratory," *American Quarterly* 29 (1977): 77–101. I would like to thank Dr. Reingold for letting me see this work prior to publication.

9. Lindemann, *Wealth and Culture*, n. 1.

10. Fosdick, *Rockefeller Foundation*, n. 2. RF.800.22.167. Transcript of conference, 29 October 1930, pp. 102–6.

11. Stanley Coben, "The scientific establishment and the transmission of quantum mechanics to the United States, 1919–1932," *American Historical Review* 76 (1971): 442–66.

12. RF.915.22.168. Agenda for meeting, 11 April 1933, pp. 32 ff.

13. Ibid., pp. 28–32.

14. Ibid., pp. 19–20, 24–25.

15. Fosdick, *Rockefeller Foundation*, n. 2. On the reorganization of 1928, see Robert E. Kohler, "A new policy for the patronage of scientific research: the reorganization of the Rockefeller Foundation, 1921–1930," *Minerva* 14 (1976), 279–306.

16. See Abraham Flexner, *Funds and Foundations* (New York: Harper, 1952), pp. 77–100.

17. RF.915.22.168. Agenda for meeting, 11 April 1933, p. 31.

18. Ibid., pp. 36–37.

19. RF.900.21.160. Staff conference, 2–3 October 1930.

20. RF.900.22.167. Transcript of conference, 29 October 1930, pp. 82–83.

21. RF files on Division of Studies and Edwin Embree, passim.

22. RF.915.22.168. Agenda, 11 April 1933.

23. RF.900.17.125. Gunn to George Vincent, 17 April 1928; Vincent to Gunn, 27 April 1928. RF.900.19.139. "Report on Reorganization," 22 May 1928. RF.915.3.19. "Agriculture, 1927–1930." RF.915.3.22. Alan Gregg, "A Division of Agricultural Science," 1 April 1929. It was apparently David Edsall's advice that tipped the balance against an agriculture division: see RF.900.17.125, Edsall to Fosdick, 29 May 1928. Vincent to Edsall, 14 June 1928.

24. RF. *Annual Reports 1930*, pp. 189–191. Grants were made to Max von Laue, William Bragg, and Linus Pauling.

25. RF.915.1.1. Norma Thompson to Mason, 31 March 1930.

26. RF.900.22.167. Conference 29 October, pp. 92–95.

27. Ibid., p. 91.

28. RF.915.1.1. Max Mason diary, 22 April and 10 June 1931.

29. Warren Weaver, *Scene of Change* (New York: Scribners, 1970), pp. 49, 56–60.

30. Coben, "Quantum mechanics," n. 11.

31. See Garland Allen, *Biology in the Twentieth Century* (New York: 1975).

32. Robert Olby, *The Path to the Double Helix* (Seattle: University of Washington Press, 1974).

33. RF.900.22.167. Conference, 29 October 1930, pp. 64–65. See also RF.915.1.2. Jerome Greene to Raymond Fosdick, 29 March 1937.

34. RF.915.1.1. Weaver to Lauder Jones, 19 November 1932.

35. Ibid., Weaver memo, 18 October 1932.

36. RF.915.21.160. Memo of staff conference, 14 March 1933.

37. RF.900.22.168. Agenda for meeting of trustees, 11 April 1933. Weaver's statement is on pp. 76–87 and includes proposed budgets.

38. RF.915.21.160. Memo of staff conference, 14 March 1933.

39. RF.915.1.1. Weaver to Lauder Jones, 16 February 1934. Weaver to Fosdick, 14 November 1934. RF.900.22.166. "Report of the Committee of Appraisal," December 1934, p. 61.

40. RF.915.1.7. "The Medical and Natural Sciences," 13 December 1933.

41. Ibid., pp. 5–6.

42. RF.915.1.1. "Report of the Committee of Appraisal," December 1934, pp. 70–75.

43. RF.915.1.7. "The Medical and Natural Sciences," 13 December 1933, pp. 5–6. See also RF.915.1.2. Weaver to W. E. Tisdale, 27 March 1935.

44. See Carroll Pursell, "The Savage Struck by Lightning: the Idea of a Research Moratorium, 1927–1937," *Lex et Scientia* 10 (1974): 146–58.

45. Cited by Weaver, RF.915.1.6. "The Science of Man," 29 November 1933.

46. RF.915.1.6. Weaver, "The Benefits from Science," 27 January 1933, pp. 9–10.

47. Ibid., p. 5. See also Mason's remarks in RF.900.22.168. "Agenda," 11 April 1933, pp. 61–65.

48. RF.915.1.7. Weaver, "Progress Report, the NS," 14 February 1934, pp. 1–3. A similar strategy in the 1960s was used by Alvin Weinberg to argue for giving molecular biology priority over high energy physics. See *Reflections on Big Science* (Cambridge: MIT Press, 1967).

49. RF.900.22.166. "Report of the Committee of Appraisal," December 1934. The other members, James Angell and Walter Stewart, were little involved. The report was conceived and executed by Fosdick.

50. Ibid., p. 30.

51. RF.915.2.16. Weaver, "Report on NS for Committee of Review," November 1938. So anxious were the trustees over the dead hand of old obligations that they appointed two committees to estimate termination costs of old programs. RF.915.1.1. Weaver to Lauder Jones, 19 January 1933.

52. RF.900.21.160. Ernest M. Hopkins to Fosdick, 16 November 1934. Hopkins to Mason, 16 November 1934.

53. Ibid., Hopkins to Fosdick, 22 November 1934.

54. RF.900.22.166. "Report of the Committee of Appraisal," December 1934, pp. 44–45. The statement that the foundation would not promote research as an end in itself was deleted before the report passed. RF.915.1.1. Weaver to W.E. Tisdale, 27 December 1934.

55. RF.915.1.1. Weaver to Tisdale, 27 December 1934.

56. RF.915.4.41. Mason's nominees for this committee were all academic biologists (Mason to Fosdick, 28 March 1934). However, Fosdick chose Simon Flexner as chairman, perhaps because he knew Flexner was not sympathetic to Foundation programs. Dakin and Howell were Flexner's choices.

57. Ibid., H. D. Dakin to Simon Flexner, 16 November 1934.

58. Ibid., p. 2. ". . . the problems worked upon are set from the outside . . . instead of arising out of the interests of the workers themselves. One outcome has been the development of a sort of competitive struggle for tangible results which gives to scientific research something of the character of a business proposition." On G. E. Hale's similar ideas, see Reingold, "Disappearing Laboratory," n. 8.

59. Ibid., W. Howell to S. Flexner, 10 November 1934, and addendum.

60. Ibid., Simon Flexner to Fosdick, 19 November 1934, pp. 2–6.

61. Ibid., S. Flexner to Fosdick, 20 November 1934.

62. Ibid. See also Fosdick to S. Flexner, 19 November 1934.

63. RF.915.4.41. W. B. Cannon to S. Flexner, 21 November 1934, pp. 4–6.

64. Ibid., F. R. Lillie to S. Flexner, 10–12 November 1934, pp. 5–6.

65. Ibid., R. Fosdick to W. Stewart, 25 November 1934.

66. RF.915.4.38. Weaver to Fosdick, 22 March 1934.

67. RF.900.22.166. "Report of the Committee of Appraisal," p. 22.

68. RF.915.4.41. David Edsall to Fosdick, 23 November 1934.

69. RF.900.22.166. "Report of the Committee of Appraisal," p. 32.

70. Diana Long Hall, personal communications.

71. RF.900.22.166. "Report of the Committee of Appraisal," pp. 58–59.

72. Ibid., pp. 36–37.

73. Ibid., pp. 46–47, 78–79, 89, 90–92.

74. RF.900.22.166. "Report of the Committee of Appraisal," p. 61.

75. Ibid.

76. RF.900.21.160. A. Gregg memo, 12 November 1940. An injunction against large projects was deleted by the trustees before being passed —as Weaver saw it, "a friendly act." RF.915.1.1. Weaver to Tisdale, 27 December 1934.

77. RF.900.22.166. "Report of the Committee of Appraisal," pp. 89–90. A resolution to appoint a committee of trustees to sit with the officers on policy questions was passed but never acted upon.

78. RF.915.1.1. Weaver to W. E. Tisdale, 27 December 1934.

79. Ibid., Tisdale to Weaver, 16 January 1935.

80. Ibid., Weaver to Tisdale, 8 February 1935.

81. RF.915.1.1. "Progress Report," 27 January 1933. RF.915.22.168. Agenda for meeting, April 1933, pp. 83–87.

82. RF.915.1.6. "Progress Report," 27 January 1933. RF.900.23.171. "Director's Report," 11 December 1934, pp. 13–14. See also Weaver to F. B. Hanson, 20 April 1933.

83. RF.915. "Progress Report," 14 February 1934, pp. 3–4.

84. RF.915.1.6. Weaver, "Science and Foundation Program," 26 January 1933, pp. 13–14. For example, Weaver pressed for a capital grant to the Institute for Organic Chemistry at Göttingen on the grounds that its weakness hindered Foundation programs in other institutes of the University.

85. RF.915.22.168. "Agenda," 11 April 1933, pp. 83–86.

86. RF. Files on "Columbia, Heavy Water" and "Columbia Biological Chemistry." R. E. Kohler, "Rudolph Schoenheimer, Isotopic Tracers and Biochemistry in the 1930's," *Historical Studies in the Physical Sciences* 8 (1977): 257–298. See also RF.915.1.8. "Progress Report," 16 May 1936, p. 49.

87. RF.915.1.7. "Medical and Natural Sciences," 13 December 1933, p. 8. The budgets were $185,000 and $154,900 for the natural science and medical projects; but for the moratorium on grants over one year, appropriations would have topped $1 million.

88. RF.900.23.171. "Director's Report," 11 December 1934, pp. 17–19. Of these thirty-nine, thirty-one were in the United States, and six of the eight European projects were in biochemistry.

89. Ibid., pp. 19–20.

90. RF.915.1.2. Weaver to Tisdale, 27 March 1934. D. P. O'Brien to Gregg, 29 October 1934. Gregg to O'Brien, 19 July 1935. Weaver diary, 10 December 1935.

91. RF.915.1.1. Weaver to Lauder Jones, 26 January 1933.

92. Ibid., Mason to Strode, 9 May 1934.

93. RF.915.1.2. Gregg to O'Brien, 17 October 1934. Weaver to Tisdale, 28 October 1934. Tisdale to Weaver, 13 November 1935.

94. Ibid., Weaver to Tisdale, 26 May 1936.

95. RF.915.1.8. Weaver, "Program and Administration," 10 October 1937, p. 18.

96. RF.915.1.2. Weaver to Tisdale, 7 April 1936.

97. RF.915.1.8. "Progress Report," 16 May 1936, pp. 48–51.

98. RF.915.1.2. Weaver to Tisdale, 19 November 1936. Weaver, "A Case for Small Projects," 18 December 1936. See also RF.900.23.172.

"Minutes of Special Trustees' Meeting," 30 November 1936, p. 1, verso.

99. RF.915.1.2. Fosdick to Greene, 25 March 1937.

100. RF.915.1.8. Weaver, "Progress Report," 16 May 1938, p. 47 and passim.

101. RF.915.1.12. "Report of the Committee of Review," November 1938, p. 29.

102. RF.915.1.8. Weaver, "Progress Report," 16 May 1938, p. 32.

103. RF.915.1.2. Weaver to Fosdick, 29 November 1937.

104. RF.915.1.8. Weaver, "Program and Administration," 1 October 1937, pp. 24 ff.

105. RF.915.1.2. Weaver to Fosdick, 20 December 1937.

106. RF.915.3.26. Weaver to Fosdick, 4 October 1938, p. 2.

107. RF.900.23.172. "Summary of Conference," 10–11 October 1938.

108. Ibid., pp. 3–4. See also RF.915.2.12. Weaver memo 27 November 1940 and Gregg memo, 12 November 1940, on the divergence of the administrative policies of the NS and MS divisions.

109. RF.900.23.172. "Minutes of Trustees Meeting," 30 November 1937, p. 4.

110. RF.915.1.2. Weaver to Tisdale, 19 November 1936.

111. Ibid., Tisdale to Weaver, 18 December 1936, p. 3.

112. RF.205D. Files for "California Institute of Technology, Biology and Chemistry." See also files on "Chicago Biology" and "Stanford Biology."

113. RF.900.23.171. Weaver, "Director's Report," 11 December 1935, pp. 19–20.

114. RF.205D. "California Institute of Technology," Mason to Gregg, 19 August 1938.

115. RF.915.1.8. Weaver, "Progress Report," 16 May 1936, pp. 11–14. Weaver thus missed the chance to fund the Kinsey Reports.

116. RF.915.2.9., 10 and 12. RF.915.3.26 and 27 contain reports and correspondence concerning the 1938 review.

117. RF.915.1.8. Weaver, "Progress Report," 16 May 1936, pp. 42–44. Research funded under this head included photosynthesis, effects of ultraviolet rays on vitamins, X-rays on tissues and Gurwitch's alleged "mitogenic radiation" from dividing cells.

118. Ibid., pp. 14–16.

119. Ibid., pp. 32–41.

120. Ibid., p. 8. He also noted that many biologists disliked the term "biophysics."
121. RF.915.1.3. Weaver to Fosdick, 17 October 1938. This memo was advice to a new foundation as to opportunities for patronage of experimental biology. Weaver recommended biophysics to those with faith and patience.
122. Ibid.
123. RF.915.1.8. Weaver, "Progress Report," 16 May 1936, pp. 17–18.
124. RF *Annual Report 1935*, pp. 150–151 stresses the importance of "physiological genetics." See also RF files on Alfred Kühn, Boris Ephrussi, George Beadle, and M. R. Erwin.
125. RF.915.1.8. Weaver, "Progress Report," 16 May 1936, pp. 19–25.
126. Ibid., pp. 8–11.
127. Ibid., pp. 10–11. For similar statements, see RF.915.1.7. "Progress Report NS," 14 February 1934, pp. 10 ff. RF.915.2.16. "Report of the Committee of Review," November 1938, p. 12.
128. RF.915.2.16. "Report of the Committee of Review," November 1938, supplement p. 12a. These figures differ slightly from those presented in table 2.
129. RF.915.1.8. Weaver, "Progress Report," May 1936, p. 52.
130. Ibid., pp. 52–53.
131. RF *Annual Report 1938*, pp. 39–40, 203–217.
132. RF.915.1.3. Weaver to Trevor Arnett, 27 Dec. 1938.
133. RF *Annual Report 1938*, p. 203.
134. A detailed analysis of the effects of Weaver's program on the discipline of biochemistry is in preparation at this writing.
135. Charles E. Rosenberg, "Science, Technology and Economic Growth: the Case of the Agricultural Experiment Station Scientist, 1875–1914," in *Nineteenth Century American Science*, ed. George H. Daniels (Evanston: Northwestern University Press, 1972), pp. 181–209.
136. Ibid., p. 190.
137. Daniel J. Kevles, "George Ellery Hale, the First World War, and the Advancement of Science in America," *Isis* 59 (1968): 427–37. Daniel J. Kevles, *The Physicists*, (New York: Knopf, 1978).
138. Lewis E. Auerbach, "Scientists in the New Deal: a Pre-war Episode in the Relations between Science and Government in the United States," *Minerva* 3 (1965): 457–82.

Acknowledgments:
Research for this paper was done with the assistance of a grant from the National Library of Medicine of the National Institutes of Health (grant no. LM-02630).

An abbreviated and revised version of this paper was published as "The Management of Science: The Experience of Warren Weaver and the Rockefeller Foundation Program in Molecular Biology," in *Minerva*, XIV, no. 3 (Autumn 1976), 279–306. Portions are reprinted here with permission of Edward Shils.

THE PHYSICS BUSINESS IN AMERICA, 1919-1940: A STATISTICAL RECONNAISSANCE

SPENCER R. WEART
American Institute of Physics

PHYSICISTS AND THEIR HISTORIANS LOOK BACK ON 1932 AS A "YEAR of miracles." In that year, the discoveries of the positron and the neutron tore down the last barriers on the way to a new understanding of the fundamental nature of matter, while the discoveries of heavy water and of two different kinds of particle accelerators handed science entirely new types of experimental tools. American physicists could be particularly proud, for their country had at last reached a level of parity with the rest of the world: three of these five revolutionary advances were made in the United States.

Other people have darker memories of 1932. What physicists recall as a year of triumph, many other Americans recall as a year of confusion and despair. What historians of science hail for a revolution in knowledge, other historians name as a time when American workers, farmers, and intellectuals flirted with an uglier sort of revolution.

This is a disturbing paradox, if we believe that scientists are not bodiless intellects but are somehow rooted in the society that raises and sustains them. The two images of 1932, which seem to refer to altogether different worlds, somehow must reflect a single reality. When I began to study the development of American physics between the two world wars it was this paradox—or rather the more general paradox of the triumphant advance of physics during a period of national disaster—that gradually drew my attention.[1]

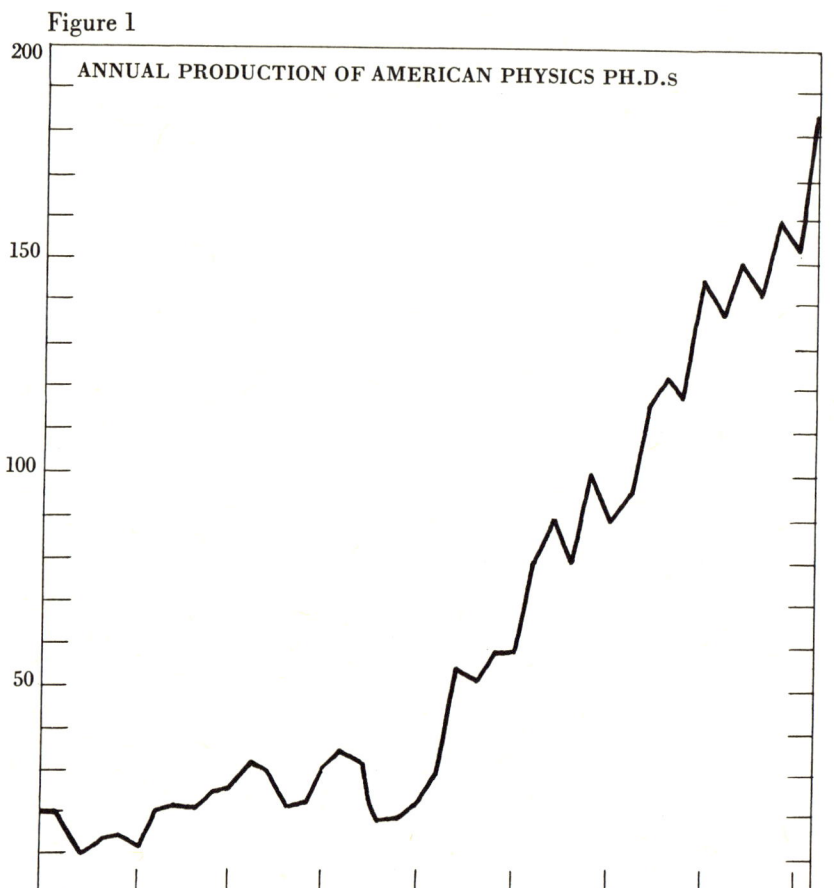

Figure 1

ANNUAL PRODUCTION OF AMERICAN PHYSICS PH.D.S

See footnote 2.

A first look at some statistics will quickly show the depth of this problem. Figure 1 gives the oldest, most complete and most unambiguous of all the relevant statistical series we possess. The shaded area shows the number of Ph.D.'s granted in physics in the United States from 1900 through 1941.[2] It shows a spectacular rise, a doubling every decade. We recognize this as only one segment of a much more extensive curve, the well-known exponential growth curve of science, extending back into the sixteenth century and forward to the 1960s.[3] The only visible interruption in the curve of figure 1 was caused by World War I, and afterward the rise continued as if the war had never been. The Great Depression of the 1930s is not visible at all.

In the natural sciences, exponential growth curves commonly point to feedback mechanisms. I am inclined to believe that science has risen exponentially because it is part of a larger system involving multiple feedbacks among technology, industry, education, government, bureaucracy, and all the other communities whose proliferation we subsume under the catchall term "modernization." If so, my specific problem becomes only one strand in a tangle of more general and less answerable problems whose theme is the question: how has science grown? The Great Depression then appears as merely incidental to this growth.

These are the issues that this essay addresses, but since the field of inquiry is wide and our knowledge of it slight, we cannot hope to squarely meet and solve the problems. I thought it best simply to reconnoiter, to locate some of the elements that may eventually be combined into an explanation of the growth and strength of American physics. Of all the various types of evidence, in this reconnaissance I have looked particularly for statistics. Certainly statistics may obscure raw human reality, but it is also precisely in statistics that we may hope to recover this reality, to comprehend in a glance something at least of the life histories and struggles of thousands of human beings and their variegated institutions.

I am not the first to want numbers on these matters. The search was started by the physicists themselves, hoping to understand the forces acting on their lives, beginning around the 1920s and continuing with increasing intensity down to the present moment. A few historians have helped them, and I have added some missing items. I will begin with comments on the 1920s, reserving until the discussion of the 1930s those statistical series that give an overview of the entire period between the wars.

The Universities in the 1920s

The most striking feature of American physics in the 1920s is of course growth. The rush of new Ph.D.'s seen in figure 1 stayed on as physicists: 90 to 95 percent of them can be turned up, years after they graduated, in the membership lists of The American Physical Society or in Cattell's *American Men of Science*.[4] Where precisely were they? Above all, they were in institutions of higher education. Of the members of the American Physical Society—a group that comes as close as any to including by definition all American physicists—some three-quarters were in education throughout the 1920s and 1930s.[5] These professors were not spread uniformly among the hundreds of colleges in the United States but were highly concentrated. Throughout the 1920s twenty leading departments of physics employed more than 40 percent of all the physics teachers in the American Physical Society. The twenty leading departments were even more important than this figure suggests, for they produced more than three-fourths of all papers published in the leading American journal, the *Physical Review*, and they trained about nine-tenths of all the new physics Ph.D.'s.[6]

The story of American physics is thus largely the story of the growth of these leading academic physics departments, and historians have naturally focused their attention on them. Such historical studies have concluded that there were qualitative changes fully as striking and significant as the quantitative growth.[7]

Around the 1920s, American physics came of age. From a state of distinct inferiority before World War I it improved until, by the early 1930s, it was the equal or superior of physics anywhere in the world. This too has its quantitative dimension. For example, an early citation study, covering all the reference citations found in papers published in a number of physics journals in 1934, turned up only 21 citations to papers published in the leading American journal, the *Physical Review*, in the interval 1895–1914, compared with 169 citations to the German *Annalen der Physik*. But in citations to papers published in 1930–1933, the *Physical Review* beat the *Annalen* three to one, and as of 1933 it had become the most-cited of all physics journals.[8]

The rapid increase in the simple numbers of American physicists no doubt played an important role in this rise by making possible a numerical increase in publication. Beyond this, the first thing to look for would be an increase in the research spirit, the desire of the physics teacher to do research and the willingness of his school to support him while he does it. We are unusually fortunate here, for we can find surveys of the amount of time spent on research in physics departments at a number of important schools, both around 1910 and in the 1930s. These surveys indicate that there was no very pronounced change. Physics teachers spent about a third of their time on research in 1910, and about a third in the 1930s.[9] At Columbia University, for example, physics professors by their own reports spent 27 percent of their working hours on research in 1932, compared with 23 percent of their time more than in 1910.[10]

From such data, as well as the historian's usual biographical and anecdotal evidence, we can be confident that the research spirit as it existed in the 1930s was already present at the leading schools before World War I. We also know that it was less strong at the hundreds of smaller schools scattered across the land.[11] It remained true, as mentioned before, that the leading departments produced the lion's share of publications, so there cannot have been any very significant diffusion of the research spirit out to the smaller colleges.

So physics teachers may not have had much more time available for research in the 1920s and 1930s than they had before. But there was a significant increase of the nonteaching staff, of purely research positions. I refer not to research professorships—although some of those were created too—but to a great increase in the numbers of postdoctoral fellowships. Among these the National Research Council Fellowships, administered by the Council and paid from Rockefeller Foundation money, have often and justly been pointed out as a main force in the improvement of American physics. In the 1920s and early 1930s they were granted to 15 percent of all physics Ph.D. recipients, giving these elite students one or two years of free time for research at an institution of their choice.[12] The effects were quickly felt. For example, of all papers from identified American sources published in the *Physical Review* in 1925, fully one-eighth acknowledged support from a National Research Fellowship.[13]

All this is important but does not get us to the center of the problem. For all the volumes of valuable work done in this country, one cannot escape the impression that even into the 1920s, many American physicists were not coming to grips effectively with the central and exciting problems of modern physics. The Americans themselves were aware of this.[14] The point was neatly put in 1927 by a young physicist who had recently taken up a post at a leading school. It was "a very fine place with a good research reputation," he wrote to his old professor, "though I confess that much of the research seems of a difficult rather than important nature."[15] This disgruntled young man was a theorist, and it was evidently the lack of theorists that he noticed. Historians looking back on the 1920s have also noticed this, and point to the introduction of modern theory as a chief factor in the improvement of the quality of American physics.[16]

Theory could be introduced only from where theory already existed, and this meant Europe. Quite suddenly, the spirit of European theoretical physics flowed into America. Some of this took place through classic channels of communication such as published papers, books and conferences; much more took place through direct contact when young Americans went to study in Europe, as most of the best new physicists did in the 1920s. Many young Europeans also came to work in America.[17] Along with the students came professors, for American physics departments invited the renowned physicists of Europe to come and lecture for a few days or a few semesters.[18] The advantages of this were widely spread. For example, in 1923 William F. Meggers invited the great theorist Arnold Sommerfeld to lecture at the National Bureau of Standards. Of course government regulations prevented Meggers from being able to pay Sommerfeld's travel expenses; but Sommerfeld came nevertheless, for he was already in the United States lecturing at Wisconsin, Caltech, and other schools. At the Bureau he gave seven lectures on quantum theory and atomic spectra, attracting more than a hundred physicists from the Washington, D.C. area.[19] The traffic in foreign lecturers became so heavy that John Slater, looking back on the 1920s, recalled that he had seen European physicists almost more at MIT than he had in Europe.[20]

Some came to stay. By 1930, Paul Epstein, Samuel Goudsmit, Otto Laporte, John von Neumann, and George Uhlenbeck had all come from Europe to teach theoretical physics as permanent residents in the United States.[21] This unprecedented movement clearly indicates that some saw better chances for advancement in America than in Europe.[22] At the same time, American students began to write theoretical theses and find jobs, so that by 1930 there were perhaps two or three native Americans teaching theoretical physics in this country for every transplanted European.[23]

There were various reasons for the new popularity of theorists in the United States, but the principal one was probably a fundamental change in the nature of physics research. A study by George Magyar indicates that in 1910 only about a fifth of the world physics literature consisted of papers that were mainly theoretical, whereas since 1930 nearly half of the papers have been mainly theoretical.[24] This remarkable shift must have had various causes, but surely one of the chief of these was the development of quantum theory, the most broadly successful intellectual revolution in the history of physics, taking place mostly in the 1920s. "Perhaps we shan't need to do any more experiments now," Robert Mulliken wrote in 1927; "the wave mechanics seems to predict and explain everything so well."[25] While things didn't go so far, already by the early 1920s quantum theory was casting a shadow over the experimentalists predominant in America. In some fields, such as spectroscopy, it was impossible to do up-to-date work in ignorance of the latest and most difficult theories.

Spectroscopy is worth a closer look, for since the nineteenth century Americans had made a specialty of manipulating light. As late as 1935 in a listing where American physics departments indicated their field of special facilities or interests, spectroscopy was in front (closely followed by the related field of X-rays).[26] Spectra, whether of light or X-rays, became in the 1920s the subject of theories that were mathematically and conceptually difficult and changing from month to month, so it is not surprising that spectroscopists called for theoretical assistance.

Such calls for help are a recurring theme in the histories of American physics departments in the 1920s.[27] For example, Robert Seidel, looking into the appointment of J. Robert Oppenheimer as theorist by the University of California at Berkeley, finds this was justified in part by the expectation that Oppenheimer would help the experimentalists on the staff with theoretical questions arising from their experimental work.[28] Another example: at the time Meggers invited Sommerfeld to lecture at the Bureau of Standards, Sommerfeld's theoretical work was the most exciting development in the field that Meggers and his section were cultivating experimentally, the field of spectroscopy. This last example may serve as an introduction to the next major topic of this essay: applied physics and its relation to the universities.

Applied Physics and the Universities

Why would Sommerfeld be invited to the Bureau of Standards? This federal agency had always been a center for practical work, and under Secretary of Commerce Herbert Hoover it was becoming more so. The research was sometimes paid for directly by the industries concerned, and applied work was threatening to crowd out more basic research.[29] But in spectroscopy there was no contradiction, for although it had its roots in the most basic atomic physics, spectroscopy had branches bearing fruit in applied areas. Meggers himself said he got interested in theory when he grew dissatisfied with the purely empirical methods used in spectroscopic chemical analyses; he hoped his research would lead to an increase in the practical value of spectroscopy, any increase in basic knowledge being welcome but incidental.[30]

What did Meggers, from his vantage point at the Bureau, think of the trends in American physics in the 1920s? Discussing the scene in a letter dated 1924, he did not seem aware of the movement historians have discerned, the advance and improvement of pure science. He feared that there was too little of this. The main trend he noticed was an enormous advance of applied physics—in government, in industry, even in the universities—a "fever of commercialized science."[31] And I believe he was not mistaken; the advance of applied physics was an obvious characteristic of American physics in the 1920s, at least as significant as the advance of theoretical physics.

In 1913 industrial physicists made up only about a tenth of the membership of the American Physical Society. In 1920, they made up a quarter of the membership—and this advance took place while the total membership of the society was doubling.[32] The proportion of papers published in the *Physical Review* that originated in industrial laboratories showed a similar fundamental increase.[33] These facts tend to confirm what many people were claiming at the time: the relations between physics and industry in America were revolutionized around the time of World War I.

The trend was already becoming clear before the war, but the war itself was widely believed to have played a central role in accelerating the transformation. Around 1917, many physicists left the campuses, but not for the Front; for them, as for most Americans, the great event of the war was a fierce intensification of industrial organization and production. Industrialists and physicists were thrust into closer contact with one another, both at the production level and at higher levels in Washington. The result was that both groups became more aware of what could be done with physics. Before the war only a few corporations thought of using the Ph.D. physicist outside his academic home, but by war's end this had changed.[34] The lesson was not lost on those who expected gunfire to be replaced by fierce postwar commercial competition. As one industrial physicist put it, "Bankers have become science-minded. They are familiar in general with the influence of scientific research upon the trend of security prices." They had come to esteem science, for in "commercial warfare . . . research supplies the ammunition."[35]

Research no longer meant simply development work aimed at cutting costs and improving an existing product. In some industries it had become essential to do the sort of research, sometimes verging on basic research, that could produce an entirely new product or could keep a company supplied with information and patents for defense against new products developed elsewhere. For example, both General Electric and the Bell Telephone Company were able, thanks chiefly to the trained physicists in their laboratories, to maintain a strong position when the new field of radio communication emerged around the time of the war. (It was such large companies that played the main role in industrial research; General Electric and Bell alone employed some 40 percent of all the American Physical Society members in industry between the wars.)[36]

The presence of trained scientists also brought a company various indirect benefits. Publicity and prestige were perhaps chief among these. As one knowledgeable observer bluntly stated, "The use of real or imaginary research laboratories to back up merchandise . . . has an importance to company prestige that cannot be exaggerated."[37] This was only true, of course, because the era of the 1920s was "the golden age of scientific faith," not only among scientists and industrialists but also for the public at large.[38]

For all these reasons, a good market for trained physicists appeared in industry. One straw in the wind was Ernest Fox Nichols, a prominent professor of physics who had worked for the Navy during the war and who in 1920 resigned from Yale to join General Electric. "The position," he informed the university as he departed, "offers complete freedom in the choice of research problems, and places at my unhampered disposal such human and material resources as no university I know of can at present afford."[39] This sort of migration may have been strengthened by the cruel inflation of 1919–1920 when college teachers' salaries, less flexible than salaries in industry, plunged in real terms. The crisis passed swiftly but probably left a strong impression, and academic pay remained notoriously lower than industrial.[40] Fears were expressed that the ranks of academic science would be decimated by the movement into industry.[41]

Some academic scientists responded by defending their old ideals of pure noncommercial research. The physicist in a college or university had his own rewards. "Financially his lot is not a wealthy one," as one of them admitted "nor socially is he high. . . . [But] close to nature and by it closer and closer every day to the Almighty God who made him, what matters it to the physicist if the days be dull or neighbouring man uncouth? His soul is more or less aloof from mortal strife."[42] In short, it was all very well to go into industry if money was your object, but there was still something superior about the life of pure research. Frank Jewett of Bell Labs recalled that his mentor Albert A. Michelson "thought that when I entered industrial life . . . I was prostituting my training and my ideals."[43]

Yet in the end the schools were not drained but strengthened by the growth of industrial physics. For it was far less the established physicist who was snatched up than the new graduate. There is some evidence that the majority of the students enrolled in basic physics courses—by far the most numerous of the physics teacher's clients—expected to make a career in industry (not necessarily as physicists, of course).[44] This meant that industry was supporting an increase in enrollments that must have had much to do with the overall rise of academic physics departments in the 1920s and 1930s, although to say just what this effect amounted to would require more sophisticated models and more detailed statistics than we can at present deploy.[45] Certainly there were physics teachers who felt that their departments would be strengthened by attracting students destined for industry. Applied physics courses or programs, sometimes labelled "engineering physics," were set up by physicists at a number of schools in the early and mid-1920s.[46]

Industry took not only undergraduates trained in physics but also fully formed Ph.D.'s. Back in 1900 only a tenth of the few Ph.D.'s who graduated each year landed in industry within a few years after graduation; in the 1920s about a fifth of the far more numerous Ph.D.'s were going into industry.[47] For research, where the quality of personnel is everything, the leaders of industrial laboratories had concluded that the best recruits were those with graduate training.[48]

To select and attract these recruits, they had to have close contacts with the professors who were training students. This made for strong ties between industry and the campuses.[49] This also gave industry a new motive for supporting research in the universities. Through such support, a company could keep its eye on promising young people while at the same time staying abreast of scientific developments. These were years when industries were glad to pay outside laboratories like the National Bureau of Standards and the Mellon Institute for Industrial Research to hire scientists to do research of specific interest.[50] So it was not surprising that the companies were also willing to pay for research at universities. For example, Corning Glass sponsored fellowships and research at Cornell, Illinois, MIT, and Purdue on problems of chemical physics directly related to glassmaking. Such arrangements were not uncommon.[51] If it didn't wish to give outright fellowship or grant support, a company could still contribute in smaller ways and keep up close relations; reading the acknowledgments in papers published in the *Physical Review* in the 1920s, I find frequent thanks given to companies for their loan or construction of pieces of apparatus or for the use of their laboratories.

All this led some to feel that the universities were "beginning to be led by the industries, instead of vice versa."[52] But insofar as the point was to train researchers, the exact nature of the research was secondary.[53] I do not believe that freedom of inquiry in universities was seriously infringed upon by the support industry gave but rather that the possibilities for inquiry were broadened all around.

I am even willing to entertain the hypothesis that the close ties with industry gave American physics a peculiar national advantage. The Europeans who came to the United States in the 1920s and 1930s noticed things that were unusual in the Old World: close cooperation, ample funding, large and technically difficult experiments. Industry may have been a source of these tendencies and of the related tendency, still embryonic, toward the differentiation of research managers and the separation of research workers from proprietorship of their tools. Even the style of research may have been subtly affected. For example, one European made a nice distinction between European physicists' *"simplicité"* and *"improvisation"* and the Americans' beloved *"élégances techniques."*[54] Without going further into this foggy question, let us note that there does seem to have emerged—in America before anywhere else—a new breed of scientist, the combination of physicist and industrial engineer.

I am not claiming that the contributions from industry, intellectual or material, were more than supplementary so far as academic physics was concerned. The main intellectual trend in the leading schools was the advance in the quality of pure physics described before. And the department budgets continued to come chiefly from traditional sources: endowment, direct philanthropy, state government appropriations, and above all tuition. All of these were growing vigorously in the 1920s, and all of these will have to be studied in detail before we will fully understand the growth of American physics.[55] Lacking such information, I will try to press on with the inquiry in another way, by crossing the watershed that the year 1932 marked in the life of the nation.

The Effects of the Depression

The Great Depression gives us a peculiar opportunity, the opportunity to study a system struck by a great shock. By listening for the reflected seismic waves we can hope to learn something about the internal structure of the system. Of course, we expect ambiguity and obscurity in such echoes. I therefore present such statistics as I have been able to discover or create so that readers, if they find my interpretations implausible, will have some materials to construct their own.

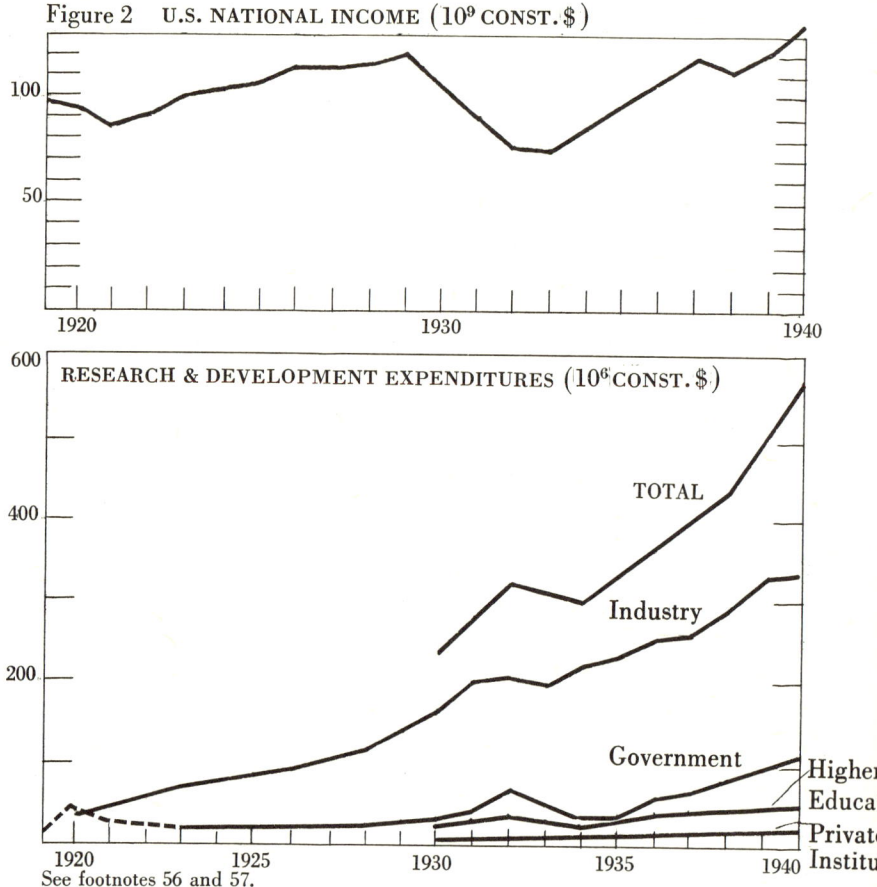

Figure 2 U.S. NATIONAL INCOME (10^9 CONST. $)

RESEARCH & DEVELOPMENT EXPENDITURES (10^6 CONST. $)

See footnotes 56 and 57.

We may begin with the most general and therefore most crude series (figure 2). The shaded area of the top graph gives the backdrop to everything else that happened, the trend of national income in constant dollars from 1920 to 1940.[56] Below are some rough figures (from studies made for the 1945 report *Science, the Endless Frontier*) suggesting the trend in funding for all U.S. research and development.[57] The top line, giving the total, apparently shows some effects of the Depression, but these are delayed and much softened in comparison with the monstrous collapse of the national income. The next line gives the investment of industry, by far the biggest component of the total when we consider not only pure research but also all types of development work; next is government expenditures, which shows a definite delayed drop during the Depression; at the bottom are higher education and private institutes. In these figures, even those for higher education, items such as engineering studies and agricultural experiment stations are strongly present, so we haven't learned much about physics as such. What we do learn is that the average scientist and engineer may not have had exactly the same experience in the 1930s as the average American.

It will become evident below that pure research in physics was done chiefly in the universities, so those must be the first places to look for more detail. Figure 3 indicates how college and university teachers fared from 1914 through 1940. The shaded area of the upper graph shows the rise over this period in the number of teachers in higher education in the U.S.[58] Only in a single year, 1934, was there even a gentle decline.

The total sum invested in academic staff depends not only on how many there were but also on how well they were paid. The median pay of teachers in large public institutions is shown in the lower graph of figure 3, where the higher line is for full professors and the one beneath for instructors.[59] The worst period was apparently the inflation of 1919, not at all the Depression. The graph, like all graphs in this paper, is given in terms of constant (1947–1949) dollars rather than in current dollars, which would show a very different picture. Many professors took a cut of 10 percent or more in pay during the Depression. But deflation—a phenomenon almost incredible to the modern reader—cancelled out the effects of the cuts in terms of purchasing power.[60] College and university teachers fared better than people in most fields, even other professions: while the average pay loss for academic people in current dollars was perhaps 15 percent, lawyers' incomes dropped 30 percent, physicians' 40 percent, consulting engineers' 60 percent.[61] Of course some schools were affected more than others, and there was a real decline in some professors' standard of living. Complaints were heard that they were forced to make their own repairs around the house and even live without a maid.[62]

All these statistics describe average faculty in higher education, but such hard evidence as I have found for physicists indicates that their experience did not deviate much from the norm, either in numbers employed or in salaries.[63] Leaving numbers aside for a moment, the relatively strong position of academic physics in America is also suggested by the welcome given to refugees from fascism. Something like a hundred physicists came to the United States from fascist countries, chiefly Germany and Austria. At the Spring meeting of the American Physical Society in 1937, one physicist remarked that there were so many Germans that it bore a striking resemblance to the old German Physical Society.[64] The

Figure 3

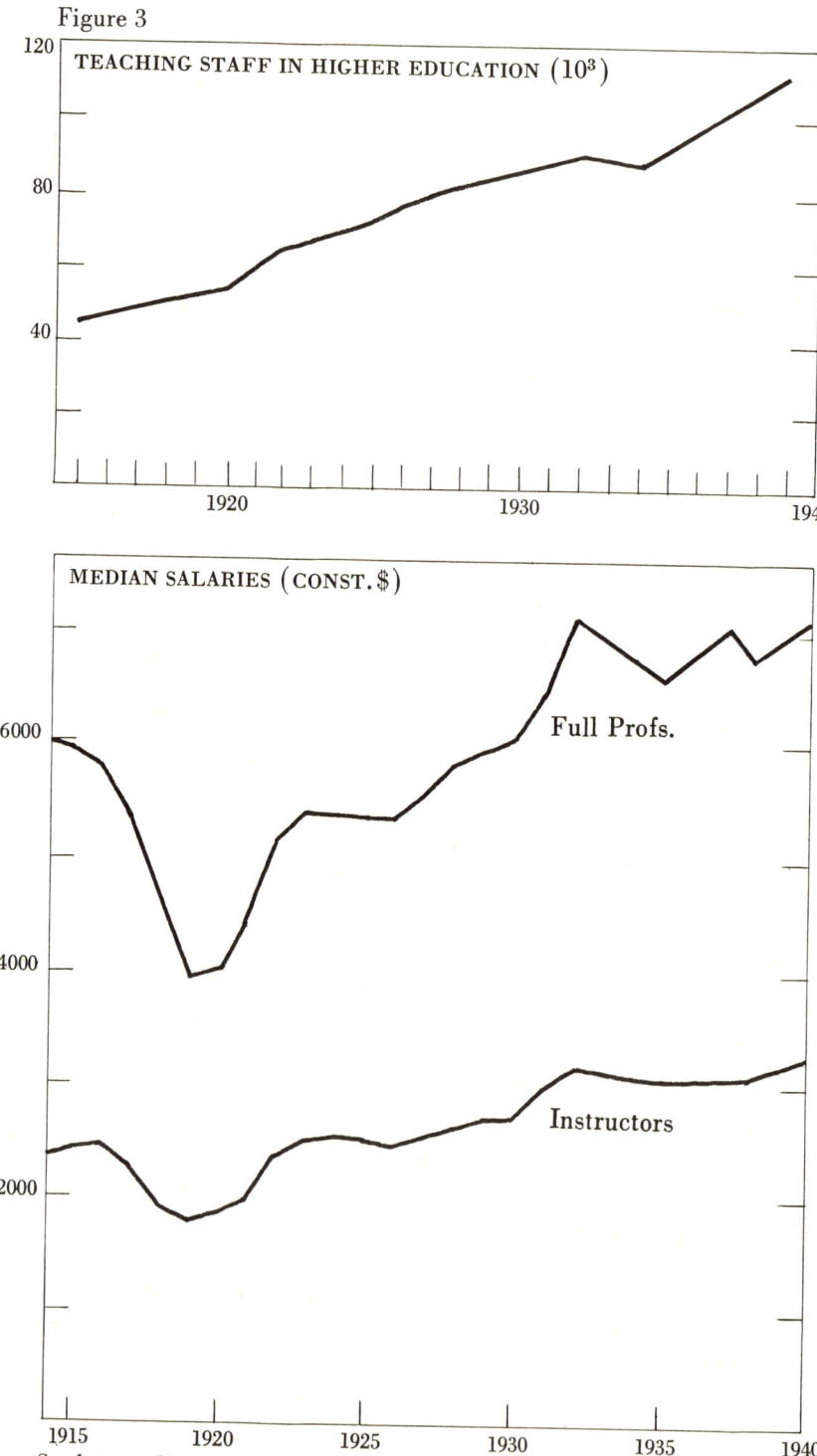

See footnote 58.

refugees encountered some resistance, partly because they were competing with Americans for jobs and partly because they were often Jews (exclusion of Jews had been the rule in some American physics departments in the 1920s).[65] That these resentments were overcome, and that space was made for most of the exiles, spoke both for the generosity of Americans and for the continuing demand in America for good physicists. Conditions in the United States were better than in Europe not only politically but economically; this had been clear when Europeans were first attracted in the 1920s and was still more clear from 1939 on, when more than half the refugees arrived.[66] In return for the gift of refuge, the brilliant group of emigré physicists greatly enriched America. The transfer of the cream of Central European physics to this country meant that the United States, already the peer of any country in the world by 1932, became predominant in physics.

American academic physicists were generally aware of the strength of their profession. In reading their correspondence of the time and their retrospective accounts, I find few signs of concern about the Depression before 1932 or after 1935.[67] During the few intervening years, they confidently expected that the "emergency" would soon pass. Nevertheless they did have some bad moments. As university budgets were cut and cut again around 1932 and 1933, many physics departments were invaded by uncertainty or even fear. For example, in 1933 one noted physicist wrote another, "They pay our salaries very irregularly at present and nobody knows whether we shall receive what is due us. Instead of dismissing people they simply keep them on but don't pay them, at least those are the expectations for next year."[68] Moreover, a decline in the numbers of junior faculty and teaching assistants and the threat to personal income probably made for an increase in the teaching load.[69] And there was a group which experienced not only anxiety and salary arrears but acute suffering: the new Ph.D.'s.

For two or three years around 1933 universities froze employment or sometimes laid off junior faculty. Even a gentle dip in employment can mean a great deal to those who are just entering the job market, particularly when their numbers keep increasing.[70] To make matters worse, the number of National Research Fellowships awarded dropped abruptly.[71] The new physics Ph.D.'s had a hard time finding positions. For example, in 1933 Prof. Samuel Allison of Chicago wrote Prof. E. C. Kemble of Harvard on the "very slim chance" that Kemble might have work for someone who knew X-rays; Allison had no less than three "exceptional men" looking for a job. "As you know," Allison continued, "conditions are such that there is no longer any question as to what salary is acceptable, as long as it is possible to eat and live on the income." Kemble promptly replied that there was no hope for a job at Harvard, adding, "The situation is a sickening one . . . I could give you a list of theoretical men to parallel yours in the X-ray field, but I guess every university is choked with its own graduates . . ." The only thing to do, Kemble felt, was "for every university to keep as many as possible of its fledgling Ph.D.'s as assistants at nominal salaries."[72] This was the solution most departments did adopt; part-time appointments and fellowships managed to keep the young physicists going while they waited for a job to come along. The most delicate sort of paternalism had to be exercised. At one school where a few half-time assistantships had to be doled out among a larger number of "lame duck" Ph.D.'s, two of the graduates were excluded because their wives had jobs in the university library.[73]

The extent of the new Ph.D.s' problem is indicated by some statistics. If we take the group of physicists who graduated in 1934 and look them up in the membership list of the American Physical Society for the following year, we find close to half of them still at the school where they got their Ph.D.—twice the proportion that stayed on at alma mater in normal years.[74] I will return later to the question of where these lame ducks finally went. While they were staying on, they continued their physics research, their labor providing an important supplement to their departments' strained resources.

Funds for research were a source of anxiety to all physicists. Even the professor who could be fairly confident of keeping his job and pay often did not know whether he would be able to support his research, at least for the few years he expected the crisis to last.[75] Support funds were cut at some departments, but the overall picture did not turn out to be too bad.

It was generally true for all fields that universities did not reduce their research activities even during the depths of the Depression.[76] Where income fell, as it often did, great efforts were made to cut corners in order to keep the research going. In two cases I know of, physics departments simply ran through their reduced funds before the end of the year but kept going on emergency grants from their universities' general funds.[77] The most serious consequence of tight budgets was probably the almost complete stoppage of planning and construction of new buildings, including laboratories, in the universities. But since some departments had moved into handsome new quarters during the 1920s, and since capital expenditures in higher education recovered fairly quickly, the problem of crowding in outgrown quarters probably was not widespread.[78]

Figure 4

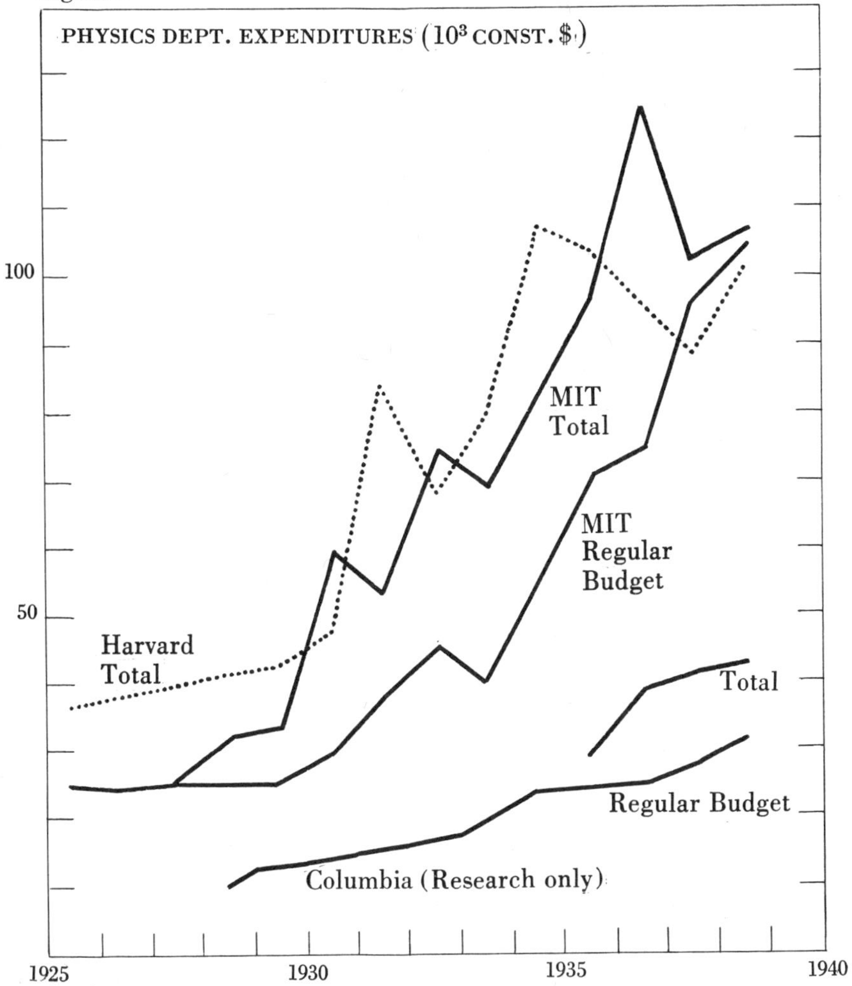

PHYSICS DEPT. EXPENDITURES (10^3 CONST. $)

See footnote 80.

In a number of leading schools the physics departments flourished as never before during the 1930s.[79] Figure 4 shows budgets, aside from salaries, at three important physics departments from the late 1920s through 1939.[80] The Depression makes no imprint on these curves. The dotted line is the Harvard equipment budget. The solid lines paralleling it are the same for MIT (the lower of this pair of lines is the regular departmental appropriation, and the upper line gives the total equipment expenditure, including grants, particularly for the construction of a cyclotron). The lowest lines are for the Columbia physics department and display rare and precious data: expenditures for research alone, excluding teaching supplies and general overhead (again the lower curve gives university appropriations and the upper gives the total spent, including grants for a cyclotron).

Certainly these three departments may not have been typical of all academic physics. Before we can reach final conclusions we should find full information on the amounts, sources and disbursements of all funds spent in a large sample of physics departments—a task beyond the limits of this essay.[81] Lacking this, we must fall back on such general data as are available on the overall funding of research.

The university budgets came first and foremost from student fees and tuitions in private schools and from state appropriations (with considerable help from fees) in public ones; by and large these funds continued to flow.[82] Endowment, though badly hurt by the Depression, remained another important source of income, particularly for research.[83] Besides these traditional sources, academic science found a new patron in the 1930s: the federal government. Beginning around 1933, many scientists, led by physicists, made strenuous efforts to found a centralized funding agency. Candidates for this were the new Science Advisory Board, set up by President Franklin D. Roosevelt, or the National Bureau of Standards. In either case, the plan was to pour copious federal funds into coordinated research contracts with industry, private research institutes, and of course the colleges and universities. These ambitious schemes failed, not least because many scientists were wary of centralized bureaucratic control.[84]

But failure to establish a federal science funding agency did not mean failure to get federal funds. The New Deal provided various sources for these. Money to help college students find work was lavished by the Federal Emergency Relief Administration and later by the National Youth Administration; in December 1936 nearly 113,000 undergraduate and 5,000 graduate students in all fields were holding jobs paid from these funds. This not only underwrote university finances in general but provided research assistance for specific jobs.[85] The Civil Works Administration and its successor, the Works Projects Administration (WPA), were even more important in paying research assistants and the researchers themselves. In 1938 the machine shop at the famous Berkeley Radiation Laboratory was manned entirely by WPA employees, while at MIT a battery of WPA helpers was working on a quarter-million dollar, four-year project to produce tables of spectroscopic wavelengths.[86] The federal government was not yet the sovereign force in research it became after 1940, but the way funds appeared, almost before scientists demanded them, showed that when the scientists did feel a pressing need for much more money, there would be a place to find it. Central governments were in fact already playing a dominant role in research in Britain, France, Germany, Italy, Japan, and the Soviet Union, among other countries; American scientists entered the game very late.[87]

One reason the physics professors were not yet pressing the government hard for money may have been that adequate outside funds were still flowing from the philanthropic organizations. Up to about 1920 the number of papers in the *Physical Review* which acknowledged such outside support ran around 10 percent of the total. The proportion rose from 1920 to 1930, chiefly because of the many papers written by National Research fellows, and from 1930 to 1940 the proportion continued to rise even while the National Research Fellowships fell off. By 1940 some 25 percent of the papers were acknowledging outside support, whether fellowships or grants, chiefly from foundations.[88] It was such outside support, for example, that permitted American physicists to go on a spree of cyclotron-building in the later 1930s, constructing two or three times as many as the rest of the world put together.[89] The usual source of these funds, sometimes a generation back, was industry. The root support, then, was the belief of industrial leaders and their lieutenants in the prestige value, public utility, and other advantages of donations to science.

Government and philanthropic funding, major sources of strength though they were, would not have saved academic physics research if the universities and colleges themselves had not been sound. The growth of physics departments from 1930 to 1940 was in parallel with the growth of college and university faculties overall, and the increased numbers of Ph.D.'s also simply paralleled the increase of Ph.D.'s in all fields.[90] These monotonously rising exponential curves were linked to a similar rise in university enrollments. In 1939 the faculty/student ratio in physics departments was about the same as it was in 1925.[91] Aside from a brief drop at the nadir of the Depression, people managed to go to school in ever-increasing numbers, aided by a great increase in part-time jobs, loans and grants from foundations, the universities, and government.[92] The increase in enrollments was a burden in the short term but a boon in the end, making a direct contribution in tuition and fees and presumably exerting indirect pressure on state appropriations.

At the root of this we must again look to an attitude, a faith that led people to contribute to education or sacrifice for it. As a com-

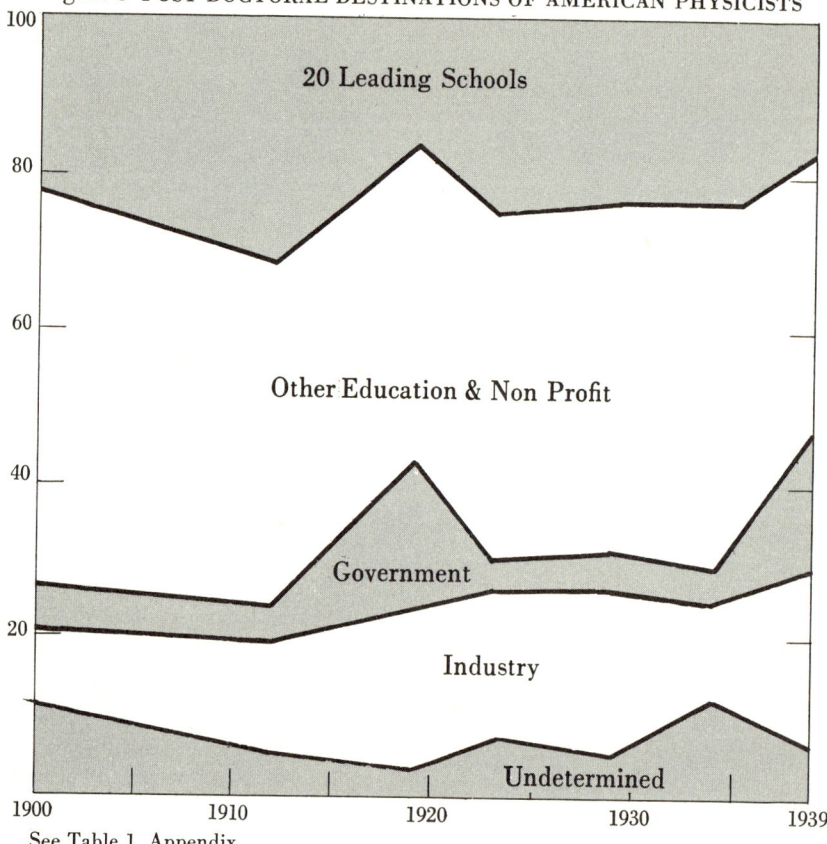

Figure 5 POST-DOCTORAL DESTINATIONS OF AMERICAN PHYSICISTS

See Table 1, Appendix.

mittee of the American Association of University Professors put it in 1937, this source of support could be traced to the students' motives, which "reduced themselves to the basic considerations of enhanced prestige and vocational advancement. These two are the marrow in the bone which gives strength to America's faith in education."[93]

Were these motives realistic? I will pass over the question of the employment of Bachelors in order to study in detail the destination of physics Ph.D.'s. Did many of them eventually have to drop out of physics? To find out we must take a sample and follow their careers for a few years. Figure 5 gives the results of the study.[94] It shows the first job held by physics Ph.D.'s after graduation, eliminating postdoctoral fellowships and similar temporary positions, for cohorts graduating in selected years. The figure is a breakdown of the total by percentages; the overall number of physics Ph.D.'s rose as shown in figure 1.

The white area along the bottom of figure 5 represents the physics Ph.D.'s who could not be found at all up to around ten years after graduation in either the membership lists of The American Physical Society or in Cattell's *American Men of Science*. This area therefore represents in a crude way mortality—whether because the physicist died or because he or she dropped out of physics. It appears that this mortality was significantly higher for the 1934 sample than for the two samples taken in the 1920s; the Depression did cause a real loss of physicists from the profession.[95] It also appears that at no time, even in 1934, was the mortality very high. Reducing percentages to numbers, we find that the Depression may have caused a permanent loss to physics of perhaps 8 of the 117 physicists graduated in 1934.[96]

The other parts of the figure suggest other interesting conclusions. Just above the bottom is a solid area representing those physics Ph.D.'s who found jobs in industry within a few years of leaving school. As I mentioned earlier, there were a good many of these in the 1920s. This was not as strong as the movement into education and private institutes (which in this and subsequent figures are merged) or even into the twenty leading physics departments (the top shaded area on the chart). Nevertheless, we see a strong movement into industry beginning around the time of World War I, which the Depression seems to have checked. Movement into government work was low except around the end of World War I and the start of World War II.

These are crude figures, only barely sufficient to give us a hint that something happened to industrial physics during the Depression. The next chart (figure 6) is less ambiguous. At the top is another background statistical series. The shaded area shows total nonfarm employment in the United States, displaying the Depression collapse and subsequent stagnation.[97] The graph below this shows employment in industrial research laboratories.[98] The solid line is for all research personnel; the dotted line is for a sample of physicists, scaled up by a factor of fifteen for clarity. These lines tell a tale of swift growth during the 1920s, a terrifying plunge during the Depression—as bad as the general collapse of employment in the United States—and then recovery.[99]

Data on budgets are harder to come by than data on personnel. One would guess that the trend in funds closely paralleled the trend in numbers of employees, and in fact this has been true for every institution, industrial or otherwise, for which I have found data. In particular, the most important single employer of physicists in the United States, the American Telephone and Telegraph

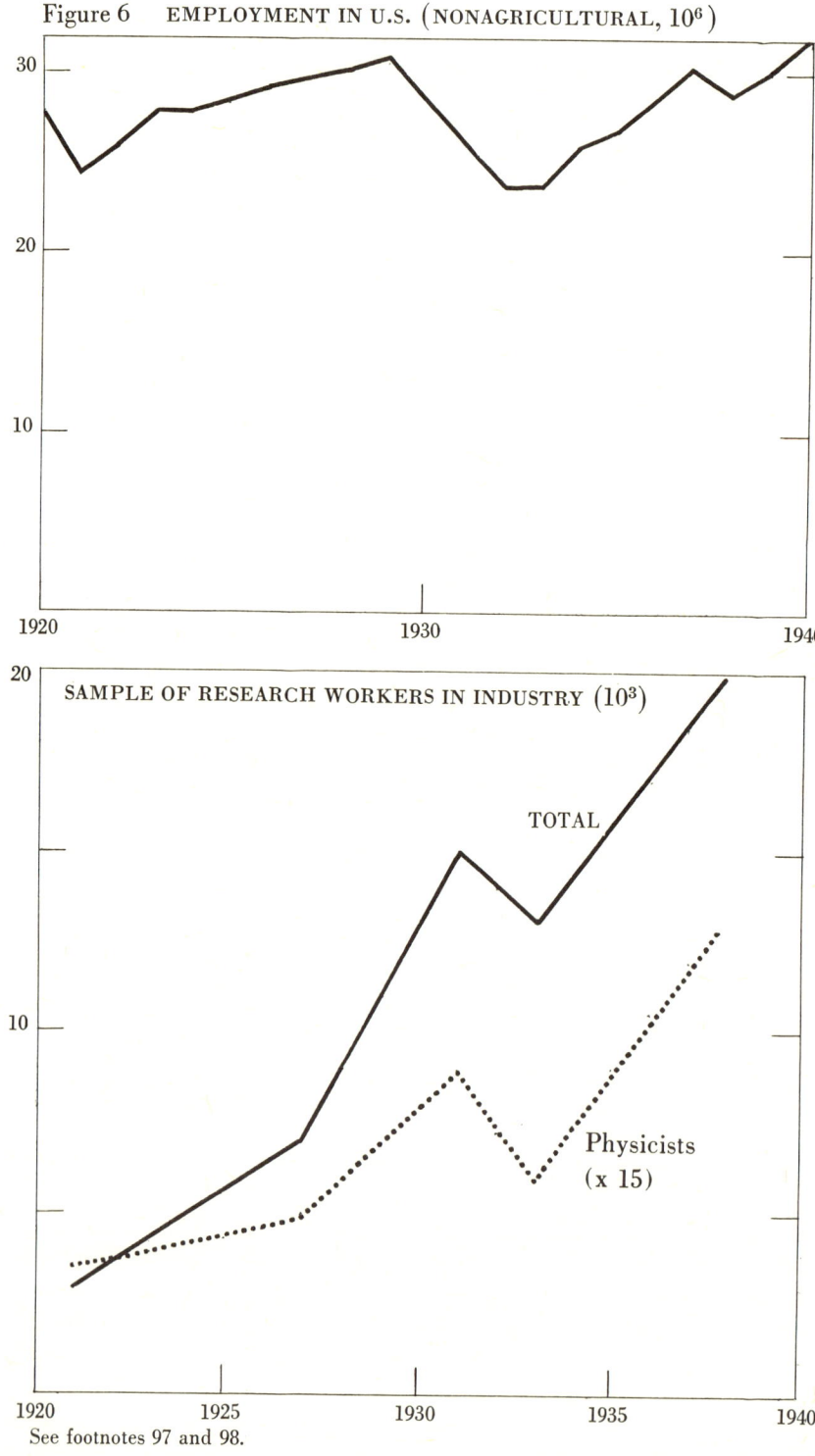

Figure 6 EMPLOYMENT IN U.S. (NONAGRICULTURAL, 10^6)

SAMPLE OF RESEARCH WORKERS IN INDUSTRY (10^3)

TOTAL

Physicists (x 15)

See footnotes 97 and 98.

Weart

Company, cut its research and development budget by over 25 percent in constant dollars from 1930 to 1933, but by 1939 this had recovered and was more than 10 percent higher than the 1930 figure.[100]

Applied science fared as poorly in government as in industry. For example, the largest government employer of physicists, the National Bureau of Standards, dropped some 20 percent of its professional staff from 1933 to 1934. The survivors had to accept pay cuts and budget cuts far worse than their colleagues in academia were taking. The younger employees were the first to go, usually accepting furloughs without pay. In the late 1930s, as New Deal programs took hold, many were rehired and government physics made a satisfactory recovery.[101]

In both government and industry cuts were sometimes made by reducing the work week; this happened at the Bureau of Standards, at Bell Labs, and at General Electric.[102] Suffering from direct and indirect pay cuts, watching friends being laid off, these applied physicists had a hard time. The director of the General Electric laboratories had a nervous breakdown that the Depression was said to have helped bring on.[103] For a few years it seemed, as one industrial physicist said in 1935, that "the era, or rather wave of industrial physics [had] subsided."[104]

It is difficult to determine whether the cutback affected fundamental research more than engineering work. There were claims that laboratory directors were ordered to "cut out work on all projects not producing a profit."[105] And certainly the Depression affected some laboratories, such as those of Westinghouse, so severely that pure research was temporarily crippled. But it was probably more common for research to decline only in proportion to the general contraction of industrial activity, and in some important laboratories research may have been specially protected.[106] For example, at Bell labs in the depths of the Depression, work was begun on transmitting signals along waveguides, a curious and unexpected development of electromagnetic theory. There were no economic benefits that could be predicted, short-term or long-term. But the telephone company could not afford to ignore a fundamental development that might on some distant day revolutionize the communications industry.[107] Lillian Hoddeson has suggested that the Depression may even have had some beneficial effects on research at Bell Labs, for with the free time in their truncated work week some staff took to studying the quantum mechanics of solids, a topic remote from immediate practical concerns. Something of the sort may have happened at the Bureau of Standards too.[108]

In short, there are scattered hints that applied physicists may have been spared the worst. The recovery shown in figure 6, so much stronger than the recovery of employment generally, also indicates a special position for industrial research. The net rise of physics employment in industry from 1930 to 1938, ignoring what happened in between, was proportionally comparable to the rise in academic employment over the same interval. A large part of this swift recovery was due to the exploitation of a new application for physics: the use of geophysics in the oil industry. Gravimeters and seismographs greatly aided geological prospecting, and some schools were quick to train their students in the new field and to carry out related research. Thus physics, as part of a feedback system, helped enlarge its own field of employment.[109]

Nevertheless for a few years industrial physicists did suffer. Since their experience was quite different from what the physics professors met, we might imagine some strains would have appeared in the physics community. Strains are helpful in understanding structure, so for the last major topic of this essay, I will look into the relations between industrial and academic physics in America in the 1930s.

The Community of Physicists in the 1930s

In the 1930s as in the 1920s, the employment of new physics Ph.D.'s provided a vital connection between the campus and industry. The job problems of new physicists were in large part due to hiring freezes in the academic world, but the collapse of industrial employment exacerbated their difficulties. Let us consider not their eventual employment (figure 5) but their location one or two years after graduation. Of those physicists who got their Ph.D.'s in 1928 and 1929 and whose location I could determine from the 1930 membership list of The American Physical Society, 22 percent were in industry and 4 percent more in government; but of those who got Ph.D.'s in 1934 and who are found in the 1935 list, only 6 percent were in industry and none in government.[110] This decline did not escape the attention of the professors who sent the students forth.

Some academic physicists pointed out that there were fields where industry had neglected to exploit the talents of scientists. They began working to persuade industrialists to employ physicists and meanwhile tried to make physics students more tempting to the industrialists. "A going concern must find a market for its product," one professor pointed out, "and this is true of the profession of physics."[111] A leader of the physics community was equally blunt: "The number of teaching jobs available for physicists will always depend on the number of students taking physics. We know that the number of these students will increase if it becomes clear that there are industrial opportunities for them."[112] So the academic branch of the physics business determined to improve the salability of its product.

Critics, industrialists and professors alike, charged that teaching was directed too much toward producing a new generation of teachers. The introduction of advanced theoretical work into American physics did not impress these critics; what the graduate needed, they said, was less work with quantum mechanics and more with classical physics and a slide rule.[113] Physicists feared that this sort of training was being usurped by the engineering faculties, whose graduates more and more resembled the product of the physics departments.[114]

To see how physicists were meeting these problems, Henry Barton wrote letters to a number of departments at the end of 1935. He found that about half of them had taken or were seriously planning to take steps to increase the attractiveness of their students to industry.[115] Not all schools joined this movement, and we have noted that the majority of students continued to end up in academic jobs. But at certain schools a majority of the physics Ph.D.'s graduated in the 1930s found their homes in industry. For example, one department toward the bottom of the list of the top twenty reported to Barton that it had not been able to place a single Ph.D. in teaching since around 1930; all the good jobs were being snatched up by graduates of the top half-dozen departments. But this school had some limited success placing its Ph.D.'s in the paper industry, oil exploration, and so forth. Hard experience was compelling the department to consider industry as the ultimate destination of its students, even though all its present graduate students, following tradition, hoped to somehow get academic jobs.[116] Some other departments told similar tales. In their recovery from the Depression industrial laboratories were pulling in many new Ph.D.'s.

Besides this indirect assistance to the schools, industry continued as in the 1920s to provide fellowships, grants for research projects, and other sorts of direct aid. I have not been able to find a quantitative measure for such assistance, but it was probably higher in 1940 than in 1930.[117]

I have noted that the movement to teach applied physics in the

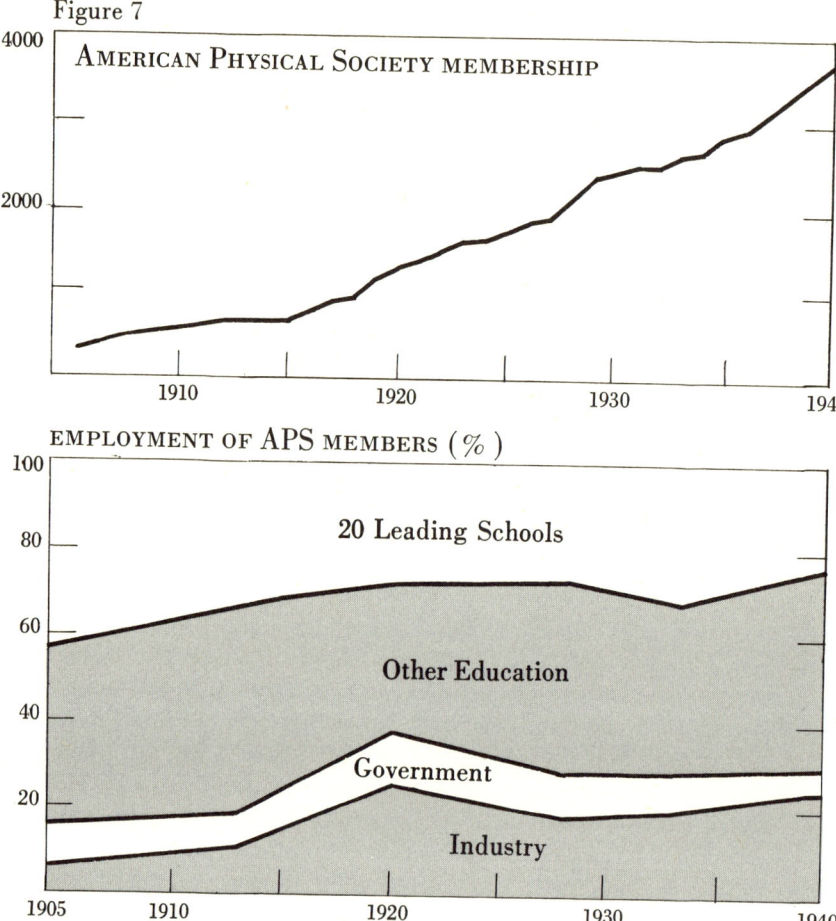

Figure 7

See footnote 118 and Table 2, Appendix.

universities included some elements of a reaction, or counter-reaction, to the famous growth of pure and theoretical physics. The conflict also came to the fore in another theater, the physics societies. Figure 7 analyzes the chief of these, The American Physical Society.[118] The upper graph shows another face of our familiar exponential, in this case the growth of Physical Society membership from 1905 to 1940. The effects of the Depression, or even World War I, on this curve would be difficult to prove. (This is partly because members who could not pay their dues were kept on the rolls for three years, by which time many were able to resume payments.)

The lower graph gives a breakdown of the membership, showing what percentages were in various locations. The two shaded areas at the top, "education, including private institutes," dominate throughout. But there is a significant change in the proportions employed by industry (the black area) and government (the white). Their combined total grows strongly into the 1920s and then drops off. Note that because of the rapid growth of the Physical Society as a whole, the absolute number of applied physicists in it was higher in 1940 than in 1920. But they were not so large a proportion of the total as they were in the 1920s, the golden age of applied physics research.[119]

People noticed that not all those who might be called physicists were members of The American Physical Society. Many belonged to the engineering societies and called themselves engineers, even though their work was in direct application of physics research.[120] Of the 1,500 or 2,000 so-called physicists in industry in 1938, only about half were in the Physical Society.[121] The respected physicist John Tate explained part of the reason: "To many physicists the meetings of the American Physical Society became less and less interesting ... those applied physicists have come to feel themselves as set apart." Another scientist confirmed that "We did not feel at home in the Physical Society Meetings. I once read a paper before the Physical Society, and I am sure the Physical Society was not interested in the paper."[122]

This feeling had gained strength through the 1920s as the numbers of applied physicists grew and as they came to feel that the Physical Society was chiefly concerned with quantum mechanics and atomic spectra rather than with physics in the broadest sense. Already in 1916 the Optical Society of America was founded by a group at Eastman Kodak; in 1929, the Acoustical Society of America was founded at Bell Labs and the Society of Rheology was founded by people interested in the study of plastic materials. In 1930, there were rumors that a Society of Applied Physics would be created.[123] It is unclear how many people actually left the Physical Society as a result of all this, but the trend seemed to be toward a complete schism in the physics community.[124]

This was avoided by the founding of the American Institute of Physics to undertake operations, such as publishing and public relations, which were most efficiently done jointly. Through the mid-1930s this organization played a leading role in bringing industrial and academic leaders together, in trying to explain the value of industry to physicists, and the value of physicists to industry, and in working to persuade applied physicists to stay within the fold.[125] The Physical Society meanwhile formed a Committee on Applied Physics which also sponsored joint industrial-academic symposia and the like. "To those of us who have been engaged in industrial research," one physicist wrote in 1937, "there appears to have been a social upheaval in the realm of the physical sciences in which the researcher in practical or applied fields has been lifted from the gutter of so-called 'prostituted' science and at least his better qualities held up as ideals for aspiring young physicists to emulate."[126]

Publications were, along with meetings, the societies' main activity and showed the same tensions the societies experienced. The increasing strength of American physics in the 1920s naturally meant a rapid increase in the number of papers, pure and applied. The traditional journal of the Physical Society, the *Physical Review*, although it doubled and redoubled its volume from 1920 to 1930, could not keep up with all areas of physics. It was increasingly seen as an outlet for papers on quantum mechanics, atomic physics, and the like. "The Physical Society," Barton recalled, "didn't want to publish [applied] papers in the *Physical Review*; they wanted it to be more pure and fundamental."[127] Thus the founding of new societies was closely tied with the creation of new journals. The *Journal of the Optical Society of America* was first printed in 1917 and was joined in 1929 by the *Journal of the Acoustical Society of America* and the *Journal of Rheology*.

Academic physics was also expanding so rapidly that it had to create new outlets, not only in the expansion of the *Physical Review* but also in new, predominantly academic journals such as the *Reviews of Modern Physics* (1928), the *Review of Scientific Instruments* (1930) and the *American Physics Teacher* (1933). But it was the upstart industry-oriented journals, threatening to institutionalize further the split between pure and applied physics, that provoked concern. The Physical Society responded by founding its own *Journal of Applied Physics* (1931) as companion to the *Physical Review*, while the American Institute of Physics produced *Physics* (1933) and the *Journal of Chemical Physics* (1931) in a combined appeal to both branches of the physics community.[128]

Some of the results of all this can be seen in figure 8.[129] In the top half I give the numbers of papers printed in various journals, which jumps with the introduction of new outlets. Also given is the number of papers published in the *Physical Review* alone—a number that declined in the 1930s, alone among physics indicators. One might expect this to be some side-effect of the Depression, but on close inspection this turns out to be unlikely. Rather, in expanding through the 1920s the *Physical Review* seems to have outrun the size that was practicable for a physics journal at that time. Already in 1928 it was running into financial difficulties, and in 1929, before physicists had felt the least shadow of economic difficulty, the journal lost money despite severe efforts to economize. The editors began to toy with the idea of page charges—the beginning of a concept that now supports much of the country's scientific publication.[130]

The Depression certainly made the situation worse; from 1931 to 1933 the number of members of the Physical Society whose dues were in arrears jumped from 8 to 19 percent of the membership, and most gave "economic conditions" as the reason.[131] But the Society, even if it had enjoyed a more prosperous membership, could not have kept expanding its journal indefinitely under the prevailing constraints on publishing. Nevertheless the *Physical Review* not only maintained itself as the world's leading physics journal but also, in an important sense, expanded. Although the total number of papers it published did not rise, the number of papers in the most exciting, important, and "pure" field, nuclear physics, increased quickly. In 1930, papers in this field numbered 8 percent of the total published in the journal, in 1935 they were 28 percent, and in 1940, 44 percent.[132]

The lower part of figure 8 shows the proportions of papers contributed to the *Physical Review* from various sources. Here the top twenty academic physics departments greatly dominated, contributing about three-fourths of the total in 1940, almost the same proportion as they had held in 1910. The production of papers by a few schools out of all proportion to their size no doubt reflected the quality of their faculties but was also related to the exponential rise of the numbers of physicists and to the concentration of advanced training. The heavy production of Ph.D.'s implicit in the rise required the few schools doing the training to have on the average three or four advanced students for every trained physicist.[133] The papers all these students wrote in connection with their theses or postdoctoral work swelled the output of the schools that did the training.

Aside from the great strength of the leading schools, the most notable feature of figure 8 is the black area at the bottom, which shows the coming and passing of the wave of industrial physics. The drop in this proportion after 1920 does not imply a drop in the absolute number of papers published out of industrial laboratories, for in the 1920s total publication in the *Physical Review* was soaring, and around 1930 many industrial papers were simply displaced to the *Journal of Applied Physics* and its kin. But in fact the absolute number of these applied papers did not rise much after the mid-1920s.[134] What this graph therefore shows, and shows more

Figure 8

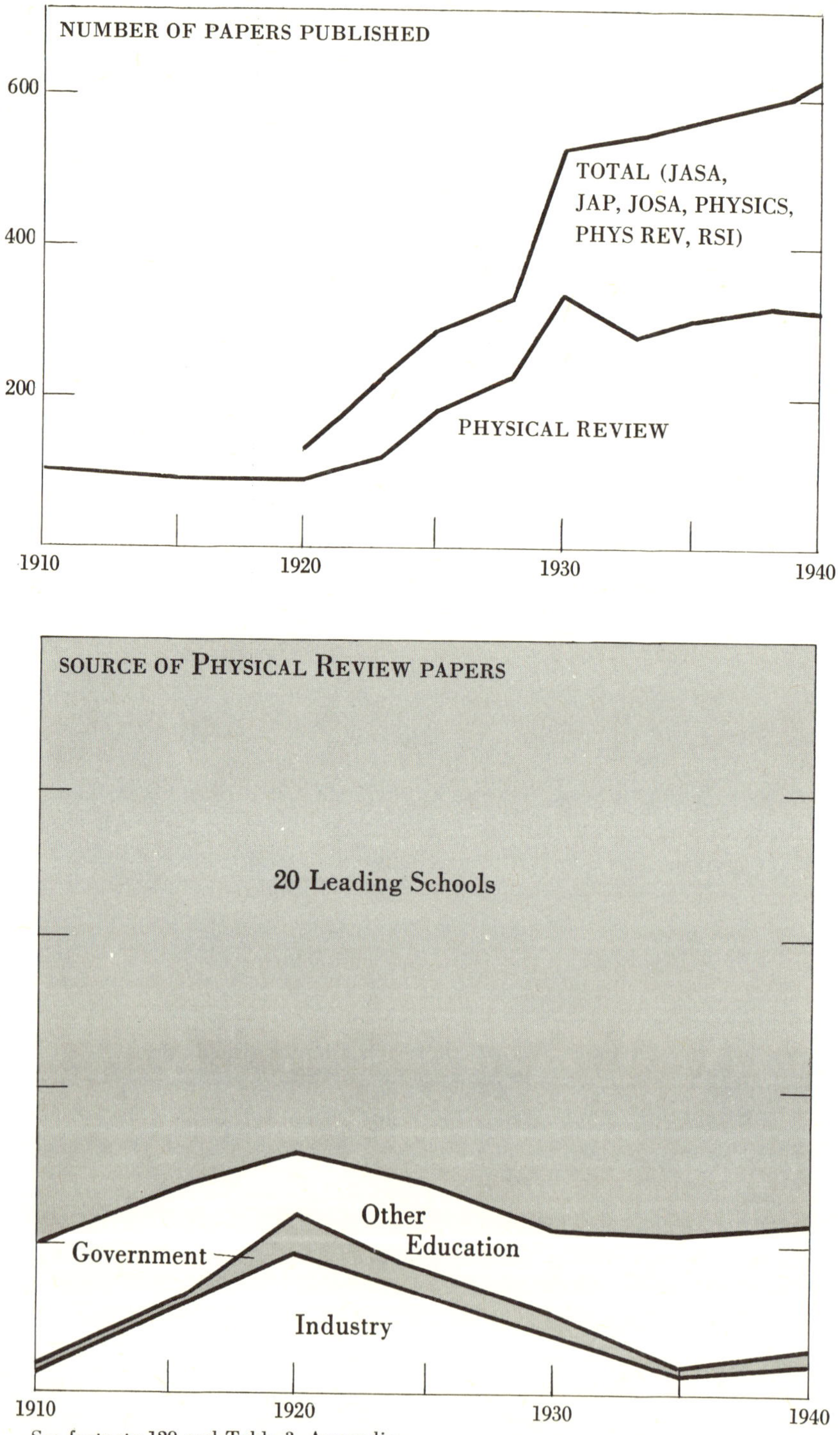

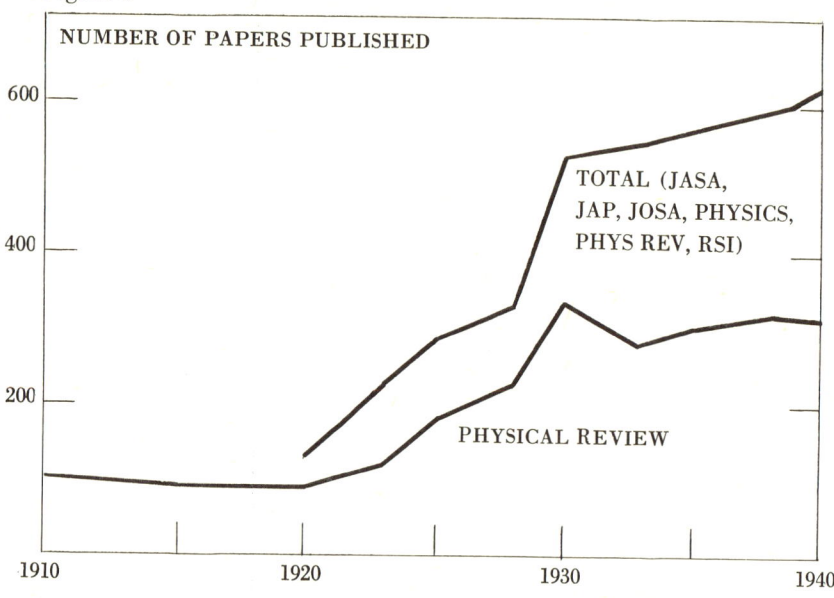

See footnote 129 and Table 3, Appendix.

clearly than any other measure I have found, is a rise of physics research in industry around the time of World War I, and its setback and partial separation from academic physics before and during the Great Depression.

Setback and separation—these are distinct and perhaps not even compatible trends. Clearly the separation was taking place already in the 1920s before any economic problems appeared, and we may surmise that the growth and strength of applied physics helped promote its independence. It is also clear that the setback came as a straightforward result of the contraction of industrial corporations in the Depression. But to the academic physicists of the 1930s these distinctions were not important, and everything merged into a single stark threat to the employment of physicists.

The physicists' anxiety about industrial physics was part of a broader worry that mushroomed during the Depression. In 1934 Karl Compton said, "The idea that science takes away jobs, or in general is at the root of our economic and social ills, . . . has taken an insidious hold on the minds of many people. . . . The spread of this idea is threatening to reduce public support of scientific work, and . . . to stifle further improvements in our manufacturing processes. Either of these results would be nothing short of a national calamity."[135] Fears of science and technology have been with us for centuries and it is not surprising that economic crisis led to new attacks. Even such a liberal as Stuart Chase, whose articles gave Roosevelt the phrase "A New Deal," worried that technological obsolescence was trampling the social order. "These damned inventors," he wrote, "are destined to throw us into one financial paroxysm after another" unless they were brought under social control and led "back into the stalls where they belong." Equally upsetting to scientists was the suggestion, originating with a British bishop in 1927, that a holiday should be declared for invention.[136] The physicists, often acting as spokesmen for all science, reacted vehemently with speeches, articles, and conferences to explain to the public that "science makes more jobs."[137] With hindsight, it appears that the physicists were overreacting. The critics were greatly outnumbered, and in any case aimed their barbs more at reckless invention than at pure science. If Stuart Chase, for example, felt invention should be regulated, he would have it regulated by physical scientists and the like, "that class most able, most clearheaded of all," aided by "magnificent laboratories and research bureaus." The belief in science that had sustained physics through the 1920s had not failed, and the physicists' public pronouncements were not needed to keep this sustenance through the 1930s.[138]

The most interesting point in all this, in the response of academic physicists through new curricula, new organizations, new publications, and new modes of addressing the public, is their extreme sensitivity in the area of applied physics. If they were not all wrong, then this sensitivity is itself evidence that any weakening of applied physics or of its ties to the academic world would be a heavy blow to all physics.

Fortunately, industrial physics recovered, as we have seen. Toward the later 1930s there were even hints that the most abstract physics of the universities might have some value for industry after all. General Electric advertised for a quantum theorist to look into the fundamental processes involved in fluorescence, as in fluorescent lights; people at Bell Labs, from their studies of the quantum theory of solids, set out on the path that would lead to the transistor; Westinghouse went into nuclear physics, a subject that later became one of its major fields of industrial interest.[139]

Conclusions

I began by asking why physicists seem to recall the year 1932 differently from the way most other Americans recall it. Part of the answer, surely, is that physicists are less concerned than most with the passing pageant of the world; but I have deliberately set out to find other sorts of answers. We have seen that academic physicists did not have too rough a time during the Depression. There were temporary job problems for many graduating students and loss of pay for some faculty, but in general the professors continued up the exponential, riding on rising enrollments, rising outside funding from foundations, and the beginnings of government grant support. In fact, by comparing the figures I have presented with similar figures from the 1960s and 1970s, one can easily demonstrate a capital fact: the Great Depression was by no means as damaging to pure physics as the past ten years have been (see Epilogue).

On the other hand, one branch of the profession, applied physicists in industry and government, did have a hard time in the Depression. This was masked by their swift recovery. Also, their rapid growth during the 1920s had brought them to a relatively independent position, so they have mostly become invisible to those who do not look beyond the boundaries of the *Physical Review*.

What can we say about the larger and less well-posed question of how American physics came to grow so greatly? I insist with regret that I have not really been able to attack the problem; at most I have made a reconnaissance in force. I can therefore offer only some tentative suggestions. The great surge of applied physics in the 1920s points to one element of strength, but the fact that the physics community in general maintained its progress in the mid-1930s, despite the setback the Depression gave applied research, warns us that physics did not depend directly and simply on industry. Apparently the real strength lay in the academic sector. Yet this sector does seem to have drawn over the longer run upon industry. This is suggested not least by the speed and vehemence with which leaders of physics reacted to any threat of a decline or isolation of applied research. The relation between industrial and academic physics is a topic worth further study by those who wish to understand American science.

Epilogue: The Situation Since 1940.

Once we see statistics for the past, it is natural to look for similar indicators in our own times, in hopes of better understanding the past and present by the comparison. Unfortunately, it is rare for even the most straightforward indicators to be unaffected by the flux of social conditions; the dollar and the Ph.D. transmute with the years. Nevertheless, I have assembled some indicators of the current situation similar in a rough way to those I found or constructed for the period between the two world wars.

Funding. Science funds continued their exponential rise after 1940, reached a peak in the late 1960s, and have since dropped off. Total basic research expenditures in the United States reached nearly $4.0 billion in 1968 and then declined, the 1976 level being lower by one-tenth (all figures in this section are in constant 1972 dollars). Industry played a major part in the early stages of the rise, but then fell behind: from contributing 28 percent of the total in 1960, it dropped to half this in 1976. I have not compiled pre-1960 figures, but I suspect that the 1950s, like the post-World War I years, were a period of particular strength for industrial research. Industrial firms spent about $76 million in 1974 for basic research in physics and astronomy, down by a factor of two from the ca. 1966 peak, and have not since regained enthusiasm for funding science.

The most important change since the prewar years has been the tremendous new rush of money from the federal government, which provided 59 percent of all basic research funds in 1960 and 67 percent in 1976. Federal support for physics peaked at about $350 million in 1967 and was only 70 percent of this in 1975, rising only a little thereafter. A substantial part of this basic research is now done in government-contract laboratories, which appeared in strength during World War II. Since these laboratories are most often run by universities, and since much of the remaining government money is spent in universities, federal funding has probably not greatly altered the general balance between university, in-house government, and industrial laboratories.[140]

The immediate reasons for the fall in funding are generally known. The number of people of college age, and particularly the proportion of these who go to college, have ceased their headlong rise, so faculties no longer need to expand. Unfortunately, federal funding peaked at the same time, with the two chief supporters of research and development in the physical sciences (the Department of Defense and the National Aeronautics and Space Administration) cutting back support.

Personnel. As in all periods I have studied, the number of workers changed in close parallel with changes in the supply of money. Since 1940 the number of physicists has continued its exponential rise, but this shows signs of leveling off. There are now some 20,000 Ph.D. physicists employed in the United States. The balance between industrial, academic, and other types of employment has not changed much since the 1920s except for a shift from strictly academic employment to federally-funded research centers. (In 1973, 53 percent of Ph.D. physicists worked primarily in universities, 24 percent in industry, 11 percent in government and 12 percent in federally-funded centers.) This shift, however, corresponds to a shift from teaching to research positions. There are also signs that within the universities themselves, scientists have had more time for research than before the war. In 1975 more than half of the physicists in leading departments spent more than half their time on research, a great advance from the earlier situation.[141]

New Ph.D.'s. These grew exponentially after a hiatus in World War II, reaching a peak of over 1,500 awarded in 1969–70. The drop since then has been abrupt: under 1,000 in 1977–78 with a fairly firm projection of about 1,000 through 1983.[142] There has been a marked change in the distribution of the origins of these

Ph.D.'s. Where before the war three-fourths or more came from twenty leading schools, in the 1970s less than 40 percent came from a rather different list of twenty leaders.[143] As for their destinations, around 1960 almost half of them were going into industry—another indication of a surge of industrial physics in the postwar period. The universities soon caught up, helped by a rapid and novel increase in temporary postdoctoral positions toward the end of the 1960s. A 1977 survey of 1976 Ph.D.'s found 44 percent of them holding post-doctoral fellowships, mostly in universities, a situation quite different from the more stable employment of new physicists at most times before World War II.[144] A number of people are going from one postdoctoral position to another, year after year. The final employment after any such position (for those who found final employment in physics) is not very different from the 1920s: around 1972 nearly half of this employment was academic, a quarter industrial, and the rest evenly divided between government and miscellaneous.[145]

But not everyone finds final employment, and the dimensions of the problem are only beginning to become apparent. According to the most recent study, of the physicists who entered the physics labor market in 1960, Ph.D. in hand, about half attained tenure in academia; but of those who entered in 1970, less than a tenth are so fortunate. From 1969 through 1974, "at least 6,000 physicists left the employment of physics." (Many of these, it should be noted, found satisfactory jobs in areas not too far from physics.) Projections for the near future show only small signs of hope, and by the mid-1980s "we will have lost a generation of physicists."[146] The 1930s Depression was a temporary inconvenience and was perceived as such; the climate today is otherwise.

Societies and Publications. The volume of publication in the *Physical Review*, despite a steep drop during the Second World War, was in 1950 more than double the 1940 value, in 1960 about double the 1950 value, and in 1970, at over 20 million words, slightly more than double the 1960 value. Other physics journals grew in parallel.[147] But since 1970, physics publication has leveled off.

Meanwhile, there have been deep qualitative changes. One measure of this is the number of papers published per faculty member in leading physics departments. In 1900 this publication rate was 1.1 per year for the United States (compared with 2.2 to 3.2 for other leading countries); in 1958 it was 2.0 and in 1972 it reached 4.2.[148] An increase in multiple authorship might account for some of this change.

Among papers of U.S. origin in the entire physics literature, the proportion coming from industrial laboratories averaged 29 percent as recently as 1966, then dropped to 19 percent in 1969 and has stayed there. The latter figure is about the same as the proportion of industrial papers in the 1920–21 *Physical Review*. Papers originating in government laboratories averaged 12 percent over 1960–75, more than double the 1920–21 proportion; this probably reflects the rise of federally-funded research centers.[149]

As for the modern *Physical Review* alone, I find that twenty leading schools (not entirely the same set that led in earlier years) accounted for 40 percent of the U.S. authorship of all the articles published in December, 1975. Federally-funded centers—operated by exactly these same schools—accounted for another 21 percent. Thus, elite institutions continued to take the lion's share of the space in this journal, although the concentration was a bit less extreme than it was before World War II. Industrial and in-house government labs contributed 15 percent and 6 percent respectively of the authorship, similar to the situation prevailing in the 1920s.[150]

I get the same impression of continuity with the past when I look into the locations of a sample of the 1975 membership of The American Physical Society. Twenty leading schools (if we include their associated federally-funded centers) accounted for 30 percent of the members, other educational institutions for 43 percent, industry for 20 percent, and in-house government for 7 percent—very nearly the same as the situation through the 1920s and 1930s. This continuity is not as straightforward as it might seem, for it conceals a diffusion of industrial physicists (who in the 1920s were highly concentrated in a few companies) into a variety of locations, including entirely new major industries such as computers and spacecraft.[151]

The total membership of the Physical Society presents a more disturbing picture. It rose smoothly, even through World War II, to over 28,000 in 1970; but in 1978 it is only slightly above that level. The 1970s mark the first extended period of stagnation in membership since the Physical Society was founded in 1899.

Turning to more subjective matters, some feelings that were strong in the 1930s seem to have reappeared in the 1970s. Physicists are concerned about finding employment for their graduating students outside the academic world; they have revised course offerings towards "practical" subjects, held industry-oriented conferences, and complained about being displaced by engineering graduates.[152]

In conclusion, although there were signs of a temporary surge of industrial research to the early 1960s, and although the federal government has entered physics in a big way, the general distribution of physics research in the 1970s is about the same as that established in the 1920s, provided one considers federally funded research centers as belonging to the university environment. But the collapse in rates of growth that physics has seen in the past dozen years is longer lasting, and for many indicators greater, than anything experienced even during the worst wars and depressions during the three-quarters of a century for which we have some statistics. If the exponential rise has permanently halted (and whether or not it has done so now, sooner or later it must), profound structural changes must follow. For example, physics education and related publication patterns must be different if, on the average, each physicist during his career trains only one graduate student, his replacement (see n. 133). It appears that the shape of the physics community and the relationship between physics and society in the United States are now going through a historic change, whose outcome is beyond guessing.

Appendix: Tables.

TABLE 1. Postdoctoral Destinations of American Physicists (see figure 5)

Year of Ph.D.	1895–1904	1910–1914	1918–1921	1923–1924	1928–1929	1934	1939
Size of Sample	103	105	71	92	159	96	133
Never determined	11	5	2	6	7	11	10
20 leading schools	23	33	12	23	37	23	23
Other higher education	48	45	27	36	68	41	45
Secondary schools	2	0	0	0	1	0	0
Observatories	3	1	0	3	2	1	0
Private Institutes	0	1	1	2	2	3	2
Total "other education"	53	47	28	41	73	45	47
AT&T, GE, NELA	3	6	10	5	5	2	7
Other industry, business & misc.	7	9	5	14	30	11	22
Nat'l Bur. Standards	4	4	9	2	3	1	2
Other Government	2	1	5	1	4	3	22

Notes to Table 1

Table enumerates first identified job, after temporary postdoctoral positions, of physics Ph.D.'s. The sample consists of all the physicists graduated in selected years and identified in M. Lois Marckworth, *Dissertations in Physics. An Indexed Bibliography of All Doctoral Theses Accepted by American Universities, 1861–1959* (Stanford: Stanford University Press, 1961). For convenience, we used not the published work but the original cards (giving name, date and title of thesis, and institution) deposited by Marckworth at the American Institute of Physics (AIP). These cards cover most of the Ph.D.'s awarded (from 73 percent to 87 percent for the different years, by comparison with the numbers given in Scates, op. cit. n. 2).

The names in the samples were then checked against the membership lists in the *Bulletin* of the American Physical Society and against Cattell, op. cit. n. 4, for periods ranging from the year after graduation up to about ten years later. The first position listed is tabulated, except where we know this was a fellowship, or where it was at either the same institution from which the Ph.D. was received or at one of the twenty leading departments, in which cases a fellowship or similar temporary position was suspected. For these cases the next position taken was tabulated (often this was still at the same school).

In some cases the position could be inferred from an address, e.g., Princeton, N.J. or Kodak Park, N.Y. Sometimes the immediate postdoctoral position could not be determined but a later one could, and we assumed this was in the same category of employment (National Research Council, op. cit. n. 40, suggests no great net migration between categories, and we have the same impression from our survey).

Twenty Leading Schools: the University of California (Berkeley), California Institute of Technology, University of Chicago, Columbia University (including Barnard), Cornell University, Harvard University, State University of Iowa, University of Illinois, Johns Hopkins University, Massachusetts Institute of Technology, University of Michigan, University of Minnesota, New York University, Northwestern University, University of Pennsylvania, Princeton University, Stanford University, Washington University (St. Louis), University of Wisconsin, Yale University; before World War I Clark University was substituted for Caltech. There was no clear cutoff, and a few substitutions among the lower-ranked schools would make little difference.

Other education: Includes other higher education (the overwhelming majority of this category), secondary schools, and also nonprofit institutes such as observatories and the Carnegie Institution of Washington.

Government: Both state and federal, overwhelmingly the latter.

Industry: Chiefly the communications and electrical industries but also many others, and including miscellaneous such as radiologists in hospitals and the self-employed. American Telephone and Telegraph (AT&T), General Electric (GE), and the National Electric Lamp Association (NELA) are separated.

Undetermined: These include those who never appeared in the Physical Society list or Cattell, or who appeared in the Physical Society list without an address and later disappeared from both lists. The presumed causes include death, illness, or movement to another occupation. An exceptionally high fraction (16 percent) of the undetermined Ph.D.'s were women; some may have been missed because they married and took a different name rather than because they left physics. Those known to have moved abroad while continuing as physicists—5 in the 1895–1905 cohort, fewer in later years—were not included in any of the totals.

TABLE 2. Employment of American Physical Society Members (see figure 7)

Year	1905	1913	1920	1928	1933	1939	1975[a]
Official APS membership (incl. foreign)	257	682	1296	1969	2673	3565	28,889
My sample	241	619	497	602	1160	798	810
Situation undetermined	5	89	12	215	441	51	331
20 leading schools	104	181	131	105	223	167	146
Other higher education	88	216	150	157	263	320	200
Secondary schools	3	19	12	2	2	9	0
Observatories	2	5	4	3	7	8	3
Private institutes	2	8	8	11	20	19	2
Total "other education"	95	248	174	173	292	356	205
AT&T, GE, NELA	4	20	50	44	65	51	12
Other industrial co.s	8	24	57	25	80	133	81
Independent engineers & inventors	2	10	13	0	0	0	0
Misc. business etc.	1	2	7	5	5	3	3
Total "industry"	15	56	127	74	150	187	96
Nat'l Bur. Standards	14	34	27	21	26	10	10
Other government	8	11	26	14	28	27	22

Notes to Table 2

Locations of Physical Society members whose situation is given and is in the United States. Categories defined as in table 1 above. For 1920-24 we used a 50 percent sample (every second page of the list), for 1928-33 a 33-1/3 percent sample, for 1939 a 25 percent sample, and for 1975 a 3-1/8 percent sample. The number whose situation was undetermined was below 15 percent for each year except 1928, when it was 36 percent, and 1933, when it was 38 percent, presumably because of procedures involving reporting of address (a check of 1924 shows a still higher percentage). There is good reason to believe that the restricted sample for these two years does not seriously distort my results: First, a questionnaire survey of Physical Society members in 1930, which got a 74 percent response, found 25 percent of them in industry or business and 66-1/2 percent in "educational or individual professional" employment (the latter presumably including the small number of editors, patent lawyers, physicians, etc.). Minutes of the Council, November 28-29, 1930, cited by Henry Barton, "Draft 5/28/63" (AIP history),

Barton Collection, AIP. Second, Hull, op. cit. n. 104, p. 380, counted, in an 88 percent sample of Physical Society members of 1934, 78 percent in education and the like and 20 percent in industry; his very low government rate (2 percent) may reflect at least in part a real drop from my 1933 figures, since this was a time of heavy layoffs.

Note a: In 1975, the fifty-seven members in the sample at federally funded research centers operated by major universities are included in the "twenty leading schools" category. The twenty leaders are merely those with the most physics faculty as listed in the AIP *1973–1974 Directory*, n. 143. There are now a significant number of foreign members, 12 percent of my total sample of 922.

TABLE 3. Sources of *Physical Review* Papers (see figure 8)

Year of Publication	1910	1915–16[a]	1920–21[a]	1925	1930	1935	1940	1975[b]
Total articles	104	144[a]	126[a]	175	335	302	317	318
Foreign or unidentified source	22	20	13	2	24	20	27	103
20 leading schools	67	92	77	123	241	221	222	131
Other education & private institutes	12	17	9	21	35	51	49	91
Industry & business	2	14	22	24	24	9	16	33
Government	1	1	5	5	11	1	3	13

Notes to Table 3

Sources of papers in *Physical Review* whose source could be identified and was in the United States. Except for 1975, papers with multiple authorship are counted under one institution only (these were not numerous; e.g., in 1925 there were only 4 percent more total authors than total papers). My counts are through 1925; afterward the figures are extracted from an AIP nuclear physics study, op. cit. n. 12. Categories are as in table 1, except that for leading schools I use the twenty schools publishing the most papers in *Physical Review* for the given year; schools which appeared in all samples 1910–1940 (except perhaps one of the pre-1920 samples, indicated by*) were: the University of California (Berkeley), University of Chicago, Columbia University, Cornell University, Harvard University, State University of Iowa, Johns Hopkins University, *Massachusetts Institute of Technology, University of Minnesota, Princeton University, University of Wisconsin, *Yale University.

Note a: To secure a larger sample, I counted 1-½ years, series 2, vol. 5–7, 15–17. Total number of articles in 1915 alone was eighty-five and in 1920 alone, eighty-four.

Note b: Physical Review *A,B,C,D* together, December 1975 only. This sample is treated differently from the others, for the top two figures are still numbers of articles, but for the rest I have counted each U.S. institution separately, since there was often more than one institution per article or even per author. The forty-five authorships from federally-funded research centers are included in the "twenty leading schools" category. The twenty leaders were those publishing the most in this one month, led by University of Arizona, University of California (Berkeley), University of California (Los Angeles), Caltech, Cornell, Harvard, MIT, State University of New York (Stony Brook), University of Pennsylvania, Purdue, and University of Washington—a very different list from the prewar leaders. Industrial labs were led by Bell and IBM.

1. In this study I have used chiefly the resources of the Niels Bohr Library of the Center for History of Physics, American Institute of Physics, New York City (hereafter AIP). I owe thanks to Charles Weiner and Joan Warnow for assembling these books and archives and introducing me to them. I have drawn extensively on the oral history collections at AIP, although in most cases I do not cite these for specific points. My thanks, too, to Tessie Liu for much bibliographical and statistical assistance, and to those who read and commented on a draft, in particular Beverly Fearn Porter. Weiner's article, "Physics in the Great Depression," *Physics Today* 23 (October 1970): 31–37, opened up many of the problems discussed here. Some of this material has been published in my article "The Rise of 'Prostituted' Physics," *Nature* 262 (1976): 13–17; I thank *Nature* for permission.

2. Figure 1: Douglas E. Scates, Bernard C. Murdoch, and Alice V. Yeomans, *The Production of Doctorates in the Sciences: 1936–1948*. A Report of a project sponsored by the Manpower Branch, Human Resources Division, Office for Naval Research (Washington, D.C.. American Council on Education, 1951), p. 113. Another series for the same quantity, different in minor details but identical in all important respects, is in National Research Council, *Doctorate Production in United States Universities, 1920–1962*. Comp. Lindsey R. Harmon and Herbert Soldz (Washington, D.C.: National Academy of Sciences, 1963).

3. Derek J. de Solla Price, *Little Science, Big Science* (New York: Columbia University Press, 1963).

4. My counts, table 1, appendix. J. McKeen Cattell, *American Men of Science* (New York: Science Press, 1906[1]–1944[7]).

5. My counts, table 2, appendix. The APS list includes not only most of the Ph.D.'s but a number of others: if we add the numbers of all American Ph.D.'s graduated to 1940 and apply mortality tables, we find roughly 2,200 should have been alive in 1940, whereas there were 3,751 APS members (only a relatively small fraction of them foreign-trained). In 1927 there were 1,920 APS members but only 991 physicists listed in Cattell, *American Men of Science* (1927[4]). See also below, n. 121.

6. Employment: my counts, table 2, appendix, omitting nonprofit institutes. Publications: my counts, table 3, appendix. Training: the

twenty leading schools (same list as for table 1, appendix) produced 91.4 percent of all American Ph.D.'s graduated in 1920–29 and 75.4 percent of all graduated in 1930–39. My counts from National Research Council, *Doctorate Production*, table 3, p. 10 and table 8, pp. 20–26.

7. Stanley Coben, "The Scientific Establishment and the Transmission of Quantum Mechanics to the United States, 1919–1932," *American Historical Review* 76 (1971): 442–66; John C. Slater, "Quantum Physics in America Between the Wars," *Physics Today* 21 (January 1968): 43–51; Katherine Sopka, paper at Joint Atlantic Seminar, Yale University, March 1975; John H. Van Vleck, "American Physics Comes of Age," *Physics Today* 17 (June 1964): 21–26; Charles Weiner, "A New Site for the Seminar: The Refugees and American Physics in the Thirties," pp. 190–234 in *The Intellectual Migration: Europe and America, 1930–1960*, ed. Donald Fleming and Bernard Bailyn (Cambridge: Harvard University Press, 1969).

8. "A Study of Scientific Periodicals" (by the U.S. Navy), *Review of Scientific Instruments* 6 (1935): 333. The most-cited 1920–32 was the *Zeitschrift für Physik*, which did not exist in the 1895–1914 period. For the status of the *Physical Review* in 1922, see Van Vleck, "American Physics," p. 22. A better measure, which I have not constructed, would be citations per paper published; one should also factor in the rate of publication per physicist, making international comparisons, along the lines of Paul Forman, John L. Heilbron, and Spencer Weart, *Physics circa 1900: Personnel, Funding, and Productivity of the Academic Establishments. Historical Studies in the Physical Sciences* 5 (Princeton: Princeton University Press, 1975), table E.3, p. 119.

9. Morris Llewellyn Cooke, *Academic and Industrial Efficiency*. Carnegie Foundation for the Advancement of Teaching, bulletin no. 5 (New York: CFAT, 1910), table 5, pp. 92–93; National Research Council, *Research—A National Resource. 1. Relation of the Federal Government to Research* (Washington, D.C.: U.S. Government Printing Office, 1939), pp. 119–20, 177.

10. 1910: Cooke, *Academic and Industrial Efficiency*, combining professors and assistant professors. Cooke's figures are for 8 A.M. to 6 P.M. only, whereas considerable research was done at night to avoid electrical interference from streetcars, ibid., p. 31. 1932: Full, associate and assistant professors from folder "1932–1933 Statement of Work. Members of Physics Department," in drawer "Budget. General 1930–1958." Pupin Hall 207, Columbia University (these papers will be added to the Pegram Collection). For instructors and lecturers the figures are 1910: 32 percent, and 1932: 37 percent. For time spent by new Ph.D.'s on research (36 percent as recalled by all people who got science Ph.D.'s in 1935), see also Lindsey R. Harmon, *Profiles of Ph.D.'s in the Sciences: Summary Report on Follow-Up of Doctorate Cohorts, 1935–1960*. Career Patterns Report no. 1 for the National Institutes of Health (Washington, D.C.: National Academy of Sciences – National Research Council, 1965), table 8, p. 16. N.B. throughout this paper a school year such as 1932/33 is treated as 1932.

11. For a complaint about this see F. K. Richtmyer, "Opportunities for Research," *Journal of the Optical Society of America* 1 (1917): 1–3.

12. Fifteen percent is the average of the fractions of each year's class of physics Ph.D.'s; for individual years the fraction varied from ½ to 1-½ times this. National Research Council, *Privileged Preliminary Statistical Data from the Study of Post-Doctoral Education in the United States. Preliminary Report No. 1* (Washington, D.C.: National Academy of Sciences, April 1968), p. 56. The Fellows gravitated to the leading schools; for the years 1926 through 1938, the top seven universities were chosen by 64 percent of the 192 National Research Fellows who were physicists. Counts by Beverly Fearn Porter for a study of the rise of nuclear physics in America directed by Charles Weiner and conducted by Henry Small, Porter and Weiner at the Center for History of Physics, American Institute of Physics; data available at AIP. For an inventory of the study see Weiner, ed., *Exploring the History of Nuclear Physics*. AIP Conference Proceedings 7 (New York: AIP, 1972), app. II. The counts were made from the National Research Council fellowship listings, 1919–38 and 1939–42. For the origins of the fellowships, see Robert A. Millikan, *Autobiography* (New York: Prentice Hall, 1950), pp. 180–83, 213. See also the chapter by Stanley Coben, this volume, and n. 88, below.

13. My counts from vols. 25 and 26, counting only papers whose institutional source could be identified and was in the United States. This 1925 count is skewed in the direction of a few fellows at Harvard. In 1930 there were forty-five fellowships acknowledged among the 335 *Physical Review* papers counted, twenty-seven of them National Research Fellowships, according to an AIP nuclear physics study.

14. See Coben, "Quantum Mechanics," 446, 454–56 and passim; Jeremy Bernstein, "Physicist" [I. I. Rabi], *New Yorker* 13 October 1975): 47 ff., p. 70, 87–88, 92 (continuation is in 20 October 1975, 47 ff.). R. A. Millikan felt outstanding American physicists were too few in comparison with other countries' and in proportion to the population; address in *Centennial Memorial Volume* (Indiana University, 1920), quoted in Duane Roller, "Millikan's Influence on Undergraduate Teaching," *Reviews of Modern Physics* 20 (1948): 26 (I owe this reference to Robert Seidel). The Board awarding the fellowships deliberately favored mathematical physics "in order to overcome America's initial deficiency in this field," according to Karl T. Compton, "Physics in National Planning," *Review of Scientific Instruments* 5 (1934): 236. Coben, "Quantum Mechanics" p. 448, says seventeen of the NRC Fellowships "were for study solely in aspects of quantum theory, others for related work."

15. David G. Bourgin to E. C. Kemble, 16 October 1927, Archive for History of Quantum Physics, film 51. The Archive (hereafter AHQP) is stored at the American Institute of Physics, New York; American Philosophical Society, Philadelphia; Niels Bohr Institute, Copenhagen; and Office for History of Science and Technology, University of California (Berkeley). See Thomas S. Kuhn, John L. Heilbron, Paul L. Forman,

and Lini Allen, *Sources for History of Quantum Physics: An Inventory and Report* (Philadelphia: American Philosophical Society, 1967).

16. See articles cited in n. 7 and interviews of J. H. Van Vleck, AHQP, II, 4–5 and E. C. Kemble, AHQP, I, 10, 12, both by T. S. Kuhn. Note that it was *modern* theory that was wanting; there was probably adequate competence in classical mechanics, electromagnetic theory, etc.

17. "Of the elite group of 135 European physicists who, from 1924 to 1930, received international predoctoral fellowships from the Rockefeller Foundation, one third chose to study at U.S. institutions; more of them were attracted to the U.S. than any other country." Weiner, "Physics in the Great Depression," p. 32.

18. The practice probably began at the physics department of Columbia University, which in the decade before World War I hosted Larmor, Lorentz, Planck, etc. Notable in the 1920s and 1930s for their visitors were Caltech and the University of Michigan Summer School. Millikan, *Autobiography* p. 221; Charles F. Meyer, George A. Lindsay, Daniel L. Rich, Ernest F. Barker, and David M. Dennison, "The Department of Physics," pp. 680–701 in *The University of Michigan—An Encyclopedic Survey. Part IV* (1944), pp. 697–98.

19. Meggers to Sommerfeld, 31 October 1922; Sommerfeld to Meggers, 6 November 1922, Meggers collection, AIP/NBS 1922. Educational Committee, National Bureau of Standards, report for 1922–23, with letter, G. K. Burgess to Vernon Kellogg, 25 January 1924, Physics department chairman's correspondence (Karl T. Compton papers), Princeton University Archives/National Research Council, 1925.

20. Interview of John C. Slater by T. S. Kuhn, AHQP, II, 16.

21. See Weiner, "New Site for the Seminar," pp. 196–200. Several senior European theorists turned down invitations to move to America.

22. See S. Ulam, H. W. Kuhn, A. W. Tucker, and Claude E. Shannon, "John von Neumann, 1903–1957," pp. 235–69 in Fleming and Bailyn, *Intellectual Migration*, p. 237. For information on this and a number of other points, I am greatly indebted to Robert Seidel for letting me read his unpublished paper, "The Origins of Academic Research in California. A Study of Interdisciplinary Dynamics in Institutional Growth," 1975; a short version is in the *Journal of College Science Teaching* 6 (1976). I am also indebted to Lillian Hartmann Hoddeson and David Allison for letting me read their unpublished Princeton University seminar paper, "Physics in America in the 1920's and 1930's," 1975. A good source, available too late for me to use, is Seidel's "Physics Research in California," Ph.D. dissertation, University of California at Berkeley, 1978.

23. My estimates based on an incomplete survey. American theoretical physicists teaching about 1930 included G. Breit, A. H. Compton, E. U. Condon, D. M. Dennison, C. Eckart, N. H. Frank, W. V. Houston, E. C. Kemble, R. C. Mulliken, J. R. Oppenheimer, H. P. Robertson, J. C. Slater, and J. H. Van Vleck.

24. "Typology of Research in Physics," *Social Studies of Science* 5 (1975): 81.

25. Mulliken to Kemble, 14 March 1927, AHQP film 51.

26. My counts from "Appointments Available in Various Universities and Colleges to Graduate Students Majoring in Physics," *American Physics Teacher* 3 (1935): 42–43, 58, 194–95. See also ibid. 4 (1936): 53; 6 (1938): 342–44; 7 (1939): 74–75. Spectroscopy was listed for nineteen departments, X-rays for seventeen, nuclear physics for twelve, and theoretical, quantum or relativity physics for ten. (The statement for any given department in these listings should not be considered reliable, but in the aggregate the statements indicate something about what the departments had available and what they thought would attract students.) Other signs of the continuing importance of spectroscopy are the strong funding it got from the Carnegie Foundation (comp. from *Yearbook of the Carnegie Institution of Washington*, 1920–35, by Hoddeson and Allison, "Physics in America") and the MIT Summer Conferences in Spectroscopy that began in 1933 (see Meggers Collection, AIP/NBS 1932–33).

27. Meyer et al., "Department of Physics"; Alpheus W. Smith, "Nine Decades of Physics and Astronomy at the Ohio State University," typescript, AIP, p. 57; interview of J. H. Slater by T. S. Kuhn and J. H. Van Vleck, AHQP, I, 12. Compare Allen G. Shenstone, "Princeton and Physics. A Remarkable History," *Princeton Alumni Weekly* 61 (February 24, 1961): 6–12, and interview of Henry A. Barton by Charles Weiner, AIP, p. 7. The point has been made already by Coben, "Quantum Mechanics," p. 446. An attempt at experimental work in spectroscopy led Slater toward quantum theory: Slater interview I, p. 7.

28. Hall to Campbell, 28 November 1928, University of California (Berkeley) archives, cited in Seidel, "Physics Research in California," p. 64.

29. Rexmond C. Cochrane, *Measures for Progress: A History of the National Bureau of Standards* (Washington, D.C.: Department of Commerce, 1966), pp. 225–28, 232, 261 n. 118.

30. Meggers to Henry Crew, 8 May 1933, Meggers Collection, AIP/NBS 1933.

31. Meggers to Fred Hartman, 28 January 1924, Meggers Collection, AIP/NBS 1924.

32. Figure 7 and table 2, appendix.

33. Table 3, appendix.

34. L. O. Grondahl, "The Role of Physics in Modern Industry," *Science* n.s. 70 (1929): 177. Also Lewis to Wheeler, 1919, in *University of California President's Report 1918–19*, p. 43, and R. A. Millikan, "The New Opportunity in Science," *Science* n.s. 50 (1919): 292–97, both given in Seidel, "Academic Research in California," pp. 29–30. Also National Research Council, *Research—A National Resource.*

2. *Industrial Research* (Washington, D.C.: U.S. Government Printing Office, 1941), p. 35; Ronald C. Tobey, *The American Ideology of National Science, 1919–1930* (Pittsburgh: University of Pittsburgh Press, 1971), p. 55; chapter by Kendall Birr, this volume. The movement was international. In Britain, Germany, and the Soviet Union, in particular, the governments were closely connected with industrial research efforts, coordinating trade associations and contributing their own funds. Thus the American movement was not an isolated happenstance but presumably reflected general developments in the world economy.

35. Edward R. Weidlin and William A. Hamor, *Science in Action: A Sketch of the Value of Scientific Research in American Industries* (New York: McGraw-Hill, 1931), p. 241.

36. Average of the fractions for five years chosen at intervals over 1920–1939; see table 2, appendix. For concentration of industrial research, see National Research Council, *Research—A National Resource.* 2, passim, and Carroll W. Pursell, Jr., this volume. On patents see "Statement of William D. Coolidge," pp. 911–24 and "Statement of Frank B. Jewett," pp. 948–79 in United States Temporary National Economic Committee, *Investigation of Concentration of Economic Power: Hearings . . . Part 3. Patents*, 16–20 January 1939 (Washington, D.C.: U.S. Government Printing Office, 1939). I am grateful to Nathan Reingold for this reference.

    For more on the value of research to Bell see U.S. Federal Communications Commission (hereafter: USFCC), *Proposed Report, Telephone Investigation*, by Paul A. Walker (Washington, D.C.: U.S. Government Printing Office, 1938), pp. 231, 706 and passim, and the rebuttal (which does not dispute my point) in USFCC, *Brief of Bell System Companies on Commissioner Walker's Proposed Report on the Telephone Investigation* (New York: 1938), p. 46 and passim. I have also benefited from conversations with Lillian Hoddeson. For General Electric, see Kendall Birr, *Pioneering in Industrial Research: The Story of the General Electric Research Laboratory* (Washington, D.C.: Public Affairs Press, 1957). For the reasons Corning Glass supported research see J. T. Littleton, "The Value of a Laboratory in a Corporation," *Glass Industry* 8 (1927): 40, cited p. 221 of William A. Liddell, Jr., *The Development of Science in the American Glass Industry, 1880–1940.* Yale University Ph.D. thesis, 1953 (Ann Arbor: University Microfilms no. 70-2840).

37. F. Russell Bichowsky, *Industrial Research* (Brooklyn: Chemical Publishing Co., 1942), pp. 121–22. Similarly, Weidlin and Hamor, *Science in Action*, pp. 61–62.

38. Weidlin and Hamor, *Science in Action*, p. 54. The *Nation* remarked in 1928 that "A sentence which begins 'Science says' will generally be found to settle any argument in a social gathering, or sell any article from tooth-paste to a refrigerator." Quoted by Paul A. Carter, "Science and the Common Man," *The American Scholar* 45 (Winter 1975/76): 785; see the whole article for the "Gospel of Science" in America in

the 1920s. Although it is true that pure science was often, and sometimes deliberately, confused with technology, I doubt that any antirational movement, as referred to by Tobey, *American Ideology of National Science*, p. 75, had much effect on belief in pure science.

39. Nichols to the Corporation of Yale University, 21 April 1920, quoted in *Science* n.s. 51 (1920): 458.

40. See *Science* n.s. 51 (1920): 115–16, quoting from the *Scientific Monthly*; Trevor Arnett, *Teachers' Salaries in Certain Endowed and State Supported Colleges and Universities in the United States . . . 1926–27*. Publications of the General Education Board, Occasional Papers, no. 8 (New York: GEB, 1928), p. 31; figure 3 and n. 59, below. Data on industrial salaries are scanty. For physical scientists (including chemists and geologists) who got their Ph.D.'s in 1935, in their first postdoctoral jobs those who went into academia and stayed there got on the average only 86 percent of the salary of those who went into industry and stayed. (Of course, they also got their summers off.) National Research Council, Office of Scientific Personnel (Lindsey R. Harmon), *Careers of PhD's, Academic versus Nonacademic: A Second Report on Follow-up of Doctorate Cohorts 1935–1960*. Career Patterns report no. 2 for the National Institutes of Health (Washington, D.C.: National Academy of Sciences, 1968), pp. 29–32. Some idea of the very high rewards collected by a few men may be found in E. L. Thorndike, "The Salaries of Men of Science Employed in Industry," *Science* n.s. 88 (1938): 327.

41. Weidlin and Hamor, *Science in Action*, p. 279; Herbert Hoover, quoted in Carter, "Science and the Common Man," p. 790. "In the end, I fear, the trusts . . . will have absorbed and assimilated *l'élan vital*, the soul of the university." Barus, quoted in *Science* n.s. 57 (1923): 446; cf. Mark H. Liddell, "The Endowment of Scientific Research," *Science* 57 (1923): 612–13.

42. Richard Hamer, "The Romantic and Idealistic Appeal of Physics," *Science* n.s. 61 (1925): 109–10. Similar sentiments are not difficult to find in *Science* in the 1920s. The *locus classicus* is Henry Rowland, "The Highest Aim of the Physicist," in *Bulletin of the American Physical Society* 1 (1899).

43. Jewett in U.S. Temporary National Economic Commission, *Investigation of Economic Power*, p. 949. For regrets at the prospective loss of a good theorist to a lucrative industrial post, see Birge to Kemble, 27 October 1927, and Kemble to Birge, 13 January 1928, AHQP, film 50.

44. In surveys made 1935–37 in which these students reported their professional goals, we find about half were intending engineers and a fourth premedical students. American Association of Physics Teachers, Committee on Tests (C. J. Lapp, Chairman), "The 1934–1935 College Physics Testing Program," *American Physics Teacher* 3 (supplement) (1935): 155; ibid. 4 (1936): 112; 6 (1938): 95. I suspect that the main trend from ca. 1900 to 1930 was a rise in engineering students compared with intending teachers, the latter being only about one-seventh of the 1935–37 students. See U.S. Bureau of the Census, *His-*

torical Statistics of the United States: Colonial Times to 1957 (Washington, D.C.: U.S. Government Printing Office, 1960), p. 210, from which I calculate that the enrollment in physics courses in secondary schools, grades 9–12, rose from 98,000 in 1900 to 192,000 in 1922, and 197,000 in 1928—a small rise relative to overall secondary school enrollments, which multiplied 5-½ times from 1900 to 1928.

45. One task would be to compile, from university alumnus office records, information on the jobs taken by physics majors on graduation.

46. Some dozen are listed by Homer L. Dodge, "Training Physicists for Industry," *American Physics Teacher* 4 (1936): 174. In creating applied physics courses physicists were competing with the mushrooming engineering schools, where physics instruction was only ancillary and was largely taught by engineering graduates. Society for the Promotion of Engineering Education (American Society for Eng. Educ.), *Report of the Investigation of Engineering Education, 1923–1929*, vol. 1 (Pittsburgh: SPEE, 1930), pp. 424–25. William S. Franklin asked, "Why the widespread contempt of physics teaching among engineering faculties?"; Franklin, "What is the Matter with Physics Teaching?" *Science* n.s. 54 (1921): 475. See also *The Teaching of Physics: With Especial Reference to the Teaching of Physics to Students of Engineering. Report Number 1 of the Educational Committee of the American Physical Society*, George V. Wendell, Chairman (APS, 1922), APS Archives, AIP. For the primacy of engineers over scientists in the 1920s, see Tobey, *American Ideology of National Science*, p. 217. For the rise of engineering education see Earl F. Cheit, *The Useful Arts and the Liberal Tradition*. Report for the Carnegie Commission on Higher Education (New York: McGraw-Hill, 1975). It would be important, as we seek a better understanding of physics in America, to know more about the balance between pure and applied physics within the universities, but in this essay I consider chiefly relations between physics departments and outside applied work.

47. Figure 5 and table 1, appendix.

48. E. C. Sullivan, "Accomplishments of the Industrial Physicist in the Glass Industry," *Journal of Applied Physics* 8 (1937): 19; F. Jewett, "Finding and Encouragement of Competent Men," at a symposium on "The Organization of Scientific Research in Industry," *Science* n.s. 69 (1929): 310; Birr, *Pioneering in Industrial Research*, pp. 75–76; Liddell, *Science in American Glass Industry*, p. 234.

49. Jewett, "Competent Men," p. 311; H. M. Goodwin, "Physics at MIT: A History of the Department from 1895–1933," MIT *Technology Review* (May 1933): 312. Professors had long served as consultants to American industry, but their role as purveyors of trained personnel was more recent. On all these issues see David F. Noble, *America by Design: Science, Technology, and the Rise of Corporate Capitalism* (New York: Knopf, 1977) and Michael Sanderson, *The Universities and British Industry, 1850–1970* (London: Routledge and Kegan Paul, 1972).

50. In 1931, the Mellon Institute was running sixty-four fellowships sponsored by nine industrial associations and fifty-five different companies, which set the problems and got exclusive use of the results; 140 scientists and engineers were involved. Weidlin and Hamor, *Science in Action*, p. 57. In 1925, the Bureau hosted sixty-one men maintained by thirty-six organizations, mostly trade associations; the work was to be published. Cochrane, *Measures for Progress*, p. 225.

51. Liddell, *Science in American Glass Industry*, p. 208. Another example: General Electric proposed an arrangement to pay for a graduate fellowship at Princeton in the general field of ionization and radiation; in order to protect patent rights the work was not to be published without GE's approval. Physics department chairman's correspondence (Karl T. Compton Papers), Princeton University Archives/General Electric . . . 1922–1923. Seidel, "Academic Research in California," p. 45, reports a number of industrial fellowships at Caltech. See also National Research Council, *Research—A National Resource*. 2, p. 53. This sort of support was not often very visible. Where specific research or contacts were not offered, as in the proposed National Research Fund, industrialists showed less enthusiasm. For all these matters see Lance E. Davis and Daniel J. Kevles, "The National Research Fund: A Case Study in the Industrial Support of Academic Science," *Minerva* 12 (1974): 207–20.

52. M. I. Pupin, "The Meaning of Scientific Research," *Science* n.s. 61 (1925): 30. Pupin apparently regarded this trend favorably, as did Millikan, *Autobiography*, pp. 218–19. See George Perazich and P. M. Field, *Industrial Research and Changing Technology*. Works Projects Administration, National Research Project, Report No. M-4 (Philadelphia: WPA, 1940), p. 44; Tobey, *American Ideology of National Science*, p. 217 and passim.

53. National Research Council, *Research—A National Resource*. 1, p. 176n., suggests that since universities regard the formal granting of the Ph.D. as one of their important responsibilities, this was much more closely supervised than was the actual content of the thesis research. See Birr, *Pioneering in Industrial Research*, p. 85; Sullivan, "Physicist in Glass Industry," p. 19.

54. W. Elsasser to F. Joliot, 13 November 1936, Joliot-Curie Papers, Radium Institute, Paris/F28. See interview of Paul P. Ewald by Weiner, AIP, p. 39; Weiner, *History of Nuclear Physics*, pp. 30–40; Weiner, "New Site for the Seminar," pp. 224–26. For the American tradition of efficiency rather than craftsmanship, see Daniel J. Boorstin, *The Americans. 3. The Democratic Experience* (New York: Random House, 1973), pp. 194 ff. Sharing of tools: Luis W. Alvarez, "Berkeley—A Lab Like No Other," *Bulletin of the Atomic Scientists* 30 (April 1974): 21–22. For the proletarianization of young German research workers ca. 1900 along bureaucratic rather than entrepreneurial lines, see Forman, Heilbron, and Weart, *Physics ca. 1900*, pp. 52–53.

55. For the type of study I refer to, see Forman, Heilbron, and Weart, *Physics ca. 1900*.

56. Figure 2 (top). United States national income in 10^9 constant (1947–49) dollars, from U.S. Bureau of the Census, *Historical Statistics*, p. 139, using the Consumer Price Index, pp. 125–26, to reduce to constant dollars. The use of national income, the Consumer Price Index as a deflator, and 1947–49 dollars are all somewhat arbitrary; the latter two are used throughout for consistency. Note that the main component of research expense (even "equipment") was salaries and wages, whether of professors or mechanics.

57. Figure 2 (bottom). Scientific research and development expenditures in 10^6 constant (1947–49) dollars, from Vannevar Bush, *Science, the Endless Frontier: A Report to the President* (Washington, D.C.: U.S. Government Printing Office, 1945), and the Consumer Price Index. The dashed line for earlier government expenditures is from approximate U.S. Bureau of the Budget figures given in National Research Council, *Research—A National Resource*, 1, p. 91, adjusted to equal Bush's figures over 1923–27.

58. Figure 3 (top). Teaching staff in higher education (1,000s). George J. Stigler, *Employment and Compensation in Higher Education*. National Bureau of Economic Research, Occasional Paper No. 33 (New York: NBER, 1950), p. 210.

59. Figure 3 (bottom). Median salaries of college teachers in large public institutions (1947–49 dollars), for full professors and instructors. Ibid., p. 42.

60. Ibid., passim; American Association of University Professors (Hereafter: AAUP), *Depression, Recovery and Higher Education*. By Malcolm L. Willey (New York: McGraw-Hill, 1937), pp. 41–47, 59–62. See also reports on "The Economic Status of Scientific Men and Women," *Science* n.s. 70 (1929): 19 ff., 47 ff.

61. Loss from 1928–29 to 1933, rough figures from AAUP, *Depression, Recovery, and Higher Education*, p. 51. For more complete figures see Stigler, *Employment and Compensation*, p. 60; U.S. Bureau of the Census, *Historical Statistics*, p. 97.

62. AAUP, *Depression, Recovery, and Higher Education*, pp. 134–35.

63. Overall employment of physicists: my counts from catalogs of University of California (Berkeley), Caltech, University of Chicago, Cornell University, Iowa State College, MIT, University of Michigan, Ohio State University, Princeton University, University of Wisconsin, show that overall teaching staff in physics departments, instructors through full professors, was roughly 122 for this sample in 1925, 154 in 1930, 153 in 1933, 154 in 1936, and 163 in 1939. This parallels the rise in figure 2 through 1935 but rises a little more slowly thereafter. See also below, n. 91. The number of full professors in my sample rose steadily; assistant and associate professors rose except for a slight drop from 1930 to 1933; instructors and lecturers dropped steadily after 1930,

probably reflecting not only the Depression but also increasing disuse of the position. Salaries: I lack full evidence, but university pay cuts were generally made across-the-board, at least among the teaching staff. Anecdotal evidence from many sources follows the lines of figure 3.

64. Meggers to H. Kayser, 4 June 1937, Meggers Collection, AIP/NBS 1937.

65. Competition: see Laura Fermi, *Illustrious Immigrants: The Intellectual Migration from Europe 1930–41* (Chicago: University of Chicago Press, 1971^2), p. 72. Antisemitism: There is substantial archival evidence, but it would be invidious to cite particular cases when the problem was widespread.

66. The ninety emigrés Weiner clearly identified as physicists were scattered among forty-three universities and at least six private companies. Fourteen percent came over before 1932; the peak year was 1939. Beverly Porter for the AIP nuclear physics study; Weiner, "New Site for the Seminar," p. 226, n. 71.

67. For example, Herbert Childs, *An American Genius: The Life of Ernest Orlando Lawrence* (New York: E. P. Dutton, 1968), pp. 145, 178.

68. Samuel Goudsmit to Gregory Breit, 11 April 1933, Goudsmit Papers, AIP. Other letters complaining of salary cuts, nonpayment, etc., are in the American Physical Society Archives, AIP.

69. For example, Charles W. Edwards to Kemble, 28 September 1933, AHQP film 53; Birge to Kemble, 10 May 1932, AHQP film 50.

70. The position of junior faculty and new Ph.D.'s is studied in detail in AAUP, *Depression, Recovery, and Higher Education*, pp. 23, 26, 30, 75–77, 121.

71. Whereas in 1920 through 1932 NRC fellowships were awarded to an average of 15 percent of each year's class of Ph.D.'s, in 1933 through 1941 they went to only 3 percent of each class on the average. National Research Council, *Post-doctoral Education*, n. 12. For the enormous rise in applications in the early 1930s, see Myron J. Rand, "The National Research Fellowships," *The Scientific Monthly* 73 (1951): 74–75. For the Ph.D. class of 1935, of seventy-one physicists surveyed thirteen had postdoctoral fellowships, of which eight were from universities and only three from private foundations; for the class of 1940, eight of sixty-nine had fellowships, of which five were from foundations. Harmon, *Profiles of PhDs*.

72. Allison to Kemble, 23 January 1933; Kemble to Allison, 28 January 1933, AHQP film 53. For the lot of graduates see also F. K. Richtmyer and Malcolm M. Willey, "The Young College Instructor and the Depression," American Association of University Professors *Bulletin* 22 (1934): 507–9; G. M. Almy, "Life with Wheeler [Loomis] in the physics department [University of Illinois] 1929–1940," talk given 24 May 1957, typescript at AIP, p. 17; Raymond T. Birge, *History of the Physics Department* [University of California (Berkeley)]. 4. The

Decade 1932–1942. Copies at AIP and Bancroft Library, University of California (Berkeley), pp. IX.2,7–8, 18–19, 26–27, 30, XII.6–8, 24–25, XIII.11 ff., XIV.23.

73. Wheeler Loomis to E. H. Williams, June 1936, quoted in Almy, "Life with Wheeler," p. 4.

74. My counts, table 1, appendix. The number staying on at the Ph.D.-granting institution (nearly always one of the top twenty) for a year or two, as a percentage of all those with a location listed in the APS *Bulletin,* was 21.7 percent (1895–1904), 14.0 percent (1910–14), 7.1 percent (1918–21), 24.7 percent (1923–24), 17.1 percent (1928–29), 41.2 percent (1934), 12.2 percent (1939).

75. Physics department chairman's correspondence (Karl T. Compton papers), Princeton University Archives/Graduate School 1931–1932; Birge, *Physics Department,* pp. XII.2–3; Almy, "Life with Wheeler," pp. 3–4; Childs, *An American Genius,* p. 175; Smith, "Nine Decades of Physics and Astronomy," pp. 58–59.

76. A 9 percent drop in current-dollar research expenditures from 1929/30 to 1933/34 was probably more than offset by deflation, although the deflator for such expenditures is not obvious. (Using the Consumer Price Index gives a 21 percent rise in constant-dollar expenditures.) AAUP, *Depression, Recovery, and Higher Education,* pp. 190–92, 208–9.

77. Almy, "Life with Wheeler," p. 16, quotes F. W. Loomis' report for the Illinois University Physics department for 1934/35: "The department, whose operating expenses have been reduced to a starvation point for over three years, suffered a financial crisis this winter and pretty nearly had to close up. It was rescued, temporarily, by the allotment of $2,200 from general and engineering funds ... It is almost impossible to convey an adequate idea of the extent to which our work, both in teaching and research, has been hampered and made inefficient ... We should have had pretty nearly to cease activity in research if it had not been for the equipment which was bought in our three boom years, 1929–32." At Columbia University George Pegram wrote President Nicholas M. Butler, 19 March 1935, stating that the volume of research work was much larger than ever before, but that the appropriation for the school year was already used up. Another $3,000 was granted (Frank Fackenthal to Pegram, 23 March 1935). Folder "Budget General 1930–1939" in drawer "Budget—General 1930–1958," Pupin Hall 207, Columbia University.

78. Bush, *Science, the Endless Frontier,* p. 81, from U.S. Office of Education, *Biennial Survey,* shows that by 1939/40 the expenditures for capital outlays in higher education had recovered to the 1931/32 level —almost three times the 1933/34 low. See AAUP, *Depression, Recovery, and Higher Education,* p. 198. I lack adequate data for physics laboratories themselves, but the pattern presumably was similar to the overall university pattern.

79. Columbia: see n. 77 above; Bernstein, "Physicist," p. 105; Sidney Millman, "Recollections of a Rabi Student of the Early Years in the Molecular Beam Laboratory," *Transactions of the New York Academy of Sciences* 38, ser. 2 (1977): 87–105. Berkeley: Birge, *Physics Department*, pp. XI.24–25, XII.28. Caltech: Millikan, *Autobiography*, pp. 247–50.

80. Figure 4. Physics department expenses, omitting salaries of teaching staff, 10^3 constant (1947–49) dollars. Dotted: Harvard University, from *Statement of the Treasurer of Harvard College* (Cambridge: Harvard University, annual). Upper solid lines: Massachusetts Institute of Technology, from MIT, *President's Report* (Cambridge: MIT, annual), in AIP nuclear physics study. Lower solid lines: Columbia University, from various items in drawer "Budget—General, 1930–1958" in Pupin Hall 207.

81. Again I have in mind work along the lines of Forman, Heilbron, and Weart, *Physics ca. 1900*.

82. In 1940, student fees provided some $200 million of the total $570 million income for all institutions of higher education, private and public together; next came state government with $150 million, and earnings from endowment, $70 million, the former mainly for public and the latter for private schools. Stigler, *Employment and Compensation*, p. 35, citing U.S. Office of Education, *Biennial Survey*. In a study for 1935–36 covering all fields of university research, 37 percent of funds came from the state and federal governments, 35 percent from general endowment income, 16 percent from foundation grants and 8 percent from other gifts (the remaining 4 percent came from "sales and other income"). Apparently tuition etc. was not regarded in this study as contributing directly to research. National Research Council, *Research—A National Resource*, 1, p. 178. The records of state legislatures may be an important, unexamined source for attitudes toward and funding of universities.

83. For drop in endowment income through 1940 see Trevor Arnett, *Recent Trends in Higher Education in the United States: With Special Reference to Financial Support for Private Colleges and Universities*. General Education Board Publications, Occasional Papers, no. 13 (New York: GEB, 1940), p. 61; Stigler, *Employment and Compensation*, p. 35 n. 7, citing J. Harvey Cain, *College Investment under War Conditions* (Washington, D.C.: American Council on Education, September 1944).

84. For government funding in general, see A. Hunter Dupree, *Science in the Federal Government: A History of Policies and Activities to 1940* (Cambridge: Harvard University Press, 1957). Science Advisory Board: Lewis E. Auerbach, "Scientists in the New Deal: A Pre-war Episode in the Relations Between Science and Government in the United States," *Minerva* 4 (1965): 457–82; Carroll W. Pursell, Jr., "Anatomy of a Failure: the Science Advisory Board, 1933–1935," *Proceedings of the American Philosophical Society* 109 (1965): 342–51. NBS: Pursell, "A Preface to Government Support of Research and Development: Research Legislation and the National Bureau of Stan-

dards, 1935–41," *Technology and Culture* 9 (1968): 145–64. See also Daniel Kevles, "George Ellery Hale, the First World War, and the Advancement of Science in America," *Isis* 59 (1968): 427–37.

85. AAUP, *Depression, Recovery, and Higher Education*, p. 379.

86. Birge, *Physics Department*, p. XIII.33, see XII.29; John C. Slater, "History of the MIT Physics Department, 1930–1948," typescript, AIP, pp. 10–11. See U.S. Works Projects Administration, *Index of Research Projects* (Washington, D.C.: U.S. Government Printing Office, 1938–39).

87. For Britain one might start with Ian Varcoe, *Organizing for Science in Britain: A Case-Study* (London: Oxford University Press, 1974), and Great Britain, Department of Scientific and Industrial Research, *Scientific Research in British Universities* (London: H.M.S.O., annual). For Germany: Brigitte Schroeder-Gudehus, "The Argument for the Self-Government and Public Support of Science in Weimar Germany," *Minerva* 10 (1972): 537–70; Paul Forman, *The Environment and Practice of Atomic Physics in Weimar Germany*. Ph.D. thesis, University of California (Berkeley), 1968 (Ann Arbor, Mich.: University Microfilms no. 68-10322). For France: Jean Perrin, *L'Organisation de la recherche scientifique en France*. Discours prononcé au conseil supérieure de la recherche scientifique (Paris: Hermann, 1938); Spencer Weart, *Scientists in Power* (Cambridge: Harvard University Press, 1979).

88. My counts through 1925; later counts from the AIP nuclear physics study.

	1910	1915	1920	1925	1930	1935	1940
Percentage of papers acknowledging support:	11.0	5.7	6.3	6.6	9.6	17.2	26.5
Ditto, disregarding NRC fellowships:	11.0	5.7	6.3	19.8	17.9	23.2	27.4

89. I am grateful to Charles Weiner for discussions of his work on the cyclotron boom. The growth of American nuclear physics is covered by the AIP nuclear physics study.

90. From 1900 through 1924 the ratio of physics to all Ph.D.'s granted in the U.S. ran around 5.9 percent. The ratio was 5.3 percent for 1925–29, 4.5 percent for 1930–34, and 5.3 percent for 1935–40. For the constant ratio (about 11 percent) of physics to all science and engineering degrees see H. William Koch, "On Physics and Employment of Physicists," *Physics Today* 24 (June 1971): 24. For discussions of Ph.D. ratios and other matters, I am grateful to Arnold Thackray and his students. Faculty: see above, n. 63, and below, n. 91.

91. Ratio of physics faculty to total university or college enrollment. There was a slight rise from 2.57 to 2.71 per 1,000 students in a sample of thirty-five departments—statistically significant but not large enough to affect the argument. David M. Blank and George J. Stigler, *The De-*

mand and Supply of Scientific Personnel. National Bureau of Economic Research, General Series, no. 62 (New York: NBER, 1957), p. 95. Since the overall faculty/student ratio for the schools did not vary significantly in the 1920s and 1930s, fluctuating around 76 per 1,000 (U.S. Bureau of the Census, *Historical Statistics*, p. 210), we see again that physicists' employment paralleled that of other faculty (see n. 63 above).

92. There was a mild drop in undergraduate enrollment and a rise in graduate enrollment ca. 1933. For rise in tuitions, see Stigler, *Employment and Compensation*, p. 34. Graduate students in physics: for seventy-one who got their degrees in 1935, according to their recollections 52 percent of their financial support came from the university and 46 percent from self or family (about half of this from the graduate student's earnings). Harmon, *Profiles of PhDs*, p. 92. A partial survey of physics departments listed 104 graduate teaching appointments, some already filled, at thirteen top schools in 1935/36, and 267, 59 percent of them filled, at twenty top schools in 1939/40; the gain in numbers of positions, for a uniform sample of eleven top schools, was 38 percent. This supports Harmon's figures in showing much university employment of physics graduate students. My counts from *American Physics Teacher*, "Appointments Available to Graduate Students."

Year	1935/36		1939/40		
	No. Places Covered by Survey	No. of Positions	No. Places Covered by Survey	No. of Positions	Fraction of Positions Vacant
Appointments					
20 top depts.	13	104	20	267	41%
Other	24	84	67	286	55%
Fellowships and/or Tuition Waivers					
20 top depts.	8	13	11	45	69%
Other	14	24	19	35	83%

Concerning the cost of training these students, note that the mean time lapse between baccalaureate and doctoral degrees for American physicists was 7.4 years for the decade 1920–29 and 7.1 years for 1930–39. National Research Council, *Doctorate Production*, pp. 40–41. (There was also an average one-half year of postdoctoral work.)

93. AAUP, *Depression, Recovery, and Higher Education*, p. 276. See Christopher Jencks and David Riesman, *The Academic Revolution* (Garden City, N.Y.: Doubleday, 1968), pp. 93–94, 108.

94. Figure 5. See table 1, appendix.

95. Statistical significance: the "unknown" rates for 1918–21, 1923–24, and 1928–29, equally weighted, average 4.6 percent. If for 1934 we

therefore assume an a priori probability of 0.046 for a Ph.D. to end up "unknown," the probability that ninety-six Bernoulli trials will yield eleven or more "unknowns" is, from the binomial distribution, only about 0.005. Harvard University Computation Laboratory, *Tables of the Cumulative Binomial Probability Distribution* (Cambridge: Harvard University Press, 1955). Some of the assumptions in such a calculation are overly simple as a description of the job market; the point is that the excess of "unknowns" is most unlikely to be due to a simple random process.

96. Note that if a person was located by my procedure, even working as a chemist or engineer or administrator, I do not consider him "lost" as a physicist; only those who disappeared from both Cattell and the Physical Society are so considered. Assuming as in n. 95 a normal loss rate of 4.6 percent, there would have been four lost from my 1934 sample, and the actual loss exceeded this by seven people. My sample numbered 96 while Scates et al., in *Production of Doctorates*, gives 117 Ph.D.'s for 1934, so by extrapolation the actual excess loss would be about eight people. This is a flimsy calculation, of course, and the number could have been, say, four or twelve for all we know. Scates claims no excess loss of physicists due to the Depression, although his charts do show an excess loss in this period for the sciences as a whole, which he estimates as 1,000 Ph.D.'s never awarded (p. 37). Also worth noting is the job stability of successful science Ph.D.'s: 53 percent of the 1935 Ph.D. cohort had no job change during the period 1936–40, and this was about the same as the experience of the 1950 and 1955 cohorts after graduation in more prosperous times. Harmon, *Profiles of PhDs*, p. 47.

97. Figure 6 (top). All nonagricultural employees in the United States ($\times 10^6$). U.S. Bureau of the Census, *Historical Statistics*, p. 73.

98. Figure 6 (bottom). *Solid line:* Total research workers in a sample of industrial companies, from Perazich and Field, *Industrial Research*, p. 78. Because of poor coverage for 1921, I raise their figure for this year by 23 percent as they recommend, p. 60. *Dotted:* Physicist research workers in a sample of industrial companies ($\times 15$). Generally the largest component of this is from ibid., p. 78, back-calculated from the percentages given. But these figures are so small that evidently the three largest physics-based companies—Bell Labs, General Electric, and Westinghouse, in that order—are not in the Perazich and Field sample. Because of their importance I have added them in. Employment figures for GE and Westinghouse are from National Research Council, *Industrial Research Laboratories of the United States.* Comp. Callie Hull (Washington, D.C.: National Academy of Sciences, 1921^2–1938^6), also published as *Bulletin of the National Research Council* nos. 16, 60, 81, 91, 102, and 104. For Bell, where the NRC directory lacks full information, I use my counts from the membership lists in the *Bulletin* of the American Physical Society for nearby years (table 2, appendix). Comparison of Physical Society counts with NRC figures for GE and Westinghouse shows a general, although very rough, agreement for numbers of physicists.

Year	1921	1927	1931	1933	1938
Perazich & Field sample	38	156	376	196	693
My total sample	159	332	598	394	915

99. See Perazich and Field, *Industrial Research*. For other details on the drop from 1930 to 1933 in industrial research employment, see C. J. West and Callie Hull, "Survey of Personnel Changes in Industrial Research Laboratories—1930–1933," *Research Laboratory Record* 2 (1933): 154–58. They find that the 33 labs reporting 100 or more research workers in 1930 showed a 23 percent drop in employment, including many scientists. For 150 companies in physics-oriented industries (aeronautics, automotive, electrical, iron and steel, nonferrous metals, optics, radio, and scientific apparatus) the drop was 36 percent.

100. American Telephone and Telegraph Co. research and development expenditures, reduced to constant (1947–49) dollars by the consumer price index ($\times 10^6$):

Year	1916	1918	1920	1922	1924	1926	1928	1930	1932	1933	1935	1939
$	6.1	9.4	10.8	11.5	14.3	16.6	20.6	31.7	27.2	22.9	28.2	35.4

The 1916–35 figures are from USFCC, *Proposed Report*, p. 221; 1939 figure from "Statement of Frank B. Jewett," U.S. Temporary National Economic Commission, *Investigation of Economic Powers*, pp. 974–75. Jewett estimated expenditures at $20 to 22 million current per year, and $7 to 9 million for Bell Labs alone. At this time the GE research labs, the next largest employer of physicists, were spending about $1 million a year: "Statement of William D. Coolidge," ibid., p. 915. For post-1939 AT&T figures see U.S. Senate, 79th Congress, 1st Session, Military Affairs Committee, Subcommittee on War Mobilization, *The Government's Wartime Research and Development, 1940–1944: Report*... (Washington, D.C.: U.S. Government Printing Office, 1945), p. 9; I owe this reference to Kendall Birr.

101. [United States, National Bureau of Standards,] *Report of the [President's] Science Advisory Board*. 1 (Washington, D.C.: U.S. Government Printing Office, 1934–35), App. 2, pp. 65, 72; Cochrane, *Measures For Progress*, pp. 317–19, 322. See Meggers to Capt. A. W. Stevens, 20 December 1933, Meggers Collection, AIP/NBS 1933. For government research in general in the 1930s see National Research Council, *Research—A National Resource*. 1.

102. Cochrane, *Measures for Progress*; Lillian Hartman Hoddeson, paper delivered to History of Science Society, Atlanta, 30 December 1975; Birr, *Pioneering in Industrial Research*, p. 120.

103. Birr, *Pioneering in Industrial Research*.

104. Albert W. Hull, "Putting Physics to Work," *Review of Scientific Instruments* 6 (1935): 379.

105. Bichowsky, *Industrial Research*, p. 113.

106. David Deitz, "Science, Uncle Sam, and the Future," *Review of Scientific Instruments* 7 (1936): 2; Birr, *Pioneering in Industrial Research*, pp. 120–21; Cochrane, *Measures for Progress*, pp. 335–44, 350 and passim; USFCC, *Proposed Report*, p. 219, n. 39. See quotes from electrical industry labs in West and Hull, "Personnel Changes in Industrial Research Laboratories," p. 158.

107. "Ultra High Frequency Transmission by Wave Guides," Case Survey Report, 25 June 1934, in notebook, "Appendix I to History of Waveguides," G. C. Southworth Collection, AIP/Box 3.

108. Hoddeson, personal communication and paper delivered to History of Science Society, Atlanta, 30 December 1975; Cochrane, *Measures for Progress*, pp. 335, 381 and passim.

109. In the sample of Perazich and Field, *Industrial Research*, there was an increase of some 500 industrial physicists from 1933 to 1938. (For the larger sample I use for figure 6, n. 98 above, the increase was about 520.) For a different sample, 1.35 times larger, Perazich and Field find a growth of 2,100 researchers of all kinds in the oil industry; in 1938, 9.3 percent of all researchers in this industry were physicists (ibid., p. 79), which implies a growth of 215 for this sample even if the proportion of physicists in this industry was not increasing. That reduces to $215 \div 1.35 = 160$ for the first-mentioned sample, suggesting that the oil industry alone accounted for over 30 percent of the recovery of industrial physics; the argument is of course very approximate. For background on a key instrument, see J. Clarence Karcher, "The Reflection Seismograph: Its Invention and use in the Discovery of Oil and Gas Fields," typescript, AIP.

110. The rest were virtually all in academic institutions. A proof of significance could be made along the lines of n. 95 above. My counts (Table 1, appendix). The sample numbered 110 for 1930 and 62 for 1935. The figures I give here differ from those displayed in figure 5 and those given in n. 74 by excluding followups of people whose locations were not found from the first look at the Physical Society membership list (forty-nine of these in 1930 and thirty-three in 1935) or who stayed on at the institution where they got their Ph.D. (twenty-three in 1930 and thirty-one in 1935). For the Ph.D. cohort of 1923–24, thirty-two were not found in the 1926 list and fifty-nine were; of the latter twenty stayed on, eight went into industry and two into government. For 1939, forty-six were not found in the 1941 list and eighty-eight were; of the latter eleven stayed on, nineteen went into industry and eleven into government.

111. Daniel S. Elliott, "Some Economic Aspects of Physics," *American Physics Teacher* 6 (1938): 198. Similarly: "We folks in the teaching business have a product to sell . . ." Wheeler P. Davey in "Report of Conference on Applied Physics," *Review of Scientific Instruments* 7 (1936): 119.

112. Henry A. Barton, "Annual Report of Director" [of the American Institute of Physics], 23 February 1940, Barton Collection, AIP.

113. Paul D. Foote, "Industrial Physics," Address as Retiring President of The American Physical Society, Boston, 29 December 1933, *Review of Scientific Instruments* 5 (1934): 63–64; Saul Dushman in Davey, "Report of Conference," p. 119; George R. Harrison in Davey, "Report of Conference," p. 120; O. E. Buckley in "A Conference on Applied Physics," *Review of Scientific Instruments* 6 (1935): 32; "AIP Anniversary Meeting," *Review of Scientific Instruments* 7 (1936): 504.

114. For example Elliott, "Economic Aspects of Physics," p. 199; see also above, n. 46. For the increased interest in research and well-rounded training in engineering education in the 1930s see Dugald C. Jackson, *Present Status and Trends of Engineering Education in the United States* (New York: Engineers' Council for Professional Development, 1939), pp. 133, 136, and passim; physicists noted this trend, e.g., Karl T. Compton, "Engineering in an American Program for Social Progress," *Science* 85 (1937): 277.

115. Of fifteen department leaders giving clear responses to Barton, about four said their departments had made changes in the curriculum along the lines desired by industry (e.g., more chemistry instruction); four were seriously considering changes in this direction; four were undertaking no changes, arguing that they were already giving training adequate for either applied or academic physics; three were not interested, feeling that academic-oriented training was sufficient. My "content analysis" of letters in Barton Collection, AIP/Box I, Committee on Training.

116. Letter to Barton, 8 November 1935, ibid.

117. A blue-ribbon panel felt that since 1929 "the science departments of universities have found it necessary, in view of the decrease in gifts by individuals, to rely more upon industrial corporations for research. This may imply the distortion of university research in the direction of shortrange problems . . ." Isaiah Bowman et al., "Report of the Committee on Science and the Public Welfare," App. 3 in Bush, *Science, the Endless Frontier*, p. 85.

118. Figure 7. The top graph shows total American Physical Society membership, from the Society's *Bulletin*. For the lower graph see table 2, appendix. The academic character of the Physical Society was stamped from the beginning—at its organizational meeting in 1899, of thirty-eight people present only two were from industry and one from government, while twenty-five were from ten leading universities. Editorial note to Ernest Merritt, "Early Days of the Physical Society," *Review of Scientific Instruments* 5 (1934): 148.

119. The same held for the elite group of "starred" physicists in Cattell, *American Men of Science*. Those in "Applied Science" and "Government" were respectively 15 percent and 12 percent of the whole group of starred physicists in 1906; 24 percent and 14 percent in 1921–27; 12 percent and 4 percent in 1933–38. Stephen S. Visher, *Scientists*

Starred 1903–1943 in "American Men of Science" (Baltimore: Johns Hopkins Press, 1947), p. 482. Of the 132 physicists starred in 1938, 53 percent were in twenty leading schools, 17 percent in industry, 9 percent in government, and 6 percent in private research foundations. National Research Council, *Research—A National Resource*. 1, pp. 174, 192–93. For the sharp drop in the "birth rate" of industrial laboratories after 1930, see National Research Council, *Research—A National Resource*. 2, p. 176.

120. Karl T. Compton, "What's Ahead in Physics," *Review of Scientific Instruments* 8 (1937): 43; Foote, *Industrial Physics*, p. 61; Davey, "Report of Conference," p. 117.

121. Perazich and Field, *Industrial Research*, p. 12, give 1,600 in 1938; my counts of a sample of The American Physical Society for 1939 (table 2, appendix) give 25 percent, or 890, in industry. There were 2,030 professionally trained physicists in industry in 1940 according to National Research Council, *Research—A National Resource*. 2, p. 176. Most of the physicists outside the Physical Society must have lacked the Ph.D.

122. Tate in Davey, "Report of Conference," pp. 115–16; L. A. Jones in Davey, "Report of Conference," p. 118.

123. "Proceedings of the American Physical Society, Minutes of the Council," 28–29 November 1930, APS Archives, AIP/Box 6. In 1930 the American Association of Physics Teachers also was formed. See *Physics Today* 4 (October 1951): 20–25, and AAPT interviews (Paul Klopsteg, etc.), AIP.

124. Many people held multiple memberships. In 1930 about half of the members of the Optical Society and one-third of the members of the Acoustical Society were also Physical Society members, although many other members of these societies would never have been likely to join the Physical Society. Survey in Minutes of American Physical Society Council, 28–29 November 1930, cited in Barton, "Annual Report of the Director," AIP. The Optical and Acoustical Societies each lost around 15 percent of their members from 1931 to 1936, recovering only after 1940—possibly another sign of the stagnation of industrial physics.

125. Karl T. Compton, "The American Institute of Physics," *Review of Scientific Instruments* 4 (1933): 57–58; K. T. Compton, "The Founding of the American Institute of Physics," *Physics Today* 5 (February 1952): 4–7; Barton, "Annual Report of the Director," AIP, and "The Story of AIP," *Physics Today* 9 (January 1956): 56–66. For AIP efforts see Weiner, "Physics in the Great Depression." Some examples: The AIP and the National Association of Manufacturers (NAM) formed a joint "Science Committee" which published a Science Supplement to the NAM newsletter, aimed at promoting research. In 1938 the AIP and the physics department at the University of Michigan sponsored a symposium on "Physics in the Automotive Industry," and in 1939 AIP held another on "Temperature and Its Measurement and Control in Science and Industry." Barton, "AIP Director's Report for

the Year 1938," Barton Collection, AIP. Besides the threatened defection of applied physicists, chief motives for founding AIP were the financial crisis in publishing and the felt need for better public relations, discussed below.

126. A. R. Olpin, "Training of Physicists for Industrial Positions," *American Physics Teacher* 5 (1937): p. 14. This change of opinion was also noted by Jewett, in U.S. Temporary National Economic Commission, *Investigation of Economic Power.*

127. Interview of Barton by Charles Weiner, AIP, p. 27. See editorial, *Journal of Applied Physics* 1 (July 1931): 69.

128. Of sixty-four papers in *Physics* in 1935, I count thirty-seven originating in colleges and universities, twenty-three in industry, and four in government. The *Review of Scientific Instruments* is a special case, beginning in the 1920s as a section of *JOSA*. Of seventy papers I counted in the *Review of Scientific Instruments* in 1935, sixty-one were from colleges or universities and only nine from industry (forty-two of the former came from sixteen top departments). The *Journal of Applied Physics* was not published 1933–36, the years when *Physics* (incorporating the *Journal of Rheology*) appeared; the *American Physics Teacher* is now the *American Journal of Physics*; the rest survive under their original names.

129. Figure 8. The top half of the figure shows the total number of articles appearing annually—regular contributed technical or scientific articles only, excluding general interest articles, letters and abstracts, etc. Upper line: Total including the *Physical Review, Journal of Applied Physics, Journal of the Acoustical Society of America, Physics,* and *Review of Scientific Instruments.* I exclude from these totals the *Reviews of Modern Physics,* a review journal; the *Journal of Chemical Physics,* with contributors who were preponderantly chemists or chemical physicists; the *American Physics Teacher,* with articles at first chiefly of educational interest; etc. Lower line: the *Physical Review* alone. For the bottom graph see table 3, appendix.

130. "Statement of Financial Situation of the Physical Review . . ." [1928], Kemble Correspondence, s.v. *Physical Review,* AHQP film 52; Proceedings of the American Physical Society, Minutes of the Council, 30 December 1929, APS Archives, AIP/Box 6. A reason, perhaps the main one, for the difficulties was that the papers had become highly specialized; each was of interest to relatively few members of the Physical Society, who therefore could not be asked to spend more to subsidize the publication. So wrote G. S. Fulcher to Dayton C. Miller, 31 January 1930, papers of the division of physical sciences, National Academy of Sciences Archives, Washington, D.C./*Physical Review* Financial Support. See also Arthur L. Foley to George Pegram, 23 November 1928, APS Archives, AIP/Selected Membership Correspondence.

131. Proceedings of the American Physical Society, Minutes of the Council, 1 December 1933, p. 8, APS Archives, AIP/Box 3. Many letters on

the subject are in ibid., Membership Correspondence. Dues of course included the price of a journal subscription.

132. The rise was smooth except for a jump from 9 percent in 1932 to 22 percent in 1933. AIP nuclear physics study; see particularly Henry Small, "Nuclear Physics in *The Physical Review* . . . The Emerging Field, 1927–1934," typescript, 1971, AIP.

133. A numerical model, however crude, can help convert vague statements into meaningful ones. The number of physicists active in year t was $N(t) \simeq N_0 10^{t/35}$. I estimate that some 15 percent of trained physicists were faculty members at the fifteen leading schools which produced about 80 percent of all physics Ph.D.'s during 1920–40. Let there be k Ph.D.'s produced by each of these elite professors per year. Then
$$N(t+35) \simeq \sum_{t=1}^{35} \frac{0.15}{0.8} N(t)k,$$
assuming that the original lot have all retired by $t+35$. Since $N(t+35) \simeq 10N_0$, we have
$$k \simeq 53 / \sum_{t=1}^{35} 10^{t/35} = 0.38.$$

For such a rough model this is not too far off the mark: in eleven leading physics departments which I could check, the annual Ph.D. production per physics faculty member averaged 0.45 in the 1920s and 1930s, with all but two schools falling in the range 0.2–0.6. Multiplying 0.45 by the average duration in graduate and postdoctoral positions, eight years (n. 92) gives about 3.6 advanced students per faculty member at a given moment. In fact, the ratio of all graduate students to faculty in five leading departments I checked averaged 3.9 in the 1930s. (For 1900 the ratio was only 1.5; Forman, Heilbron, and Weart, *Physics ca. 1900*, Table A.4, p. 24.) Over a thirty-five-year career, each of the elite professors would train some $0.45 \times 35 = 16$ students.

134. See above, n. 128, for the strong academic component in the *Review of Scientific Instruments* and *Physics*. Moreover, at first many papers in the newer journals were of a less rigorous and more popular character than papers in the *Physical Review*, so simple counts do not tell the full story.

135. K. T. Compton, "Science Makes Jobs," *The Scientific Monthly* 38 (1934): 297. Compton was referring particularly to restrictive NRA codes.

136. Stuart Chase, *A New Deal* (New York: Macmillan, 1932), pp. 151, 218. Chase's "suggestion of 1932 for a ten-year moratorium on invention seemed only half-jocular; it came after the publication of a book describing the satisfactions of the static economy of rural Mexico," writes Arthur M. Schlesinger, Jr., *The Crisis of the Old Order, 1919–1933* (Boston: Houghton Mifflin, 1957), p. 201. See Tobey, *American Ideology of National Science*, pp. 150–51; Dupree, *Science in Federal Government*, p. 347; Carroll Pursell, "'A Savage Struck by Lightning': the Idea of a Research Moratorium, 1927–37," *Lex et Scientia* 10

(1974): 146–61; William E. Akin, *Technocracy and the American Dream: The Technocrat Movement, 1900–1941* (Berkeley: University of California Press, 1977), pp. 149–152.

137. Weiner, "Physics in the Great Depression"; "Science Makes More Jobs," *Review of Scientific Instruments* 5 (1934): 139.

138. Chase, *A New Deal*, p. 218. Pursell has concluded that "the sometimes violent reactions from the scientific and technical communities were based either on a vast overestimation of the support for a moratorium or on a deliberate desire to win some advantage from beating a dead horse." In "Savage Struck by Lightning," p. 149.

139. GE: W. D. Coolidge, "The Research Laboratory of the General Electric Company," *Journal of Applied Phys.* 8 (1937): 34–39; Bell: L. Hoddeson, personal communication; Westinghouse: see announcement, *Review of Scientific Instruments* 9 (1938): 206.

140. National Science Foundation, *Science Indicators 1976: Report of the National Science Board 1977* (Washington, D.C.: U.S. Government Printing Office, 1977), table 3–3, p. 225, table 3–6, p. 229, and table 3–18, p. 243; see "Basic Research" section.

141. Beverly F. Porter, Sylvia F. Barisch, and Raymond W. Sears, "A First Look at the 1973 Register," *Physics Today* 27 (April 1974): 29; National Research Council, Physics Survey Committee, *Physics in Perspective*. 1 (Washington, D.C.: National Academy of Sciences, 1972), pp. 832–33. Similar figures (but showing 5 percent in hospitals and other nonprofit work) are in National Science Foundation, "Work Activities of Employed Doctoral Scientists and Engineers in the U.S. Labor Force, July 1973," *Reviews of Data on Science Resources* No. 24, NSF 75-310, June 1975. Time: National Science Foundation, *Science Indicators 1976*, p. 279.

142. Susanne D. Ellis, "Enrollments and Degrees," AIP pub. no. R–151.16, February 1979.

143. New leaders include University of Maryland, State University of New York (Stony Brook), Purdue, and University of Tennessee (Knoxville). American Institute of Physics, *1973–1974 Directory of Physics & Astronomy Faculties in North American Colleges and Universities* (New York: AIP, 1973), App. 5.

144. Also, 23 percent in education institutions, 18 percent in industry, 6 percent in government, 6 percent in federally-funded or private research institutions, and 3 percent in miscellaneous. Susanne D. Ellis, "1975–1976 Graduate Student Survey," AIP pub. no. R–207.9, July 1977. See also National Research Council, *Physics in Perspective*, p. 850; National Research Council, "Summary Report 1974. Doctorate Recipients from United States Universities," June 1975; National Research Council, Office of Scientific Personnel, "Employment Status of Doctorate Recipients. Tabulations by Doctoral Field," May 1970.

145. Beverly F. Porter, "Post-Doctoral Positions in Physics and Astronomy," AIP pub. no. R–270, October 1975; Koch, "Physics and Employment."

146. L. Grodzins, "Supply and Demand for Ph.D. Physicists, 1975 to 1986," pp. 52–86 in Martin L. Perl, ed., *Physics Careers, Employment and Education.* American Institute of Physics Conference Proceedings, no. 39 (New York: AIP, 1978), pp. 83, 54, 63, 67.

147. National Research Council, *Physics in Perspective,* p. 916.

148. David E. Drew, *Science Development: An Evaluation Study.* National Board on Graduate Education, Technical Report No. 4 (Washington, D.C.: National Academy of Sciences, 1974), pp. 173–74.

149. National Science Foundation, *Science Indicators 1976,* table 3–22, p. 248.

150. My counts from *Physical Review 12 A, B, C, D* (December 1975), see table 3, appendix. Note also that the proportion of foreign authorships is now much higher.

151. My counts from the APS *Bulletin,* 1975 Membership Directory. See table 2, appendix.

152. See e.g., Perl, *Physics Careers,* and the monthly "Letters" column of *Physics Today.* Also Bruce L. R. Smith and Joseph J. Karsky, *The State of Academic Science: The Universities in the Nation's Research Effort* (New York: Change Magazine Press, 1977), pp. 102–18.

SCIENCE AGENCIES IN WORLD WAR II: THE OSRD AND ITS CHALLENGERS

CARROLL PURSELL
University of California (Santa Barbara)

THE OFFICE OF SCIENTIFIC RESEARCH AND DEVELOPMENT (OSRD) was, by universal agreement, the major American science and technology agency of World War II. This distinction was won not because the office spent the most money or employed the most people (both the Army and Navy spent more for research and development), but because it was the chosen instrument of the nation's technical elite; it became the training ground for that generation of science administrators who shaped the postwar scientific establishment. However, the OSRD hegemony, which now seems so clear and so secure, was repeatedly challenged by other government agencies, and by a number of ambitious administrators who hoped to establish rival centers of scientific power.[1] Like any other bureau in Washington—in wartime or peace—the OSRD spent a significant amount of its time and effort defining and defending its own peculiar mission. Its survival proved its fitness in the jungle of science politics.

The parent organization of the OSRD, the National Defense Research Committee (NDRC), was established on 27 June 1940. The presiding genius behind the organization was Vannevar Bush—scientist, engineer, businessman, educator, and administrator extraordinary.[2] In his own person, Bush represented all the estates of science and typified both its major virtues and its several weaknesses. He had been a professor of power transmission at MIT in the early years of the Depression. Karl T. Compton, president of MIT, made him a vice president of the school, and he sat in on some of the meetings (though he was never a member) of the Science Advisory Board that Compton chaired. Toward the end of the decade, Bush moved to Washington to become president of the Carnegie Institution of Washington (CIW), the private philanthropic research establishment which was, in many ways, more important than the National Academy of Sciences as the scientific embassy in the nation's capital. In 1938, Bush was made a member of the National Advisory Committee for Aeronautics (NACA). A year later, in 1939, he was made chairman of that committee.

Through both his civilian contacts with European scientists as president of the CIW and his official cognizance of the NACA research program, Bush became concerned with the traditional but dangerous unpreparedness of the nation in the face of approaching war. As early as 1939, he helped accomplish the election of Frank B. Jewett to the presidency of the National Academy of Sciences. The move was without precedent in that Jewett, unlike his predecessors (who had been academic scientists devoted to basic research), was an industrial scientist devoted to applied research and an accomplished administrator entrusted with the directorship of the Bell Laboratories of AT&T. The evidence indicates that Jewett was chosen for the position deliberately so that the National Academy would be better prepared for the coming war.[3]

The genesis of Bush's concern seems evident. Officially, the government was handicapped in its war preparation by the strong isolationist sentiment in the country, as well as its own failure to realize that science and engineering could play a key role in the process. Bush, being out of the public limelight, could avoid the former problem and realized the importance of the latter one. As early as March 1939 he wrote Jewett: "I seem to be involved in the matter of national defense a bit on account of the work of the National Advisory Committee for Aeronautics, and also because of a private conviction that antiaircraft is not receiving the attention it should have."[4] By October he had made contact with the military on the matter.

In the spring of 1940, Bush was working closely with a group of the most important and powerful scientists in the nation: Karl T. Compton of MIT; James B. Conant, a chemist and president of Harvard University; and Jewett, president of the National Academy, director of Bell Labs, and a vice president of AT&T. Together these men planned to establish the NDRC, bypassing all existing agencies both within and outside the government. Working through Frederick Delano, uncle to President Roosevelt and a trustee of the CIW, and through Harry Hopkins, the presidential advisor, the group succeeded in winning the President's approval for an executive order establishing the committee.

The new committee was set up as an agency of the old Council of National Defense, established in 1916 as a consortium of cabinet secretaries. Bush was named chairman of the NDRC, and Compton, Conant, and Jewett were members. In addition, membership was given to Conway P. Coe, as commissioner of patents, and Richard C. Tolman, a Caltech professor who had come to Washington uninvited to offer his services to the war effort. In addition, two members were appointed from the military, one from the Army and one from the Navy.

It was a typical blue-ribbon committee, at least in concept. The members were to receive no salary but were to be given transportation expenses and per diem. Their mission, as stated in the order, was to:

> correlate and support scientific research on the mechanisms and devices of warfare, except those relating to problems of flight included in the field of activities of National Advisory Committee for Aeronautics. It shall aid and supplement the experimental and research activities of the War and Navy Departments; and may conduct research for the creation and improvement of instrumentalities, methods, and materials of warfare. In carrying out its functions, the Committee may, ... within the limits of appropriations allocated to it, ... enter into contracts and agreements with individuals, educational or scientific institutions (including the National Academy of Sciences and the National Research Council) and industrial organizations for studies, experimental investigations, and reports.[5]

It is difficult to imagine that anyone less accomplished than Bush could have been so successful. Beginning with the assumption that the military was incompetent to care for its own research needs, he won the military's enthusiastic approval for his new responsibility in that area. Fearing to get involved with Congress, he managed to spend nearly half a billion dollars on the basis of an order by the Council of National Defense. He helped select an industrial scientist for president of the jealous and conservative National Academy of Sciences (NAS), then used the president to help bypass that organization. Finally, he convinced six cabinet secretaries to give him the authority to compete with their own laboratories and responsibilities in carrying on research and development for the war. Bush accomplished all this because he had a plan when no one else even saw the problem, and he was able to keep this enormous power because he used it only sparingly.

The assignment sheet for the new NDRC was significant. Meeting in Bush's office at the Carnegie Institution even before the order was issued, preliminary duties were divided thus:

To Mr. Coe: the consideration of the evaluation of patent applications and informal suggestions relating to national defense through existing and proposed machinery in the Patent Office and the War and Navy Departments. To Dr. Conant: the determination of facilities and personnel in scientific and educational institutions which might be utilized for research for national defense. To Dr. Jewett: the determination of research facilities in industry which could be available on new problems. To Dr. Compton: the determination of (1) military developments under way in government laboratories, with especial attention to programs likely to be slowed down in the interests of immediate production; (2) developments now considered desirable by armed services, but not under way; and (3) military research programs which it would be desirable to supplement.
To Dr. Bush: general relations of the Committee with other government organizations; and special projects needing immediate attention, especially uranium-fission.[6]

Several things are immediately apparent from this listing. First, the committee was going to make significant alliances and had picked the best man to do so. Second, some of the members were obviously less important than the hard core of organizers. Thus, Coe was merely to worry about gearing in the Patent Office and reporting on new military agencies; Tolman, though present, was not on record as having any assignment at all. Compton, rather than the two military officers added later, was responsible for deciding what the Army and Navy needed and what they were doing—both right and wrong.

As Conant later reported, the individual members were left free to do their jobs in the way they thought best. Notes left from that first meeting indicate, for example, that "Dr. Jewett mentioned the existence of an informal organization of research directors of a number of industrial organizations. He is to inform them of the establishment and purpose of the [NDRC]. . . . In considering the facilities of industrial research laboratories," it was discreetly decided, "it will be necessary to avoid disturbing any existing contracts between the Army and Navy and such laboratories."[7] As Conant noted:

No laboratories were built or leased or staffed. All operational work was to be done through contract with industry or research institutes or universities. Liaison with the Army and Navy was assured by having an Army and a Navy representative on NDRC. Furthermore, officers in the various specialities were appointed to confer with the subcommittees or divisions which were being organized rapidly by the four civilian scientific members of the main committee, to whom the chairman had assigned responsibility for certain areas.[8]

Like any invading army, the NDRC was careful to secure its rear. Jewett was given the job of breaking the news to the head of the National Research Council that the chairman of the council's Engineering and Industrial Research Division (Bush) had taken over a lion's share of wartime research work and was now assigning it without direct consultation with the NAS/NRC. Conant, once a practicing chemist and respected member of the profession, called in all his debts when he convinced the American Chemical Society to cooperate with the NDRC rather than allowing the Chemical Warfare Service, its traditional governmental patron, to monopolize the society's talents.

University presidents and industrial research directors were reassured by news that money would flow to their laboratories rather than to key personnel away from those labs. And finally Bush, who had already proved himself adroit at wearing any number of hats at once, remained head of the NDRC and its successor, the CSRD, while continuing as a member of the NACA, special scientific advisor to the Manhattan District, and chairman of the Joint Committee on New Weapons and Equipment of the Joint Chiefs of Staff, in which position he saw to it that the military actually used the weapons he was developing in his other capacities.

The NDRC worked superbly well, and its very success brought about its subordination, on 28 June 1941, to the newly established OSRD. The one-year experience with the NDRC had shown two major weaknesses in the research and development effort. First, Bush needed more power to see to it that devices invented by NDRC were put into production and properly used by the military —in line with the new emphasis that appeared in the name change from research to research *and* development. Second, medical research, which had been left in the hands of the National Research Council, had been so badly bungled that it was decided to give this field also to Bush. The new OSRD was therefore to have two branches, the NDRC and a new Committee on Medical Research. Bush moved up from head of NDRC, where he was replaced by Conant, to chairman of OSRD. The medical section was never of much interest to Bush, but he did what was required to see that the physicians and research people did what they had to do.

During its first year, the NDRC had begun by using money from the President's Fund until the beginning of the new fiscal year, then supplemented $37 million from that source with $2.4 million transferred from the military for the fiscal year 1941–42. In fiscal 1942–43, the President's Fund was no longer used, but Congress began to make direct contributions, and military transfers went up. By the end of fiscal 1945–46, the OSRD (and the NDRC before it) had spent a total of $453,656,657 in what was undoubtedly the most concentrated and varied research and development effort ever undertaken up to that time.

Predictably, OSRD grants and contracts went to a relatively small number of well-established centers. The policies of buying the best and causing as little disruption to the status quo as possible both indicated such an outcome. In dollar volume, the largest single recipient of funds was MIT, with seventy-five contracts totaling $116,941,352. The University of California had the largest number of contracts (106), but its dollar volume was only $14,384,506, placing it fifth on the list of nonprofit institutions behind MIT, Caltech, Harvard, and Columbia, in that order. The largest corporate contractor was Western Electric Company, a subsidiary of AT&T, with ninety-four contracts valued at $17,091,819. The top eight contractors included such blue-chip luminaries as Du Pont, RCA, Eastman Kodak, General Electric, and so forth.[9] This concentration of research money, which so violated the time-honored concept of geographical distribution, was much criticized at the time, but such criticism was overridden by the preeminent need to stop the Axis.

The enormous success of the OSRD within its own sphere did not prevent pressure from those who thought the OSRD was either doing the wrong job or doing the right job but not well enough, or from those who were not even aware of the agency's existence. Even before the NDRC was officially underway in the spring of 1940, the idea of some sort of science mobilization was finding voice in other quarters. Jerome N. Frank, chairman of the Securities and Exchange Commission, wrote President Roosevelt on 31 May 1940, suggesting the appointment of "a central non-military (civilian) committee of engineers, chemists, physicists (and perhaps even psychiatrists) for the purpose of considering and stimulating inventions useful for war purposes, and evaluating their potentialities. The committee," he continued, "would maintain close contact with the research laboratories of the great industrial corporations and the scientific research departments of the leading universities."

This suggestion was sent to Secretary of Commerce Harry Hopkins for evaluation, and he in turn sent it to Bush on 3 July, just six days after the NDRC was set up. Replying to Frank on the eighth, Bush reported the establishment of the new NDRC and noted that "it is designed to meet to a rather extraordinary degree the very matters which you recite in that memorandum."[10]

Facing another rival idea, Jewett wrote Bush early in June of 1940 that a committee was studying the problem of mobilization but that he had his people trying to "keep things on the proper track and avoid the turmoil and confusion which would result from various Departments of Government setting up hastily conceived groups to organize science and invention."[11] This group was officially known as the Advisory Commission to the Council of National Defense but was commonly called the Defense Commission. Two of its members, William Knudson and Edward R. Stettinius, Jr., who were in charge of industrial production and industrial materials respectively, were leading the committee investigation, but their move in the direction of research was blunted by Bush.

Other camps were heard from as well. Secretary of Agriculture Henry A. Wallace wrote Bush in July 1940 that "agriculture through our programs of extension, conservation and research, can mobilize at once to meet any national emergency." Specifically, he suggested that "in view of the fact that the Department of Agriculture is one of the greatest of the Scientific Research Agencies, I think a representative of this Department might sooner or later be included on the National Research Defense Committee [sic]." Agriculture, however, did not fit into Bush's conception of modern electronic warfare.[12]

There were other areas besides agriculture with which Bush did not want to be too closely involved. One such was the flood of volunteer inventions which would sweep into defense agencies as soon as the war effort picked up momentum. During the Civil War, a permanent commission had been set up by the Navy, in part to deal with such ideas. During World War I, the Navy had again acted to establish the Naval Consulting Board to consider this problem. The dismal record of such activities (only one idea put into production in World War I) led Bush to shy away from this thankless task.[13]

At the first informal conference of the yet-unborn NDRC, Commissioner Coe "suggested that the World War experience indicated that it would be advisable to have a preliminary sifting of patent applications and informal suggestions of interest to national defense by the regular machinery of the Patent Office. Applications and suggestions showing promise of being worthwhile," he said, "would then be referred to a board, on which the War and Navy Departments would be represented." It was decided that the NDRC would "refer to Commissioner Coe for consideration all formal and informal suggestions relating to patents and inventions."

At the same meeting, Coe:

> Stated that a bill had been introduced in Congress which would authorize the Commissioner of Patents to require an inventor to keep his invention secret. The purpose ... would be to keep certain types of information which might be used contrary to the interest of the United States from being made public. The bill is expected to pass; and at the time it is signed Commissioner Coe expects that an Inventors Council will be set up as a part of the machinery for sifting suggestions relating to patents.[14]

On 11 July 1940, Harry L. Hopkins sent a letter to Charles F. Kettering, head of research at General Motors, asking him to chair such a council, and upon his acceptance the council was established in August. The real work of the group was done by a small but dedicated staff of fifty-five persons, only nine of whom had technical training. Although unsolicited inventions formed the bulk of their work, they always tried to stimulate inventors in specific directions. Two major handicaps were encountered, however. On the one hand, anyone not familiar with the conditions under which a device might be used could not gain a clear picture of how the device ought to be designed. On the other hand, the military, which understood the problems, refused until 1943 to issue any kind of list of inventions that it wanted. As a result, the independent inventor was hardly able to make a useful contribution to the war effort.[15]

The NDRC used this council as a convenient dumping ground for most volunteered ideas but, at the same time, held on to a few. In a memo to Conant dated 31 August 1940, Bush explained the agency's policy on extramural inventions:

> In the first place, I think it is quite correct for us to divide suggestions into two categories in accordance with the general scientific reputation of those involved. In the first category we shall place suggestions by men known to us. In the second category we would place the suggestions of distinguished individuals, or suggestions from individuals who are endorsed strongly by some outstanding scientist, or by some government official or Congressman who has reason to urge particular attention on the basis of competent scientific advice.
>
> In the case of the first category, I think we should promptly refer the invention to the National Inventors Council, where it

will receive careful evaluation and proper attention by the Armed Services. . . . In cases of the second category, we can sometimes still adopt the above procedure, but I think we will usually find it well to do some evaluation of our own, in conference with liaison officers.[16]

The statistics for the National Inventors Council reflect its minor role in pushing the war effort. Through the end of June 1946, it had spent only $519,779. During this same time, it received 208,975 ideas for evaluation, found that 8,615 (4.1 percent) had enough merit to warrant further work, and saw 106 into production. On balance, and in view of the small number of people involved and small amount of money spent, it appears that the council more than paid for itself, at least in terms of the time it saved other agencies with more important work to do.

Another agency established to provide for better mobilization of science and technology during World War II was the National Roster of Scientific and Specialized Personnel. The history of the roster was checkered and not unlike that of the Inventors Council. It was established by the National Resources Planning Board (NRPB) on the basis of a letter from the President dated 18 June 1940. After August of that year, the roster was administered jointly by the NRPB and the Civil Service Commission. In April 1942, it was shifted to the War Manpower Commission and to the Department of Labor in September of 1945.[17]

The roster idea, although endemic at least since World War I, seems to have been stimulated by the report, in December 1939, that the Royal Society in England had collected some 80,000 names on a register of British scientists and that the list had been taken over by the Ministry of Labor.[18] Discussions within the National Academy of Sciences and National Research Council during that winter probably led to the sending of a letter from President Roosevelt to Jewett in May 1940, asking about the feasibility of such a list being made for technical people in America. Jewett was anxious to make such a list the responsibility of the National Academy but, despite his enthusiastic answer to the President, he found himself preempted by the NRPB, which was headed by Secretary of the Interior Harold Ickes.[19]

The NRPB picked psychologist Leonard D. Carmichael, later secretary of the Smithsonian Institution, to head the roster, and began by operating through the four societies represented on its Science Committee: the National Research Council, the Social Science Research Council, the American Council of Learned Societies, and the American Council of Education. The task was an enormous one—to compile a list of all Americans with special technical competence, to record what those qualifications were, and to keep a current address and occupation for each person. The desire of each separate professional society (e.g., the American Society of Mechanical Engineers and the American Institute of Physics) to keep its own lists was resisted during the war.

Questionnaires were sent out, using the membership lists of professional societies and subscription lists of technical journals, and the data were coded and placed on punched cards for quick reference. Unfortunately, little reference was made. More often than not, technical jobs were filled from the ranks of friends or through personal recommendations. In a great chain of personal and professional patronage, the four key members of OSRD, for example, chose the heads of the various sections not from the punched cards but from among their friends. These in turn appointed their friends to head subcommittees, and so on down the line. There was little place in this system for an impersonal roster.

Early in 1941, Irvin Stewart, Bush's chief administrative officer, reported that some of Carmichael's group were "disturbed because no more use is being made of the roster."[20] After the war, the official historian of the OSRD epitomized that agency's attitude toward the roster: "Its punch cards were invaluable when one wished to know what American scientists spoke Italian, but as might be expected the Roster was used less to obtain key men than the rank and file. Those charged with recruiting chemists and physicists for OSRD and its contractors knew the outstanding men in each field already and through them got in touch with many young men of brilliant promise."[21]

Among the rival agencies that were proposed or actually sprang up in wartime Washington, two proved to be of particular potential trouble. In the executive branch, Interior Secretary Harold L. Ickes made the most concerted effort to win a piece of the research pie for his department. In part, this was merely an extension of Ickes's normal bureaucratic imperialism, evident throughout the thirties. In part, it was a direct response to the fact that the OSRD left many scientists untouched by Ickes's organization of war work. Writing directly to the President on 19 August 1940, Ickes charged that only some 500 to 600 of the 100,000 scientists and engineers in the country were available to the government through the National Academy clique. He wrote:

An office of scientific liaison is needed, which will serve as a direct contact between the scientists and the Government and permit the former to make their contributions to any defense program. It is desirable that such a clearing house, in order to

Pursell

attain maximum usefulness, should be located in a permanent administrative department of the Government where its activities would be available equally to all agencies engaged in defense work. In view of the Interior Department's identification with conservation in general, there is perhaps justification for setting up in the Department, as part of the Secretary's office, an "Office of Scientific Liaison," which would be devoted to the effective utilization of our science resources.[22]

This proposal was to plague Bush for the next few months, and it embodied assumptions and a philosophy that haunted him throughout the war and after.

Ickes's scheme lay dormant until early in 1941. Then Lyman Chalkley, who later went to work for the OSRD, published a letter in the *New York Times* in which he decried the failure of the government to use the scientific talent available in the nation and the lack of any scientific coordination in Washington. Ickes immediately sent a clipping of this letter to Roosevelt, along with a copy of his proposal of the previous August and a cover letter asking that the whole matter be reopened. The President sent the bundle to William Knudsen, by now running the Office of Production Management, who suggested that Bush be appointed to a committee to draw up a list of projects for research which would then be, in his rather vague phrase, "undertaken in connection with the [American Association for the Advancement of Science]...."

The President next referred the matter to both the Bureau of the Budget (BOB) and to Bush. Bush consulted the BOB, which was not anxious to see yet another agency set up, and then wrote to Jewett that "there is not much danger in the situation, and perhaps rather an opportunity." In his bland official prose, Bush suggested "that there is a real opportunity to clarify a difficult matter, and perhaps to stabilize some relationships in a desirable way. I have in mind particularly that something might be done about the medical research."

Jewett was not so quietly optimistic. "While there may not be much danger in what Mr. Ickes has proposed," he warned, "I think the suggestion must be looked upon as symptomatic of a desire on the part of some of the present governmental officials to take over control of things which are certainly not in the best interests of the country to have entrusted to the whims of politically appointed and temporary department officials."[23]

It was again Bush, however, who saw the situation most clearly. In his official reply to the President on Ickes's proposal, he emphasized what his own group was already doing and suggested that the problem raised by Ickes might best be solved by giving the NDRC or a successor agency more power to see to it that the military faced up to new advances in science and technology. His argument was accepted and led directly to the establishment of the OSRD. As Bush had predicted, Ickes's proposal contained more opportunity than danger.

For his part, Ickes replied to Roosevelt that the NDRC was of course doing a superb job on military weapons, the sole area claimed by Bush for his group. Indeed, said Ickes, its very success underscored the need for some similar agency which would provide the same services for the civilian bureaus of the government. He stood ready, of course, to set up such an agency if the President wished.

A new threat to the hegemony of the OSRD arose early in 1942. On January 13th, the President had set up a War Production Board (WPB) under Donald Nelson, to mobilize the nation's resources for war in the fields of production and supply. One of the first problems facing the new board was that of scarce materials such as aluminum and rubber. Efforts in the past to push this problem off onto the OSRD had been firmly resisted by Bush, and there was talk within the organization that the new WPB was a logical agency to handle the matter.

About this same time, Bush was visited by a Msgr. Miller and a Mr. Shocknessy from the Institutum Divi Thomae, a Catholic research laboratory in Cincinnati, Ohio. In early March, Msgr. Miller, this time accompanied by a Dr. Sperti, met with Maury Maverick, a former congressman from Texas now serving with the WPB. Maverick shared their belief that the scientific and technical capability of the nation was dangerously underutilized; as a result of the meeting, the gentlemen from Cincinnati left Maverick a twenty-page "Plan for the Organization of a Research Division (or bureau) of the War Production Board." Maverick was inspired and strongly urged Donald Nelson to set up such a body: *"This is an outstanding necessity.* (HITLER DID IT STARTING ABOUT EIGHT YEARS AGO). Our job is to kill Japs and Germans."[24]

Bush was still toying with the idea of doing the production research job himself, but Jewett warned him against it: "I'd shun it," he wrote, "like I would the seven-year itch." Changing the metaphor he continued, "keep out of it if you can and give the hard pressed boys all the help you can from the shore but don't go in swimming—the water is cold and there is a damnable undertow. We'll all help in the helping but most of us would fail to qualify as lifeguards." Maverick had no idea of letting Bush take over the job and pressed the bishop auxiliary of New York, Stephen J. Donahue, to write the President praising the plan. The White House sent the letter to Bush for comment, but he replied blandly that it was Nelson's problem.[25]

Nelson's solution was to appoint Maverick as chairman of a special committee to look into the matter further. This prospect was more than Bush had bargained for. "The whole subject of industrial research," he wrote Conant, "in connection with strategic materials and the like seems to be in a terrible turmoil in WPB. I will tell you about it, but it is certainly getting thoroughly out of hand and onto dangerous ground. I went so far as to put in a word of caution with Sidney Weinberg for Don Nelson's behalf a day or two ago, for I am decidedly afraid of what is there being built up."[26]

There was one man on the Maverick committee that Bush could trust, the geologist C. K. Leith, a veteran of governmental advising who had earlier been connected with the Science Advisory Board. It was reported by an observer that:

Leith has gone a long way in enlarging his ideas, but points of disagreement are what kind of man shall be chosen to head the new agency. Leith fearing Maverick, is insisting on a technical man. Maverick fearing a man of Leith's type, insists on a general executive. . . . The other point of disagreement is that Leith . . . is attempting to perpetuate a primary relationship between the National Academy of Science as the new unit for research jobs. . . . Leith undoubtedly fears that the Thomas Aquinas group of scientists may seize control of the new unit. Maverick fears that the National Academy Old Man's Club of distinguished scientists want to steal the new unit.[27]

While the Maverick committee was still wrangling, Thomas C. Blaisdell, a member of the planning committee of the WPB, went directly to Nelson with yet another proposal. "I am impressed," he wrote, "with the necessity for an organization of an agency which will be out of the regular run of WPB's activities and yet closely related to the problems which the Board is forced to face." His solution was a proposed U.S. War Research Development Corporation, the functions of which would be "(1) the testing of new industrial processes; (2) the building of pilot plants; (3) the construction up to the stage of operation of processes which it has been decided shall be carried into full-bloom activity."[28]

It was a proposal well-calculated to alarm such men as were gathered around Bush. Leith wrote with disgust and despair as follows:

> Several meetings of the [Maverick] committee have been spent largely ... [on] the contention of the chairman and others that American science and technology are not properly on the job, that organized science and technology are hidebound, that the war will be won by turning to small laboratories and to the forgotten man with an inspiration, that the direction, coordination, and "revolution" of science and technology should be directed by a committee of wise laymen with the proper social sense in W.P.B. Almost no serious attempt has been made to review or understand the broad program already underway.

To cap it off, Leith complained, Maverick had now gone over the heads of his own committee and was pushing the idea of a separate corporation with large funds.[29]

Through some leak not identified, copies of the minutes of the Maverick committee meetings were in Bush's hands within days. These were carefully screened for any explicit or implied criticism of the OSRD. Little was found, but Bush continued to worry about the intrusion of politicians with wrong ideas into the temple of science. Jewett was quite upset: "If the question is a serious matter of importance," he wrote, "(and I think it is), I can't conceive of a more maladroit way of getting a wise answer than the one Nelson adopted. This sort of thing certainly isn't subject matter for a series of indiscriminate town meetings replete with references to 'sacred cows'—'boot-licking' etc. I'm disgusted."[30]

What worried Bush was the fact that the Maverick committee persisted in raising issues that were highly divisive because they touched upon sensitive matters of corporate profit and monopoly power in the general economy. Maverick, for example, had written to Nelson about "the grave necessity that the WPB must openly, freely and honestly go into the matter of patents and new processes; that it must be done for the benefit of the public and not for any individual or special interest. We must do our duty, and NOT do anything that in the end will prove disastrous [such as if] we permitted certain special interests or individuals to get a superior interest over the public, and the single will to victory."[31] Patent reform was not a subject Bush wanted left to the likes of Maury Maverick.

The final outcome of the fight left the matter in the hands of the WPB, but largely under objective conditions Bush could live with. On 23 November, 1942, an Office of Production Research was set up in that agency, with the task of stimulating civilian rather than military technology. In 1944, it spent a mere $4.5 million, however, and was never sufficiently aggressive or well-funded to pose any threat to the industrial status quo.

Probably the most serious challenge to the assumptions and leadership of the OSRD came not from the executive but the legislative branch. In the Congress, from 1942 through 1945, Sen. Harley M. Kilgore (D-W. Va.) kept a constant surveillance on technical mobilization efforts and made sweeping suggestions for their improvement.

In 1942, Sen. Kilgore was named chairman of a special Subcommittee on War Mobilization of the Senate Committee on Military Affairs. Equipped with a staff and a small appropriation, and most importantly a conviction that the war effort was being hindered by incomplete use of the nation's technological resources, the subcommittee began hearings in November 1942. For most of the following six days, witnesses followed each other in complaining that they and the groups they represented were not being given a chance to take part in the war effort. The list of charges included the suppression of innovations that might harm the competitive positions of powerful corporations, the lack of a central office in Washington concerned with the problem and a resulting confusion, and discouragement for those who were looking for someone to accept their services.

Most of this testimony tended to alarm those who believed, in the words of the preamble to Senate Bill 2721 by Kilgore, that "the full and immediate utilization of the most effective scientific techniques for the improvement of production facilities and the maximization of military output is essential for the successful prosecution of the war to a sure and speedy victory." The various sections of the proposed bill provided for the establishment of an Office of Technological Mobilization, which was to be given broad powers to draft facilities and ideas.

Needless to say, the prospect of any such program was an immediate threat to at least three groups: those scientists who felt that their scholarly integrity might be compromised by outside political control of research; business leaders who saw the prospect of severe limits on profits and the disruption of competitive advantages; and the OSRD group, which saw a new rival for the leadership of the technical war effort.

Four days before hearings on the bill opened, Jewett visited Kilgore in his office. "I found him a man of intelligence and extremely reasonable and easy to talk to," Jewett reported with surprise and relief. "He is clearly trying to do something constructive in a sector where he thinks help is indicated. While he didn't say so in so many words, I got the distinct impression that he didn't draft S-2721 but that he is the victim of some of the starry-eyed New Deal boys, and that he knows it." Jewett was not, he said, "worried about S-2721 and the like—they are too fantastic, grandiose and regimental to go over. However, if they are allowed to run their normal course of prolonged town meeting hearings they'll waste a lot of time, muddy the situation, and prevent anything really constructive (if there is anything such) from being done."[32]

One of the problems of the OSRD group was that it was never clear in its own mind about the kind of people who were behind Kilgore. One group often blamed was the "soreheads." Lyman Briggs, head of the National Bureau of Standards, said that the Kilgore bill was "supported by a considerable group of people whose requests for government funds have been turned down. For example, the 'Engineering Colleges Research Association' sees in this bill an opportunity to obtain government grants for engineering research similar to what the land grant colleges have long enjoyed under the Hatch Act, Morrill Act, and similar laws."[33] This influence was, however, relatively innocent compared with that hinted at by Jewett when he spoke of "starry-eyed New Deal boys." As time went on, this charge that Kilgore was being used by sinister and foreign political ideologies in a deliberate attempt to revolutionize the American way of science was subtly shaded into charges of Communist influence, centered on Dr. Herbert Schimmel, chief scientific advisor to Kilgore.

Again Jewett summed it up nicely: "The basic arguments against the Bill in the fields of science and technology are identically the same as those against corresponding bills in the fields of labor and education, mainly, that it would completely revolutionize our entire American age-old concept and by placing vast powers in the hands of a federal bureaucracy would set the stage for complete domination of the life of the nation by a small group of federal officers and bureaucrats."[34] The basic point missed by Jewett, although grasped by Bush, was that American life was already being revolutionized—the only question was whether the blueprint for the future should be drawn up by politicians answerable to the people or by private enterprises answerable only to themselves.

By the summer of 1943, Kilgore's bill was going through its fourth revision, and no less a person than Bush was helping him with it. One of the reasons for this was that Kilgore was already beginning to look ahead to the end of the war when, he felt, the government would have to continue to offer massive subsidies to American science. Quite realistically, Bush realized that he would have more influence if he tried to channel rather than dam the tide of public support. Again, Jewett opposed such considerations.

"From every angle," he wrote, ". . . trouble aplenty is ahead if Congress is foolish enough in the midst of a war to pass legislation which seeks to put peacetime science and education under Federal domination beyond the narrow minimum which national safety clearly indicates as necessary."

"Everybody will be up in arms and political fights and court procedures will be the established order of the day," Jewett warned. "Industry, large and small, will marshal its vast powerful forces and institutions and individuals will have their own particular weapons of attack." He had, he claimed, received many letters from scientists opposing the Kilgore bill and, as he said, "they reflected an unalterable opposition to being made the intellectual slaves of the State and of having anyone tell the men what and how they should think."[35]

In Septermber 1943, the magazine *Modern Industry* polled its readers on the question: "should industry support the science mobilization bill?" Of those answering, 82.2 percent voted "no."[36] By June 1944, however, Kilgore had moved decisively beyond his failed effort to organize more centrally and effectively the wartime research enterprise, and he was concentrating his attention on the postwar shape of American science.

Throughout the preparedness and wartime period, the NDRC/OSRD was able to carve out for itself the major responsibility for research and development on new weapons, and to defend that responsibility against all efforts either to dilute it with tasks viewed as extraneous or submerge it within some larger, more centralized entity. In part, this defense of its position grew out of a natural bureaucratic effort to protect one's territory. In part, it grew out of an honest conviction that the agency's job was the most important, and that it was doing that job in the best possible manner.

Finally, the defense grew from a basic ideological and political commitment to the status quo antebellum arrangements of American science. The OSRD leadership could muster little enthusiasm for pulling down the mighty and raising up the weak. Except for a growing conviction that in the postwar period the federal subsidy to science should be massively increased, the OSRD saw little to blame and much to praise in the way the nation's scientific and technical resources were distributed and used. As such, the OSRD was not only a success within the narrow research limits it set for itself, it was equally successful in perpetuating the scientific leadership which, coming out of the Depression of the 1930s, built the scientific establishment of the Cold War period.

1. See Carroll Pursell, "Alternative American Science Policies during World War II," in *World War II: An Account of Its Documents*, ed. James E. O'Neill and Robert W. Krauskopf (Washington, D.C.: 1976), pp. 151–62.

2. The best history of the NDRC/OSRD is Irvin Stewart, *Organizing Scientific Research for War: The Administrative History of the Office of Scientific Research and Development* (Boston: 1948). For further studies of this and related events, see Carroll W. Pursell, Jr., "Science and Technology in the Twentieth Century," in *A Guide to the Sources of United States Military History*, ed. Robin Higham (Hamden, Conn.: 1975), pp. 269–91.

3. Telegram Gano Dunn to Vannevar Bush, 26 April 1939, Gano Dunn folder, Bush file, Carnegie Institution of Washington records; Arthur H. Compton, *Atomic Quest: A Personal Narrative* (New York: 1956), p. 34.

4. Vannevar Bush to Frank B. Jewett, 23 March 1939, Jewett folder, Bush file, Carnegie Institution of Washington records.

5. The complete text of the order is given in James Phinney Baxter III, *Scientists Against Time* (Boston: 1946), p. 451.

6. Minutes of meeting of June 18, 1940, office file of Frank B. Jewett, OSRD records, National Archives.

7. Notes on the First Informal Conference, 18 June 1940, central classified file, organization, OSRD records, National Archives.

8. James B. Conant, "The Mobilization of Science for the War Effort," *American Scientist* 35 (April 1947): 196.

9. Baxter, *Scientists Against Time*, pp. 456–57.

10. Jerome N. Frank to The President, 31 May 1940, central classified file, organization (July 1940), in OSRD records, National Archives, and Vannevar Bush to Jerome N. Frank, 8 July 1940, in OSRD records, National Archives.

11. Frank B. Jewett to Vannevar Bush, 6 June 1940, office file of F. B. Jewett (folder 49.01), OSRD records, National Archives.

12. Henry A. Wallace to Vannevar Bush, 11 July 1940, office file of V. Bush, OSRD records, National Archives.

13. See National Inventors Council, *Administrative History of the National Inventors Council* (processed, Washington, D.C., c. 1946).

14. National Archives Record Group 227, Central Classified File, Organization (NDRC May 1940–June 1940), notes on the first informal conference, 18 June, 1940.

15. See National Inventor's Council, *Administrative History*, passim.

16. National Archive Record Group 227, office file of F. B. Jewett, memorandum, V. Bush to J. B. Conant, 31 August 1940, folder 49–E.

17. *U.S. Government Organizational Manual, 1962–63*, p. 644.

18. For background see Leonard Carmichael, "The Number of Scientific Men Engaged in War Work," *Science* 98 (13 August 1943): 144–45.

19. Frank B. Jewett to Franklin D. Roosevelt, 24 May 1940, FDR papers (OF330), Roosevelt Library, Hyde Park, N.Y.

20. National Archive Record Group 227, office file of Irvin Stewart, memorandum, Stewart to Richard C. Tolman, 29 January 1941.

21. Baxter, *Scientists Against Time*, p. 127.

22. Harold L. Ickes to the President, memorandum re A proposed Office of Scientific Liaison, FDR papers (OF2240), Hyde Park.

23. Harold L. Ickes to the President, 7 February 1941, policy documentation file, file 282, War Production Board records, National Archives; William S. Knudsen to President Roosevelt, 18 February 1941, FDR papers (OF2240), Hyde Park.

24. Maury Maverick to James S. Knowlson, 31 March 1942, folder 282, policy documentation file, WPB records, National Archives. A copy of the "Plan for the Organization of a Research Division (or bureau) of the War Production Board," dated 9 March 1942, is there also.

25. Vannevar Bush to Franklin D. Roosevelt, 8 April 1942, FDR papers (OF872), Hyde Park.

26. Vannevar Bush to James B. Conant, 28 April 1942, central classified file, organization (Scientific Personnel Office), OSRD records, National Archives.

27. Robert D. Leigh to Mr. Blaisdell, 30 April 1942, policy documentation file, folder 282, WPB records, National Archives.

28. Thomas C. Blaisdell, Jr., to Donald M. Nelson, 12 May 1942, policy documentation file, folder 282, WPB records, National Archives.

29. C. K. Leith to A. I. Henderson, 15 May 1942, policy documentation file, folder 073.011, WPB records, National Archives.

30. Frank B. Jewett to W. L. Batt, 8 June 1942, policy documentation file, folder 073.011, WPB records, National Archives.

31. Maury Maverick to Donald M. Nelson, 29 May 1942, policy documentation file, folder 073.011, WPB records, National Archives.

32. Carnegie Institution of Washington Records, Bush file, Frank B. Jewett folder, letter from Jewett to Bush, 16 November 1942.

33. Lyman J. Briggs to William D. Coolidge, 8 March 1943, central classified file, cooperation (Kilgore committee, January–June 1943), OSRD records, National Archives.

34. Frank B. Jewett to Vannevar Bush, 27 May 1943, Bush file, Frank B. Jewett folder, CIW records.

35. National Archives Record Group 227, central classified file, cooperation (Kilgore Committee, July–December 1943), letter, Jewett to Bush, 2 September 1943.

36. National Archive Record Group 227, central classified file, cooperation (Kilgore Committee, July–December 1943), letter, Eldridge Haynes to Dear Sir, 11 October 1943, referring to 15 September 1943 issue of *Modern Industry*.

ACADEMIC SCIENCE AND THE MILITARY: THE YEARS SINCE THE SECOND WORLD WAR

HARVEY M. SAPOLSKY
Massachusetts Institute of Technology

NO SCIENCE POLICY ISSUE IN THE HISTORY OF THE NATION GENERATED more emotion than the controversy in the late 1960s and early 1970s over the role of the military in the support of academic science. In the high drama of the time, university researchers with military-sponsored projects often were forced to defend their work publicly against the criticisms of their colleagues and students. The complexity of the relationships that had brought the researchers and their sponsors together were ignored for simpler versions by both sides of an impassioned and occasionally violent debate. The defenders of the military-sponsored projects preferred to see themselves as the deserving beneficiaries of disinterested patrons, while their critics preferred to see them as the willing collaborators in a design to distort the direction and purposes of science.

Intertwined as this issue was with the national debate over the Vietnam War, it is not surprising that the attention directed toward the military's role in science came years after other agencies had replaced the military as the prime sponsor of university-based research, and that the issue of the propriety of military sponsorship of such research faded before it was resolved. Collectively, military agencies were the main federal sponsors of university-based research in the initial years after World War II, but after 1960 they lost this role first to the National Institutes of Health and then to the National Science Foundation as well. In absolute terms, military support for academic science peaked in the mid-1960s and has been in decline ever since then. Nevertheless, the Department of Defense (DOD) still sponsors approximately $200 million worth of basic and applied research in the universities (about 10 percent of all such support by federal agencies), and DOD contracts and grants are once again being received in the universities without much notice or controversy.

Today, in the midst of the current calm, it may be possible to begin a dispassionate analysis of relationships during the last three decades between science and the military. To the historian, there may be in this effort an element of instant history, of the type found in television documentaries and on the racks of the supermarket paperback book stands. Certainly more distance from the events is required before the full impact of the military's involvement in science can be appreciated. And yet, a generation has matured since scientists were mobilized to develop the atomic bomb. Not everyone now thinks of the same war when someone says "the war."

To keep the topic manageable, I am only concerned here with the ties linking the military and academic science. Bypassed are a consideration of the military's own research laboratories, the subject of many official studies,[1] and the Federally Funded Research and Development Centers (FFRDCs), the large military-sponsored research facilities that exist on only a few campuses.[2] Although some events that occurred before World War II can be said to have anticipated recent relationships between the military and academic science, attention is focused on the years since that conflict, when these relationships became both intense and continuing.[3]

Of necessity, there is special interest in the history of the Office of Naval Research (ONR), an agency that was extremely influential in determining the pattern of relationships that has come to exist between the universities and government. ONR's importance can be seen in a list of attendees of a meeting held among government agencies in 1957 to discuss the coordination of basic research policies. Present in addition to the then current Chief Scientist of ONR were the Director of the National Science Foundation, once a Chief Scientist of ONR; the Chief Scientist of the Office of Army Research, also once a Chief Scientist of ONR; the head of basic research of the Atomic Energy Commission, whose program was initially managed by ONR; and the Chief Scientist of the Office of Aerospace Research, an agency that explicitly, if not always very successfully, patterned itself on the model of ONR. Needless to say, ONR's policy proposals were bound to receive a sympathetic hearing in such a setting.

The Origins of the Office of Naval Research

Neither the Navy nor any of the other services was a convert to science at the end of World War II. To be sure, senior officers were extremely impressed with the rapid progress that had been achieved in weapons. But if on occasion they praised the contribution of science to the war effort, they usually meant the contribution of technology, committing the common verbal error of confusing science and technology—an error scientists are not always moved to correct.[4] The support of science, perhaps appropriately, held no special interest for these officers. The fact that ONR filled a void in the support of science at the end of the war was quite unintentional; it had been created for a different purpose.

The concept of an Office of Naval Research originated with a group of young reserve officers who served during the war in the Office of the Coordinator of Research and Development, an organization that linked the civilian-directed Office of Scientific Research and Development (OSRD) with the Navy's materiel bureaus. Known as the "Bird Dogs" because of their task of ferreting out problems in interorganizational relationships, these young officers spent much of their spare time planning the structure of the Navy's postwar research effort. In their view, the Navy needed a permanent office to coordinate research and to work with scientists in the development of weapons. As early as November 1943, they had evolved a design for a central research office for the Navy that was nearly identical to the design adopted for the Office of Naval Research when it was established at the end of the war.[5]

The Bird Dogs sought to advance their proposal by submitting it formally in the fall of 1944 to the Secretary of the Navy, who was then beginning to consider demobilization plans.[6] Knowing President Roosevelt's personal interest in the organization of the Navy, they even had scheduled a presentation for him on their proposal; it never took place because of his death.[7] But it was not the Bird Dogs' efforts, vigorous as they were, that gave the Navy an Office of Naval Research, nor was it their plan for research that establishing the office was intended to implement.

The man who founded ONR was Vice Adm. Harold G. Bowen; his interest in the organization was to gain a base from which he could promote nuclear propulsion within and outside the Navy.[8] The Navy had been essentially excluded from the Manhattan

Project during the war. Admiral Bowen was one of the few naval officers who knew about the bomb project and the potential that existed for the use of nuclear power in ships. He also knew the reason why the Navy had been excluded from the Manhattan Project.

Vannevar Bush, head of the OSRD (which had initial control of the atomic bomb program), had had a series of conflicts with the naval officers, but especially with Admiral Bowen.[9] As Technical Aide to the Secretary of the Navy in 1940, Admiral Bowen had resented the effort of outside scientists such as Bush to become involved in the management of weapon research and had blocked their attempts to gain information on these projects.[10] Bush soon had an opportunity to repay Bowen and the Navy when he gained independent authority for weapon research with the establishment of the OSRD in 1941. Bush first stripped control over the development of radar from the Naval Research Laboratory, which Admiral Bowen also headed, and gave it to the newly established Radiation Laboratory at MIT.[11] Then, when it came time to assign control of the atomic bomb project to one of the services, Bush saw to it that the task was given to the Army. As he later said, he did not think naval officers—and especially those at the Naval Research Laboratory—had sufficient respect for and an ability to work cooperatively with civilian scientists.[12]

The Naval Research Laboratory in 1939 was the first government agency to support work in the military applications of fission. Admiral Bowen, who was then Chief of the Bureau of Engineering, had jurisdiction over the Laboratory, and he had approved of this allocation of funds.[13] A pioneer in the introduction of high-pressure steam in naval ships, Admiral Bowen was an early champion of nuclear propulsion. Bush, however, was not his only enemy. In a series of reorganizations, Bowen lost control over the Bureau of Engineering and the Naval Research Laboratory and his position as Technical Aide to the Navy Secretary. Barred also from a major operational command, he found work as a labor troubleshooter for the Under Secretary of the Navy, seizing struck war plants to ensure the uninterrupted production of munitions and ships. Admiral Bowen's toughness in these difficult situations won him the gratitude of Under Secretary James Forrestal. When Forrestal became Secretary of the Navy near the end of the war, he was in a position to express his gratitude.

It was Secretary Forrestal who received the proposal from the Bird Dogs for a central research coordination office in the Navy. Within eight days of its receipt, he implemented instead a proposal for an Office of Patents and Inventions made by R. J. Dearborn, also a friend of Admiral Bowen, and appointed the Admiral as its Chief. In May 1945, after the death of President Roosevelt, the Secretary retitled Admiral Bowen's organization the Office of Research and Invention and gave it all the powers that the Bird Dogs had outlined for their office, plus jurisdiction over the Naval Research Laboratory.[14] With Admiral Bowen in command of the Navy's postwar research planning, the Bird Dogs thought that their vision of a Navy supporting the work of civilian scientists had been shattered. They were to be surprised.

Admiral Bowen apparently intended to use the Office of Research and Invention as the agency to promote the development of nuclear propulsion for the Navy. His first need in that quest was to gain access for the Navy to the Manhattan Project and current progress in nuclear physics. Barring access was the Army, pleased with its atomic bomb monopoly. Requests for atomic clearances for naval officers in the early months after the end of the war were put off by General Groves, head of the Manhattan Project.[15] However, not all the doors to nuclear information were well guarded. The Admiral discovered that General Groves had alienated many of the civilian scientists who had been mobilized for the bomb project because of the strict discipline he had imposed on their work. Dissatisfied with Army management, at the end of the war these scientists were streaming back to the universities, but they were still anxious to continue their research. Once himself abrupt with civilian scientists, Admiral Bowen now saw them as useful allies and quickly became their patron. The funds needed to finance this venture were easily obtained from Secretary Forrestal. The Bird Dogs and others knowledgeable in the ways of civilian scientists were recruited by the Office of Research and Invention to manage a basic research program in a manner acceptable to the scientists.[16]

There was much debate about how to organize the overall postwar research effort, both military and civilian. One proposal called for establishing a Research Board for National Security, to be housed in the National Academy of Sciences.[17] Another idea, put forward by Vannevar Bush, involved the creation of a National Research Foundation that would support both civilian and military research.[18] Still another proposed placing military research under the Joint Chiefs of Staff.[19] There was also conflict over the structure of the overall nuclear development program.[20]

Admiral Bowen took part in these discussions, but he also took care that his own organization base was protected. As he expressed it, "In Washington, one cannot accomplish anything without a statute in front of him and an appropriation behind him."²¹ Quietly he had a bill introduced in Congress establishing an Office of Naval Research. On 1 August 1946, the bill became Public Law 588 of the 79th Congress. Thus while the debate over organization of postwar research dragged on for four more years, and the newly established Atomic Energy Commission attempted to formulate its operating procedures, Admiral Bowen, the representative of a victorious Navy seeking to do relevant naval research, had obtained the secure base he had desired.

But one thing he could not obtain was the internal Navy jurisdiction for nuclear propulsion matters. Admiral Bowen still had enemies within the uniformed Navy. Officers on the staff of the Chief of Naval Operations and in the Bureau of Ships, the organization that had absorbed Bowen's old Bureau of Engineering, outmaneuvered him. The Bureau of Ships in 1946, and eventually Admiral Rickover, gained, as the Navy puts it, cognizance over nuclear propulsion.²² This battle lost, Admiral Bowen soon retired, but it was through his efforts and ambitions that the Navy had gained an Office of Naval Research.

The Office of National Research

In the immediate postwar years, the Office of Naval Research was the prime federal agency supporting academic science. The National Science Foundation was not established until 1950 and then did not begin receiving significant appropriations until the Sputnik crisis. Though the Office of Army Research and the Office of Aerospace Research were patterned after ONR, neither was operating before the early 1950s.²³ The National Institutes of Health, limited in interest to the support of biomedical research, did not collectively match the level of ONR appropriations until the early 1950s.²⁴ The Atomic Energy Commission then used ONR as its preferred mechanism for the support of university-based research. Both the Department of Agriculture and the National Advisory Committee for Aeronautics supported research, but their research clients worked apart from or outside of the universities.

Senior Navy officials did not choose such a mission for ONR. In fact, involved as these officials were then with the demobilization of the fleet and the battles over unification, they could have hardly been aware that ONR was still in existence. The definition of ONR's mission was literally the responsibility of its own staff. Admiral Bowen's vision for the office departed with him, for he had been secretive about his ambition to lead the Navy into the nuclear age. What he left behind were the officers and civilians he had recruited to support university-based science in a manner acceptable to scientists and their universities.

Not surprisingly, the staff found common ground in continuing a program to support academic science. Some believed it important that there be an interim basic research agency until the Congress and the President could agree on the structure for the National Science Foundation. Others believed it was in the Navy's interest to be on good terms with scientists, particularly those who had gained national prominence because of the success of the atomic bomb project. Still others wanted to preserve a mobilization base for the nation in case of a future conflict. There was no need to rank the motives, for all were well served by the same policies. Different rationales could be cited to different audiences without provoking staff disharmony.[25]

Two problems faced the organization. One was to obtain funds for science; the other was to persuade the universities to accept those funds. Neither, it turned out, was insurmountable. The Navy at the time had large budgetary surpluses due to canceled procurement contracts, and these surpluses were relatively easy to reprogram. In addition, agencies within and outside the Navy were apparently willing to pay a share of ONR's costs. The contribution of the Atomic Energy Commission to ONR's physics research program was mentioned previously. Rear Adm. Luis de Florez, the flamboyant head of the training devices center attached to ONR, was said to have persuaded the Bureau of Aeronautics, his old agency, to transfer in $32 million.[26] (As it was recently revealed that Admiral de Florez was then already or soon thereafter to become the director of technological research at the Central Intelligence Agency [CIA], it is possible now to doubt the generosity of the Bureau of Aeronautics.[27] In any event, the CIA should be added to the list of agencies in which ONR alumni came to be research managers.)

University officials and scientists worried then about the same issues they worry about today. Pres. James Conant of Harvard feared the instrusion of secret research into the universities during the peacetime.[28] Norbert Wiener, the founder of cybernetics, did not want his work to contribute to military power.[29] University comptrollers worried about the adequacy of overhead rates and the placement of social policy clauses in government contracts.[30] Members of the MIT Corporation thought that government support of research would prove to be the most unreliable source of university support.³¹ And the leaders of the National Academy of Sciences were concerned that government financing of research would lead to government control of research.³²

ONR attempted to eliminate these concerns by designing its operating policies to meet the needs of academic science. It encouraged the open publication of research results, stressed its interest in basic research, chose the most flexible possible contract forms, pledged continuing support, and involved scientists in its research planning. From the perspective of the universities, ONR offered enlightened research management, the example always to be cited.³³

The fact that the concerns of the 1940s are once again prevalent in the universities is testimony to the unpredictable nature of politics. ONR strove to keep its arrangements with universities intact, but policies, attitudes, and admirals change; there are no permanent commitments in government. ONR was publicly hailed in 1946 for its support of science,³⁴ but in 1971 ONR avoided arranging a public ceremony to celebrate its twenty-fifth anniversary, fearing to call attention to itself and its university contracts.³⁵

Pressures for Relevance

The golden age of academic science in America actually lasted only four years. Between 1946 and 1950, ONR was able to use public funds to support promising opportunities in science—opportunities that were largely free from the considerations of geography and the expected public benefits of that science. Government in America had traditionally been indifferent to the needs of science, supporting only those projects or fields where the political and economic returns were immediate and obvious.³⁶ During these four years, however, it seemed as if the discovery of new knowledge was coming to be valued in America for its own sake. But after 1950, though ONR and other agencies of government would continue to support science, this support would never be free from pressure for relevance.

Many mistakenly believe that the Mansfield Amendment, which was attached to the fiscal year 1970 Military Procurement Authorization bill and required that all defense-sponsored research have a direct relationship to a specific military function, dates the resurgence of a concern for relevancy in government-supported research.³⁷ The Mansfield Amendment, in fact, marks only the return of a congressional interest in the relevancy of research. The executive branch, at progressively earlier stages as one moves down the hierarchy, had previously indicated the same concern. In 1966 President Johnson, in a statement specifically referring to biomedical research but of general applicability, noted that society expected visible benefits from its investments in basic science.[38] More than a decade earlier, Secretary of Defense Charles E. Wilson announced that it was not the intention of the Department of Defense to support pure scientific research or, as he characteristically put it, ". . . research on why potatoes turn brown when they are fried."[39] And in 1950 the Navy rediscovered the Office of Naval Research.

The Navy in the months preceding the Korean War was feeling the effects of the cuts President Truman had imposed upon the defense budget and was seeking to eliminate nonessential functions. To senior naval officers, ONR's university research program appeared to be an obvious candidate for a budget reduction. In January 1950, the Chief of Naval Operations assigned Vice Adm. Oscar C. Badger, a Medal of Honor winner and an officer without previous experience in research, to investigate the program for possible savings.[40] Admiral Badger immediately required that ONR defend each of its academic science contracts in writing and at hearings in which he sat in judgment. Though the Admiral professed an interest in science, the ONR staff feared his presence and purpose.[41]

Testifying during the day, the staff worked each evening refining the project rationales. An argument that once aroused the admiral's anger was not used again. Soon each project had acquired a thick coating of naval justifications.[42] By the end of the two-week inquiry, the admiral was an ally, submitting a report which the ONR staff helped write that stated, "Within the limits of approved policies, directives, and availability of funds, the *research programs* in universities of the Bureaus and the Office of Naval Research *are on a sound and objective basis* and are controlled, administered and continually reviewed and adjusted with a high degree of efficiency and appreciation of the overall objectives" (italics in original).[43]

The General Board of the Navy, the service's highest policy review body, took a less sanguine view of ONR's activities than did Admiral Badger. To begin with, the board felt that ONR was not paying sufficient attention to naval needs in its research planning. One way to gain ONR's attention that was considered was to have supervisory responsibility for the office shifted from the Navy's secretariat to the Chief of Naval Operations.[44] In addition, the board thought that there was still room for savings in ONR's university research. Its opinions were expressed in a report submitted one month after Badger's. "There is disagreement as to the justification for the amounts now spent by the Navy for basic research, which is currently about 10 percent of the research budget. The Board recognizes the necessity for basic research in the overall scientific development of the United States. However, expenditures for this purpose should be assigned a relatively lower priority if further curtailment of the total research and development budget is necessary."[45]

The Chief of Naval Operations also seemed dissatisfied with Admiral Badger's assessment. Less than two weeks after receiving his report, the Chief directed Rear Adm. H. S. Kendall to head an inquiry that would seek opportunities for a closer integration of research and development activities in the Navy.[46] But before Admiral Kendall's study group could turn to a detailed examination of the basic research program in ONR, the Korean War broke out.[47] Admirals Badger and Kendall, along with the rest of the Navy, turned their attention to the Pacific and the fighting. Soon there was money for every type of research protected by a defense justification. Because of Korea, ONR survived its Badger/General Board crisis. But ONR and government-sponsored university research would never again be free from the pressures for relevance.[48]

The Permanent Mobilization of Science

It was inevitable that the ONR would be forced to give greater priority to naval interests, as compared to science interests in its university research program. The vast bulk of the government's funds for research are appropriated not to serve the needs of science, but rather to serve particular national goals such as the maintenance of the national defense, improvement in health care, and the development of the economy. The allocation of these funds to the universities most often is made not in the form of an institutional grant that permits university administrators to exercise discretion in the use of the funds, but rather in the form of project contracts or grants in which the use of the funds has to be justified in terms of the expected contribution to a specific agency mission. If basic research were to be defined in terms of the motivation of the research sponsor rather than that of the research performer, there would be little, if any, basic research being supported by government.

It was also inevitable, given the mission-related justifications that have been required to gain support for academic science, that the military research organizations, including ONR, would be quite creative in their approach to preparing justifications. These organizations have viewed their role in the military as aiding the exploration of the most advanced scientific concepts, whose implications for warfare are as yet only vaguely discernible. In order to survive, however, they have had to appear to be supporting research whose implications for warfare are both concrete and substantial.

The process of preparing justifications, or "painting projects blue" as it was once known within ONR, not surprisingly has been considered a challenge to the collective imagination. The organizational objective has been to obtain funds for research projects thought to be worth doing. The more vividly military the project rationale, the more likely it has been that the project would be approved as it is reviewed at higher levels within the bureaucracy. And since it is quite impossible to predict all the applications that will eventually flow from new knowledge once generated, nearly any justification that is persuasive in the higher level reviews can be prepared in good conscience. As a former administrator in the Air Force Office of Scientific Research recalled the process, "We sold them [in this instance the Air Staff] the sizzle, not the steak."[49]

Although academic scientists have not been directly involved in the preparation of military research justifications, they have not been unaware of the process by which the funds for academic science are obtained within government. In the initial postwar years, scientists feared that government support of science would distort the direction and purposes of science. But when the military's research support seemed free of these consequences, scientists became the leading advocates of increased public investments in academic science. However, their enthusiasm for research subsidies and their willingness to accept whatever rationales appeared to be effective in gaining such subsidies have produced precisely the impact upon science they sought to avoid—its permanent mobilization.

Once the idea of government support of science was accepted by scientists, their search for additional support was both unabashed and insatiable. The President's Science Advisory Committee in its first public report stated that ". . . it is difficult to imagine more fruitful and prudent ways to spend taxpayers' money than on basic research."[50] The Naval Research Advisory Committee initially proposed that the Navy invest between 5 and 10 percent of its total R & D budget in basic research, and when that level was achieved the committee then suggested that the level be doubled.[51] The National Academy of Sciences, advising the Congress, argued that the nation's investment in basic research should increase at an annual rate of 15 percent.[52]

The pressure for increased research subsidies stemmed not only from the aspirations of scientists, but also from the financial needs of their universities. Higher education expanded enormously in the United States in the years since World War II, but without very much direct federal aid. Contentious issues, such as the conflict between graduate and undergraduate education and the relationship between church and state, long delayed passage of subsidies for higher education. Allocations for scientific research, training, and facilities acted as a real, if neither totally satisfactory nor fully disguised, surrogate for direct federal aid to higher education.[53]

In order to gain these allocations, however, scientists have had to claim that their work serves current national interests. First it was defense, then the space race that sustained science. Now it is the cancer crusade and Project Independence. When one national crisis fades, another must be found and promoted. American science may be affluent, but as John Maddox points out, this affluence is maintained at the price of being involved in an endless search for objectives.[54]

Although science has always made utilitarian claims to gain support, the visibility and specificity of its recent promises make science unusually vulnerable.[55] Consider the Department of Defense's Project Hindsight, which, in the mid-1960s, sought to identify the sources of weapon system improvements by describing the purpose and institutional location of research done up to twenty years prior that was embodied in a sample of deployed weapon systems.[56] The study's finding that basic research activities in the postwar years contributed little to operational weapon systems was methodologically flawed but reduced the support for basic research and increased the pressure for more applied work in the military research organizations.[57]

A counterstudy sponsored by the National Science Foundation focused on civilian technology and failed to impress military budget planners in its findings, which were favorable to basic research.[58] Moreover, not all of science can be accommodated within the framework of national programs aimed at narrow objectives. Some disciplines, quite valuable to the progress of science itself, are always certain to be beyond the margin of support. Finally, as the agitation against military research in the late 1960s showed, there is the issue of the legitimacy of the objectives selected. Now public, the goals of science can be, some say must be, contested politically both within and outside of science.[59]

The Absence of Independence and Responsibility

For Vannevar Bush, the central issue in the relationship between science and the military was the need for civilian scientists to have an independent role in the formulation of defense policy. He believed that for free men to remain free, they would have to do research in the implements of war. He believed also that making such research effective would require the active participation of at least a portion of the academic community. And he was concerned that the decisions determining the type of weapons to be developed and the use of the weapons that were developed would become, after World War II as they were before that war, the monopoly of the military.[60]

At the end of World War II, Bush, the leader of the wartime research effort, tried in a number of ways to ensure that scientists would have an independent voice in future defense decisions. First, he endorsed the proposal to establish within the National Academy of Sciences a Research Board for National Security, which would manage military research. Next, Bush included in his own proposal for what was to become the National Science Foundation a provision giving the civilian-directed NSF responsibility for "long-range research on military matters." Finally, he helped organize a civilian-led Joint Research and Development Board (JRDB) in the Pentagon to coordinate postwar research.

Each of these initiatives went awry. The Research Board for National Security was never constituted because of opposition to its independent status by the Bureau of the Budget and some segments of the military. The National Science Board chose not to have the NSF become involved in military research.[61] And the JRDB failed to become an effective organization because of conflicts that developed between it and the services.[62]

But Bush did not give up. When the Korean War broke out, he suggested what was essentially the reestablishment of OSRD in a proposal to the Secretary of Defense entitled "A Few Quick."[63] The services, however, blocked the proposal by pursuing the notion of summer studies—ad hoc task forces of civilian scientists convened often, but not always during the summer academic recess, to examine major defense issues.[64] Although civilian scientists have hailed summer studies as the ideal device to deal with complex policy problems, the initiative for convening a summer study and the decision on what to do with its recommendations lay largely in the hands of the services.⁶⁵ Few have had much impact upon policy.⁶⁶

The formal science advisory committees, of which there are many in the Department of Defense, have had even less impact upon policy. The Naval Research Advisory Committee (NRAC) is a good case in point. Established to advise the Secretary by the same statute that created the Office of Naval Research, NRAC spent the first dozen years of its existence championing ONR's budget within the Navy. Because NRAC's membership consisted mainly of university presidents and leading scientists, this was not exactly a selfless activity. Moreover, in turning its attention to broader issues, NRAC immediately ran into the uniformed Navy. The time, for example, that NRAC questioned the wisdom of building large aircraft carriers, it was told simply that it had better find something else to do, for carriers were a matter of central interest to the Navy. NRAC was soon back championing the research budget.⁶⁷

The military since World War II has controlled the research support and the access to information of civilian scientists interested in defense matters. It is not surprising that the military holds the view that civilian scientists are a special interest group, since scientists must approach the military for research support. A scientist asked to consult on defense matters may be cautious in his dealings with the military not only because of fear for his own research support but also out of concern for his institution and his discipline.

Because Bush failed in his attempts to gain autonomy for civilian scientists, the recent debate over defense research was one between scientists who lacked both a responsibility for defense policy and access to defense data and scientists who, while they may have had some responsibility and some access, lacked independence. Today, both the military and academic science seek complete isolation from each other. Neither the nation nor its defense is likely to be well served by such a situation.

1. Mindak lists nineteen major management studies of naval and/or military research that were conducted between 1947 and 1972. Almost all of these studies had some interest in military laboratories and several were exclusively devoted to the subject. Robert J. Mindak, *Management Studies and Their Effect on Naval R&D*, ONR Report ACR 205, 1 November 1974 (Arlington, Va.: Office of Naval Research, 1974).

2. These organizations are described and discussed in Harold Orlans, *The Nonprofit Research Institute* (New York: McGraw Hill, 1972). For a

discussion of MIT's experience with this type of laboratory, see Dorothy Nelkin, *The University and Military Research* (Ithaca, N.Y.: Cornell University Press, 1972).

3. A useful history of the relations between science and the military in the United States which stresses the pre-World War II period is Clarence G. Lasby, "Science and the Military," in *Science and Society in the United States*, ed. D. D. Van Tassel and M. G. Hall (Homewood, Ill.: The Dorsey Press, 1966), pp. 251–82. See also Kent C. Redmond, "World War II, a Watershed in the Role of the National Government in the Advancement of Science and Technology," in *The Humanities in the Age of Science*, ed. C. Angoff (Rutherford, N.J.: Fairleigh Dickenson Press), pp. 168–80.

4. Note Fleet Admiral Ernest J. King, *U.S. Navy at War 1941–1945: Official Reports to the Secretary of the Navy* (Washington: U.S. Navy Department, 1946) issued December 8, 1945, and "Navy-Science Link Stressed as Vital," *New York Times* (8 December 1945), p. 30.

5. The Bird Dogs, "The Evolution of the Office of Naval Research," *Physics Today* (August 1961): 32.

6. Memorandum from the Coordinator of Research and Development to the Secretary of the Navy. Subject: Organization of Research in the Navy Department, 11 October 1944. Memorandum from Lt. Comdr. R. A. Krause USNR, Lt. Comdr. B. S. Old, USNR, Lt. J. T. Burwell, Jr., USNR to the Secretary of Navy via the coordinator of research and development. Subject: Suggestion for Post-War organization of Naval Research. Unless otherwise noted, all official documents cited are from the record files of the Office of Naval Research.

7. The Bird Dogs, "The Evolution of the Office of Naval Research," p. 35.

8. See Richard G. Hewlett and Francis Dunean, *Nuclear Navy 1946–1952* (Chicago: University of Chicago Press, 1974), p. 25.

9. Vannevar Bush, *Pieces of the Action* (New York: William Morrow, 1970), pp. 104–6. See also n. 12 below.

10. Harold G. Bowen, *Ships, Machinery, and Mossbacks: the Autobiography of a Naval Engineer* (Princeton: Princeton University Press, 1954), p. 178.

11. Bowen, *Ships, Machinery, and Mossbacks*, pp. 150, 177.

12. Vincent Davis, *The Politics of Innovation: Patterns in Navy Cases* (Denver: Graduate School of International Affairs, University of Denver, 1967), p. 25. Davis cites correspondence to him from Bush to this effect in Vincent Davis, *The Admirals Lobby* (Chapel Hill: University of North Carolina Press, 1967), n. 45, p. 175. See also his *Postwar Defense Policy and the U.S. Navy, 1943–1946* (Chapel Hill: University of North Carolina Press, 1966), n. 112, p. 338. This opinion was confirmed in an interview I had with Bush, 23 February 1972.

13. Bowen, *Ships, Machinery, and Mossbacks*, p. 183. See also Hewlett and Dunean, *Nuclear Navy*, pp. 17–19, and Carl O. Holmquist and Russell S.

Greenbaum, "The Development of Nuclear Propulsion in the Navy," *U.S. Naval Institute Proceedings* (September 1960), pp. 65–71.

14. The Bird Dogs, "The Evolution of the Office of Naval Research," and Bowen, *Ships, Machinery, and Mossbacks*. Memorandum from the Secretary of the Navy to all Bureaus, Boards and Offices of the Navy Department. Subject: Office of Research and Invention, 19 May 1945.

15. Holmquist and Greenbaum, "The Development of Nuclear Propulsion in the Navy," p. 68.

16. The Bird Dogs, "The Evolution of the Office of Naval Research," p. 35. It is ironic that the Atomic Energy Commission, the successor organization to the Manhattan District, would twenty years later use the same strategy of seeking out university researchers in order to overcome a bureaucratic jurisdictional conflict when it fought for access to the nuclear-powered artificial heart project being conducted by the National Heart and Lung Institute. Harvey M. Saposky, "Government and the Artificial Heart," appendix A, *The Artificial Heart Assessment Report*, June 1973.

17. Daniel J. Kevles, "Scientists, the Military, and the Control of Postwar Defense Research: the Case of the Research Board for National Security 1944–1946," *Technology and Culture* 16 (1975): 20–47.

18. Vannevar Bush, *Science, the Endless Frontier* (Washington: U.S. Government Printing Office, 1945). There was also a conflicting proposal introduced by Senator Kilkore using the eventual title "National Science Foundation." These bills and the major issues in formulating postwar research policy are well covered in Don K. Price, *Government and Science* (New York: New York University Press, 1954). Note also Detlev W. Bronk, "The National Science Foundation: Origins, Hopes, and Aspirations," *Science* 188, no. 4187 (2 May 1975): 409–414.

19. Memorandum from R. A. Bowen, USN, to Capt. A. S. McDill, USN, Army/Navy Secretariat, Joint Chiefs of Staff, 19 March 1946.

20. Price, *Government and Science*, chap. 2; Richard G. Hewlett and Oscar E. Anderson, Jr. *The New World 1939/1946* (University Park: Pennsylvania State University Press, 1962).

21. Bowen, *Ships, Machinery, and Mossbacks*, p. 351.

22. Hewlett and Dunean, *Nuclear Navy*, p. 28; Russell S. Greenbaum, "Nuclear Power for the Navy: The First Decade (1939–1949)," mimeo, Office of Naval Research, March 1955, pp. 73–74.

23. The Air Force Office of Scientific Research, the predecessor organization to the current Office of Aerospace Research was established in 1951, having for one month the dubious title of the Office of Air Research. An excellent history of the AFOSR is Nick A. Komons, *Science and the Air Force* (Arlington, Va.: Office of Aerospace Research, 1966). Additional information on the office is contained in Thomas A. Sturm, *The USAF Scientific Advisory Board: Its First Twenty Years, 1944–1964* (Washington, D.C.: U.S. Government Printing Office, 1967). The Army

established the Office of Ordnance Research in 1951. Being tied to a specific technical branch, the Ordnance Corps, the Office had more limited functions than did either the Air Force or Navy organizations. The Army Research Office was established in 1958.

24. Stephen P. Strickland, "The Integration of Medical Research and Health Policy," *Science* (17 September 1971).

25. Interviews with former ONR personnel, 1970–71. Some of the rationales are seen in the 1946–47 speeches of Capt. R. A. Conrad, USN, the Director of Planning at ONR, before such audiences as the Army and Navy Staff College, the Commonwealth Club in San Francisco, the Industrial Research Institute, and a Navy Day ceremony at the University of Illinois at Urbana. ONR files.

26. Comments of R. A. Luis de Florez. Transcript of the Ninth Meeting of the Naval Research Advisory Committee, Washington, D.C., June 1949, at p. 75. Apparently the funds were allocated in 1947.

27. Admiral de Florez's CIA position is revealed in reports on early CIA research, *New York Times* (11 January 1976).

28. Conant recalls his postwar attitudes on secret research in James B. Conant, "An Old Man Looks Back: Science and the Federal Government, 1945–1950," *Bulletin of the New York Academy of Medicine* 47, no. 11 (November 1971) pp. 1248–51. His views are also discussed in "Science Subsidies Traced by Conant," *New York Times* (9 January 1971): 31.

29. Norbert Wiener, "A Scientist Rebels," *Atlantic Monthly* 179 (January 1947): 46.

30. See *Final Report by the Advisory Committee on Contractual and Administrative Procedures for the Department of the Army*, 15 October 1948. The report is known as "The Stewart Report" after the committee's chairman, Robert B. Stewart, vice president and controller, Purdue University.

31. Comments of Karl Compton, transcript of the second meeting of the Naval Research Advisory Committee, Washingon, D.C., 15 January 1947, at p. 47.

32. Frank B. Jewett, president of the Academy, was among the many leaders of science who both before and after the war expressed reservations about the desirability of direct government support of basic research. Note Frank B. Jewett, "The Future of Scientific Research in the Postwar World," in *Science in America*, ed. J. C. Burnham (New York: Holt, Rinehart and Winston, 1971). Don K. Price has described the problems of initiating a postwar research program in several places. See particularly his "Organization of Science Here and Abroad," *Science* 129, no. 3351, (20 March 1959): 759–65 and "Science at a Policy Crossroads," *Technology Review*, 73, no. 6 (April 1971): 31–37. Eugene Wigner, a key figure in science in the postwar period, recalled in an interview that a song was composed in 1946 by Arthur Roberts, one of the physicists who was anxious to return to pure research; began

with the line "Take away your billion dollars," as expressive of the times. See *Science* (10 August 1973), p. 533.

33. ONR's early research management procedures are discussed in "Background Information on the Office of Naval Research," mimeo, ONR, (Washington, D.C., April 1966). See also the Stewart Report, p. 11. For a typical article citing ONR as model, see John E. Pfeiffer, "The Office of Naval Research," *Scientific American* 180, no. 2 (February 1949), pp. 11–15.

34. Note, "Science Dons a Uniform," *Business Week*, no. 889 (14 September 1946), pp. 19–24, and Benjamin Fine, "Navy and Colleges Cooperate in Largest Peacetime Program of Scientific Research," *New York Times* (10 February 1946), p. IV 9:1.

35. ONR's twentieth anniversary was commemorated at a public convocation, the proceedings of which were published in *Research in the Service of National Purpose*, ed. F. Joachim Wyl (Washington: U.S. Government Printing Office, 1966).

36. This attitude is well illustrated in the history of the Geological Survey, the largest scientific agency in the federal government in the nineteenth century. Thomas G. Manning, *Government in Science: The U.S. Geological Survey 1867–1964* (Lexington, Ky.: University of Kentucky Press, 1967): William Culp Darrah, *Powell of the Colorado* (Princeton: Princeton University Press, 1951). See also my "Science Policy," in *Handbook of Political Science*, ed. F. Greenstein and N. Polsby, (Reading, Mass.: Addison-Wesley, 1975). The point has been challenged in Nathan Reingold, "American Indifference to Basic Research: A Reappraisal," in *Nineteenth-Century American Science: A Reappraisal* ed. George H. Daniels, (Evanston, 1972), pp. 38–62.

37. Note J. Thomas Ratchford, "Congressional Views of Federal Research Support to Universities," a paper delivered at the 1972 meetings of the American Association for the Advancement of Science, December 1972. The Mansfield Amendment is mistitled, for it was Sen. William Fulbright rather than Senator Mansfield who actually introduced the amendment.

38. President Johnson's statements on research productivity and the politics that lay behind them are discussed in John Walsh, "NIH: Demand Increases for Applications of Research," *Science* (8 July 1966): 149–52 and Daniel S. Greenberg, "LBJ at NIH: President Offers Kind Words for Basic Research," *Science* (28 July 1967): 403–9.

39. The "fried potatoes" quote and Secretary Wilson's attitudes toward science are reported in "Wilson Hits Generals for Opposing Air Cuts," *New York Herald Tribune* (9 June 1953) and "Wilson to Oppose Military Cutbacks After Korea Truce," *New York Times* (9 June 1953). The Secretary was asked to clarify his views on the subject at his two next news conferences in Washington, D.C. on 16 June 1953 and Quantico, Va., on 21 June 1953.

40. Eleventh meeting of the Naval Research Advisory Committee, Washington, D.C., 30 January 1950.

41. Interviews with former ONR personnel, 1972.

42. Interviews with former ONR personnel, 1972.

43. The official citation for the report is V. Adm. Oscar C. Badger sec. ltr. OCB/tn Serial: 001 of 13 February 1950 to CNO. Subject: investigation of Navy Research and Development contracts with universities and colleges; report of (8 encls.). Despite extensive searches, the report has not been located in Navy files. It is, however, extensively cited in the Progress Report of the Kendall Board; see n. 47 below.

44. Notes by N. J. Frank, Secretary to the General Board, on a meeting of the Board with Capt. G. C. Wright, OP–34, Project 8–49, 3 March 1950, General Board Files, Office of Naval History.

45. P. 59, Report of the General Board on the Relationship of Various Budgetary Programs to Maintain a Most Effective Navy, 13 March 1950, General Board Files, Office of Naval History; CNO Files, Office of Naval History.

46. Letter from CNO to R. Adm. H. S. Kendall, 24 February 1950.

47. The Kendall Board, officially the U.S. Navy R & D Survey Board, did issue one progress report on 1 July 1950 before disbanding, in which it quoted approvingly from the Badger Report but in which it also noted that there should be a careful review of ONR and the Navy's work in basic research.

48. ONR never did again forget it was in Navy; note especially the Fifteenth Meeting of the Naval Research Advisory Committee, Washington, D.C., 25 June 1951. Interviews with former ONR personnel, 1972 and 1974. Note also Luther Carter, "Office of Naval Research: 20 Years Bring Changes," *Science* (22 July 1966): 397–400.

49. Nick A. Komons, *Science and the Air Force: A History of the Air Force of Scientific Research* (Arlington, Va.: Office of Aerospace Research, 1966), p. 70. Daniel S. Greenberg, "The Air Force: Study Relates Troubled Relationship with Research," *Science* (16 June 1967): 1465.

50. President's Science Advisory Committee, *Strengthening American Science* (Washington, D.C.: Government Printing Office, 1958).

51. "Scientists Urge More Navy Funds," *New York Times* (22 October 1959): 11.

52. National Academy of Sciences, *Basic Research and National Goals*, (Washington, D.C.: Government Printing Office, 1965). For a perceptive discussion of the attitude of the scientific community on the financial needs of science, see Harold Orlans, "Developments in Federal Policy Toward University Research," *Science* (11 February 1967): 665–68.

53. Don K. Price, "Science at a Policy Crossroads," *Technology Review* (April 1971): 33–34; Daniel S. Greenberg, "The New Politics of Science," *Technology Review* (February 1971): 44.

54. John Maddox, "American Science: Endless Search for Objectives," *Daedalus* 101, no. 4 (Fall 1972): 129–40.

55. The utilitarian claims of science in nineteenth century America are discussed in Manning, *Government in Science* and Harold S. Miller, *Dollars for Research* (Seattle: University of Washington Press, 1970).

56. C. W. Scherwin and R. S. Isenson, "Project Hindsight," *Science* (1967): 1571–77; R. Isenson, "Project Hindsight: An Empirical Study of the Sources of Ideas in Operational Weapon Systems," in *Factors in the Transfer of Technology* ed. W. H. Gruber and D. G. Marquis (Cambridge: MIT Press, 1969).

57. K. Krielkamp, "Hindsight and the Real World of Science Policy," *Science Studies* (1971): 43–66.

58. Illinois Institute of Technology, *Technology in Retrospect and Critical Events in Science (TRACES)* (Washington, D.C.: National Science Foundation, 1968).

59. J. R. Ravetz, *Scientific Knowledge and its Social Problems* (London: Oxford, 1971).

60. There are several interesting histories of early science advisory mechanism utilized by the Navy both in the United States and Great Britain that illustrate how resistant the services are to outside examination: Nathan Reingold, "Science in the Civil War: The Permanent Commission of the Navy Department," *Isis* 49 (September 1958): 307–18; Lloyd L. N. Scott, *Naval Consulting Board of the United States* (Washington: U.S. Government Printing Office, 1920); Roy M. MacLeod and E. Kay Andrews "Scientific Advice in the War at Sea 1915–1917: The Board of Invention and Research," *Journal of Contemporary History* 6, no. 2 (1971): 3–40.

61. Detlev W. Bronk, "The National Science Foundation: Origins, Hopes, and Aspirations," *Science* 188, no. 4187 (2 May 1975): 412.

62. See Price, *Government and Science*, 145–52.

63. Memorandum E. R. Piore to R. Adm. C. M. Bolster, 7 January 1952. Subject: Notes on V. Bush's extensive memorandum, "A Few Quick," dated 5 November 1951.

64. Interviews with former ONR personnel, 1971. Comments of Dr. F. Ridenour at the fourteenth meeting, Naval Research Advisory Committee, 19 March 1951, Washington, D.C. For discussion of summer studies see J.R. Marvin and F. J. Weyl, "The Summer Study," *Naval Research Reviews* (August 1966): 1–12.

65. For example, Jerrold Zacharias, "Scientist as Advisor," speech delivered at the Harvard Graduate School of Public Administration, Cambridge, Mass., 28 March 1961.

66. Harvey M. Sapolsky, *The Polaris System Development: Bureaucratic and Programmatic Success in Government* (Cambridge: Harvard University Press, 1972), pp. 28–31, 49–50.

67. O. W. Helm, "Genesis of the Naval Research Advisory Committee," *Naval Research Reviews* (September 1957), pp. 21–29, minutes of the fortieth meeting of the Naval Research Advisory Committee, 3 May 1960, Washington, D.C.